Information Security and Cryptography
Texts and Monographs

For further volumes:
http://www.springer.com/series/4752

Information Security and Cryptography
Texts and Monographs

Series Editors
David Basin
Ueli Maurer

Advisory Board
Martín Abadi
Ross Anderson
Michael Backes
Ronald Cramer
Virgil D. Gligor
Oded Goldreich
Richard A. Kemmerer
Ueli Maurer
John C. Mitchell
Tatsuaki Okamoto
Kenny Paterson
Bart Preneel

Carmit Hazay · Yehuda Lindell

Efficient Secure Two-Party Protocols

Techniques and Constructions

 Springer

Dr. Carmit Hazay
Department of Computer Science
and Applied Mathematics
Faculty of Mathematics and
Computer Science
Weizmann Institute
Rehovot
Israel
and
Interdisciplinary Center (IDC)
Herzliya 46150
Israel
carmit.hazay@gmail.com

Professor Yehuda Lindell
Bar Ilan University
Department of Computer Science
Ramat Gan 52900
Israel
lindell@cs.biu.ac.il

Series Editors

Prof. Dr. David Basin
Prof. Dr. Ueli Maurer
ETH Zürich
Switzerland
basin@inf.ethz.ch
maurer@inf.ethz.ch

ISSN 1619-7100
ISBN 978-3-642-26576-1   ISBN 978-3-642-14303-8 (eBook)
DOI 10.1007/978-3-642-14303-8
Springer Heidelberg Dordrecht London New York

ACM Computing Classification (1998): E.3, C.2, H.2.8

Cover design: KuenkelLopka GmbH, Heidelberg

Printed on acid-free paper

Springer is part of Springer Science+Business Media (www.springer.com)

To our families, for all their support

Preface

In the setting of multiparty computation, sets of two or more parties with private inputs wish to jointly compute some (predetermined) function of their inputs. The computation should be such that the outputs received by the parties are correctly distributed, and furthermore, that the privacy of each party's input is preserved as much as possible, even in the presence of adversarial behavior. This encompasses any distributed computing task and includes computations as simple as coin-tossing and broadcast, and as complex as electronic voting, electronic auctions, electronic cash schemes and anonymous transactions. The feasibility (and infeasibility) of multiparty computation has been extensively studied, resulting in a rather comprehensive understanding of what can and cannot be securely computed, and under what assumptions.

The theory of cryptography in general, and secure multiparty computation in particular, is rich and elegant. Indeed, the mere fact that it is possible to actually achieve the aforementioned task is both surprising and intriguing. However, the focus of this book is not on the theory of secure computation (although a number of results with theoretical importance are studied here), but rather on the question of *efficiency*. Recently, there has been increasing interest in the possibility of actually using secure multiparty computation to solve real-world problems. This poses an exciting challenge to the field of cryptography: Can we construct secure protocols (with rigorous proofs of security) that are truly efficient, and thus take the theory of secure computation to the next step towards practice. We stress that this book is not about "practical cryptography". We do not take systems considerations into account, nor how protocols should be implemented and deployed. Instead, our aim is to provide an introduction to the field of efficient protocol construction and design. We hope that this book will make the field of secure computation in general, and efficient protocol construction in particular, more accessible and will increase awareness regarding the importance of this vibrant field.

Outline. This book is divided into three distinct parts:

- **Introduction and definitions:** We begin with a general introduction and survey of secure computation, followed by definitions of security under a number of different adversary models. This part also includes important material regarding the properties of these definitions, and the relations between them.

- **General constructions:** In this part, we present secure protocols for general secure computation. That is, we present protocols that can be applied to any circuit computing any efficient function. Although this does not enable us to utilize specific properties of the function being computed, the resulting protocols can be efficient enough if the circuit and input are not too large.

- **Specific constructions:** Finally, we study secure protocols for specific problems of interest. Two of the chapters in this part consider efficient constructions of basic building blocks that are widely used in constructions of secure protocols; namely, zero-knowledge (via Σ protocols) and oblivious transfer. The last two chapters study two specific examples of higher-level protocols; specifically, the secure computation of the kth ranked element (or median) of a distributed list, and secure search operations on databases. The constructions in this part demonstrate how specific properties of a function being computed can be utilized to achieve greater efficiency.

It goes without saying that the material presented in this book is far from an exhaustive study of results in the field. There are many alternative constructions achieving some of the results presented here, and many other problems of interest for which efficient protocols have been constructed. In some places throughout, we have added pointers to additional readings of relevance.

In order to not unnecessarily complicate the constructions and models, we have focused on the *two-party* case and consider only *static adversaries* and the *stand-alone model*. We do not claim that this is the best model for constructing protocols; indeed it is arguably too weak in many cases. However, we believe that it serves as a good setting for an initial study, as it is significantly cleaner than other more complex settings.

Prerequisite knowledge. We assume that the reader is familiar with the basics of theoretical cryptography. Thus, for example, we assume that readers know what commitment schemes and zero-knowledge proofs are, and that they are comfortable with notions like pseudorandomness and computational indistinguishability. In contrast, all the relevant definitions of secure two-party computation are presented here from scratch. Thus, this book can also be used as a first introduction to secure computation.

Reading this book. Although there are advantages to reading this book in sequential order, much of the book can be read "out of order". It goes without saying that the chapter on definitions is needed for all later chapters. However, it is possible to read definitions as needed (e.g., read Section 2.2

and then Chapter 3, then Section 2.3 followed by Chapter 4, and so on). Regarding the general constructions in Part II of the book, the constructions in Chapters 4 and 5 rely in a direct way on Chapter 3, and thus it is highly recommended to read Chapter 3 first. In contrast, Chapters 4 and 5 can be read independently of each other.

The specific constructions in Part III can be read independently of the general constructions in Part II. It is preferable to read Chapters 6 and 7 first (and in order) because later protocols use the tools introduced in these chapters. In addition, some of the oblivious transfer protocols of Chapter 7 use zero-knowledge proofs that are constructed in Chapter 6. Nevertheless, if one is satisfied with referring to an arbitrary zero-knowledge proof or oblivious transfer protocol, then the chapters in Part III can be read in any order.

Book aims and its use for teaching a course. This book can be used as a textbook for an introductory course on secure computation with a focus on techniques for achieving efficiency, as an entry point for researchers in cryptography and other fields like privacy-preserving data mining who are interested in efficient protocols for secure computation, and as a reference for researchers already in the field. Regarding its use as a textbook, due to the flexibility regarding the order of reading this book (as described above), it is possible to design courses with different focuses. For example, a more theoretical course would spend considerable time on definitions and the general constructions of Part II of the book, whereas a more applied course would focus more on the specific constructions in Part III. We remark also that Chapters 6 and 7 can serve as a nice opening to a course; the material is not as heavy as general secure computation and contains many interesting ideas that can be attractive to students. When teaching a general introduction to (computational) secure computation, it is certainly possible to base much of the course on this book. However, in such a case we would also teach the GMW construction. A full treatment of this appears in [35, Chapter 7].

Comments and errata. We will be more than happy to receive any (positive or negative) feedback that you have on this book, as well as any errors that you may find. Please email us your comments and errata to lindell@cs.biu.ac.il. A list of known errata will be maintained at http://www.cs.biu.ac.il/~lindell/efficient-protocols.html.

Acknowledgements. First and foremost, we would like to thank Ivan Damgård for generously providing us with the text that formed the basis of Chapter 6 on Σ protocols. In addition, we would like to thank Oded Goldreich, Jonathan Katz and Eran Omri for providing us with constructive advice and comments on this book.

Carmit Hazay: First, I would like to thank my co-author Yehuda Lindell who was also my Ph.D. advisor. Yehuda introduced me to the area of secure computation and has greatly contributed to my academic career. He is a continuing source of inspiration and assistance, and I am grateful to him for an amazing journey which led to this book.

During my Ph.D. I had the pleasure of working with many talented people who enriched my knowledge and deepened my understanding regarding secure computation. I would like to thank Ran Canetti, Rosario Gennaro, Jonathan Katz, Hugo Krawczyk, Kobbi Nissim, Tal Rabin and Hila Zarosim for many productive discussions and a memorable time.

Yehuda Lindell: First and foremost I would like to thank Oded Goldreich. Beyond being my Ph.D. advisor, and as such of great influence on my academic career, Oded has continued to provide valuable support, advice and encouragement. I owe much to Oded and am greatly indebted to him.

The ability to write this book is due to the knowledge that I have gained over many years of research in the field of secure computation. In this time, I have worked with many different co-authors and have benefited from countless fruitful discussions with many members of our research community. I would like to thank Yonatan Aumann, Boaz Barak, Ran Canetti, Rosario Gennaro, Shafi Goldwasser, Shai Halevi, Carmit Hazay, Yuval Ishai, Yael Kalai, Jonathan Katz, Eyal Kushilevitz, Hugo Krawczyk, Tal Malkin, Moni Naor, Benny Pinkas, Tal Rabin, Alon Rosen and Adam Smith for years of joint work and cooperation in a friendly and enjoyable environment. Finally, I would like to give a special thanks to Benny Pinkas for all I have learned from him regarding topics of efficiency in secure protocols.

My work on this project was supported by the Israel Science Foundation (grant 781/07) and by a starting grant from the European Research Council.

October 2010   Carmit Hazay and Yehuda Lindell

Contents

Part II General Constructions

Part III Specific Constructions

Part I
Introduction and Definitions

In the first two chapters of this book we provide a general introduction to the field of secure computation, as well as rigorous definitions for secure two-party computation in multiple models. Specifically, we consider security in the presence of semi-honest and malicious adversaries, as well as introduce the notion of covert adversaries and security that is not based on the full simulation ideal/real-model paradigm.

Chapter 1
Introduction

The focus of this book is on constructing *efficient* secure protocols for the *two-party* setting. In this introduction, we begin with a general high-level survey of secure multiparty computation. This places the topic of this book in its larger context. Following this, we describe the basic results and techniques related to efficiency in secure computation. Finally, we conclude with a roadmap to the book.

1.1 Secure Multiparty Computation – Background

Distributed computing considers the scenario where a number of distinct, yet connected, computing devices (or parties) wish to carry out a joint computation of some function. For example, these devices may be servers that hold a distributed database system, and the function to be computed may be a database update of some kind. The aim of *secure multiparty computation* is to enable parties to carry out such distributed computing tasks in a secure manner. Whereas distributed computing classically deals with questions of computing under the threat of machine crashes and other inadvertent faults, secure multiparty computation is concerned with the possibility of deliberately malicious behavior by some adversarial entity. That is, it is assumed that a protocol execution may come under "attack" by an external entity, or even by a subset of the participating parties. The aim of this attack may be to learn private information or cause the result of the computation to be incorrect. Thus, two important requirements on any secure computation protocol are *privacy* and *correctness*. The privacy requirement states that nothing should be learned beyond what is absolutely necessary; more exactly, parties should learn their output and nothing else. The correctness requirement states that each party should receive its correct output. Therefore, the adversary must not be able to cause the result of the computation to deviate from the function that the parties had set out to compute.

C. Hazay, Y. Lindell, *Efficient Secure Two-Party Protocols*,
Information Security and Cryptography, DOI 10.1007/978-3-642-14303-8_1,
© Springer-Verlag Berlin Heidelberg 2010

The setting of secure multiparty computation encompasses tasks as simple as coin-tossing and broadcast, and as complex as electronic voting, electronic auctions, electronic cash schemes, contract signing, anonymous transactions, and private information retrieval schemes. Consider for a moment the tasks of voting and auctions. The privacy requirement for an election protocol ensures that no coalition of parties learns anything about the individual votes of other parties, and the correctness requirement ensures that no coalition of parties can influence the outcome of the election beyond just voting for their preferred candidate. Likewise, in an auction protocol, the privacy requirement ensures that only the winning bid is revealed (this may be desired), and the correctness requirement ensures that the highest bidder is indeed the party to win (and so the auctioneer, or any other party, cannot bias the outcome).

Due to its generality, the setting of secure multiparty computation can model almost every, if not every, cryptographic problem (including the classic tasks of encryption and authentication). Therefore, questions of feasibility and infeasibility for secure multiparty computation are fundamental to the theory and practice of cryptography.

Security in multiparty computation. As we have mentioned above, the model that we consider is one where an adversarial entity controls some subset of the parties and wishes to attack the protocol execution. The parties under the control of the adversary are called corrupted, and follow the adversary's instructions. Secure protocols should withstand any adversarial attack (where the exact power of the adversary will be discussed later). In order to formally claim and prove that a protocol is secure, a precise definition of security for multiparty computation is required. A number of different definitions have been proposed and these definitions aim to ensure a number of important security properties that are general enough to capture most (if not all) multiparty computation tasks. We now describe the most central of these properties:

- *Privacy:* No party should learn anything more than its prescribed output. In particular, the only information that should be learned about other parties' inputs is what can be derived from the output itself. For example, in an auction where the only bid revealed is that of the highest bidder, it is clearly possible to conclude that all other bids were lower than the winning bid. However, this should be the only information revealed about the losing bids.
- *Correctness:* Each party is guaranteed that the output that it receives is correct. To continue with the example of an auction, this implies that the party with the highest bid is guaranteed to win, and no party including the auctioneer can influence this.
- *Independence of Inputs:* Corrupted parties must choose their inputs independently of the honest parties' inputs. This property is crucial in a sealed auction, where bids are kept secret and parties must fix their bids independently of others. We note that independence of inputs is *not* implied

by privacy. For example, it may be possible to generate a higher bid without knowing the value of the original one. Such an attack can actually be carried out on some encryption schemes (i.e., given an encryption of $100, it is possible to generate a valid encryption of $101, without knowing the original encrypted value).

- *Guaranteed Output Delivery:* Corrupted parties should not be able to prevent honest parties from receiving their output. In other words, the adversary should not be able to disrupt the computation by carrying out a "denial of service" attack.

- *Fairness:* Corrupted parties should receive their outputs if and only if the honest parties also receive their outputs. The scenario where a corrupted party obtains output and an honest party does not should not be allowed to occur. This property can be crucial, for example, in the case of contract signing. Specifically, it would be very problematic if the corrupted party received the signed contract and the honest party did not.

We stress that the above list does *not* constitute a definition of security, but rather a set of requirements that should hold for any secure protocol. Indeed, one possible approach to defining security is to just generate a list of separate requirements (as above) and then say that a protocol is secure if all of these requirements are fulfilled. However, this approach is not satisfactory for the following reasons. First, it may be possible that an important requirement was missed. This is especially true because different applications have different requirements, and we would like a definition that is general enough to capture all applications. Second, the definition should be simple enough so that it is trivial to see that *all* possible adversarial attacks are prevented by the proposed definition.

The standard definition today (cf. [11] following [37, 5, 59]) therefore formalizes security in the following general way. As a mental experiment, consider an "ideal world" in which an external trusted (and incorruptible) party is willing to help the parties carry out their computation. In such a world, the parties can simply send their inputs over perfectly private channels to the trusted party, which then computes the desired function and passes each party its prescribed output. Since the only action carried out by a party is that of sending its input to the trusted party, the only freedom given to the adversary is in choosing the corrupted parties' inputs. Notice that all of the above-described security properties (and more) hold in this ideal computation. For example, privacy holds because the only message ever received by a party is its output (and so it cannot learn any more than this). Likewise, correctness holds since the trusted party cannot be corrupted and so will always compute the function correctly.

Of course, in the "real world", there is no external party that can be trusted by all parties. Rather, the parties run some protocol amongst themselves without any help. Despite this, a secure protocol should emulate the so-called "ideal world". That is, a real protocol that is run by the parties (in a world where no trusted party exists) is said to be **secure** if no adversary can do more harm in a real execution than in an execution that takes place in the ideal world. This can be formulated by saying that for any adversary carrying out a successful attack in the real world, there exists an adversary that successfully carries out the same attack in the ideal world. However, successful adversarial attacks *cannot* be carried out in the ideal world. We therefore conclude that all adversarial attacks on protocol executions in the real world must also fail.

More formally, the security of a protocol is established by comparing the outputs of the adversary and honest parties in a real protocol execution to their outputs in an ideal computation. That is, for any adversary attacking a real protocol execution, there exists an adversary attacking an ideal execution (with a trusted party) such that the input/output distributions of the adversary and the participating parties in the real and ideal executions are essentially the same. Thus a real protocol execution "emulates" the ideal world. This formulation of security is called the ideal/real simulation paradigm. In order to motivate the usefulness of this definition, we describe why all the properties described above are implied. Privacy follows from the fact that the adversary's output is the same in the real and ideal executions. Since the adversary learns nothing beyond the corrupted party's outputs in an ideal execution, the same must be true for a real execution. Correctness follows from the fact that the honest parties' outputs are the same in the real and ideal executions, and from the fact that in an ideal execution, the honest parties all receive correct outputs as computed by the trusted party. Regarding independence of inputs, notice that in an ideal execution, all inputs are sent to the trusted party before any output is received. Therefore, the corrupted parties know nothing of the honest parties' inputs at the time that they send their inputs. In other words, the corrupted parties' inputs are chosen independently of the honest parties' inputs, as required. Finally, guaranteed output delivery and fairness hold in the ideal world because the trusted party always returns all outputs. The fact that it also holds in the real world again follows from the fact that the honest parties' outputs are the same in the real and ideal executions.

We remark that the above informal definition is actually "overly ideal" and needs to be relaxed in settings where the adversary controls half or more of the participating parties (that is, in the case where there is no honest majority). When this number of parties is corrupted, it is known that it is *impossible* to obtain general protocols for secure multiparty computation that guarantee output delivery and fairness. In particular, it is impossible for two parties to toss an unbiased coin when one may be corrupt [17]. Therefore, the definition is relaxed and the adversary is allowed to abort the computation (i.e., cause it to halt before termination), meaning that "guaranteed output delivery" is not fulfilled. Furthermore, the adversary can cause this abort to take place after it has already obtained its output, but before all the honest parties receive their outputs. Thus "fairness" is not achieved. Loosely speaking, the relaxed definition is obtained by modifying the ideal execution and giving the adversary the additional capability of instructing the trusted party to not send outputs to some of the honest parties. Otherwise, the definition remains identical and thus all the other properties are still preserved.

Recently it has been shown that in the case of no honest majority, some non-trivial functions can be securely computed with complete fairness [39]. Despite this, we will forgo any attempt at achieving fairness because (a) general constructions cannot achieve fairness due to [17], and (b) we focus on efficient protocols and all currently known techniques for achieving fairness for non-trivial functions are inherently inefficient.

We note that there are works that aim to provide intermediate notions of fairness [77, 29, 6, 37, 40]. However, we limit our reference to the cases that either (complete) fairness and output delivery are guaranteed, or neither fairness (of any type) nor output delivery are guaranteed.

Adversarial power. The above informal definition of security omits one very important issue: the power of the adversary that attacks a protocol execution. As we have mentioned, the adversary controls a subset of the participating parties in the protocol. However, we have not described the corruption strategy (i.e., when or how parties come under the "control" of the adversary), the allowed adversarial behavior (i.e., does the adversary just passively gather information or can it instruct the corrupted parties to act maliciously), and what complexity the adversary is assumed to have (i.e., is it polynomial time or computationally unbounded). We now describe the main types of adversaries that have been considered:

1. **Corruption strategy:** The corruption strategy deals with the question of when and how parties are corrupted. There are two main models:

   a. Static corruption model: In this model, the adversary is given a fixed set of parties whom it controls. Honest parties remain honest throughout and corrupted parties remain corrupted.

   b. Adaptive corruption model: Rather than having a fixed set of corrupted parties, adaptive adversaries are given the capability of corrupting parties during the computation. The choice of whom to corrupt, and when, can be arbitrarily decided by the adversary and may depend on what is has seen throughout the execution (for this reason it is called adaptive). This strategy models the threat of an external "hacker" breaking into a machine during an execution. We note that in this model, once a party is corrupted, it remains corrupted from that point on.

   An additional model, called the proactive model [67, 13], considers the possibility that parties are corrupted for a certain period of time only. Thus, honest parties may become corrupted throughout the computation (as in the adaptive adversarial model), but corrupted parties may also become honest.

2. **Allowed adversarial behavior:** Another parameter that must be defined relates to the actions that corrupted parties are allowed to take. Once again, there are two main types of adversaries:

a. **Semi-honest adversaries:** In the semi-honest adversarial model, even corrupted parties correctly follow the protocol specification. However, the adversary obtains the internal state of all the corrupted parties (including the transcript of all the messages received), and attempts to use this to learn information that should remain private. This is a rather weak adversarial model. However, it does model inadvertent leakage of information by honest parties and thus is useful in some cases (e.g., where the parties essentially trust each other but want to ensure that nothing beyond the output is leaked). This model may also be of use in settings where the use of the "correct" software running the correct protocol can be enforced. Semi-honest adversaries are also called "honest-but-curious" and "passive".

b. **Malicious adversaries:** In this adversarial model, the corrupted parties can *arbitrarily* deviate from the protocol specification, according to the adversary's instructions. In general, providing security in the presence of malicious adversaries is preferred, as it ensures that no adversarial attack can succeed. However, protocols that achieve this level of security are typically much less efficient. Malicious adversaries are also called "active".

These are the classic adversarial models. However, in some cases, an intermediate adversary model may be required. This is due to the fact that the semi-honest adversary modeling is often too weak, whereas our protocols that achieve security in the presence of malicious adversary may be far too inefficient. An intermediate adversary model is that of **covert adversaries**. Loosely speaking, such an adversary may behave maliciously. However, it is guaranteed that if it does so, then it will be *caught cheating* by the honest parties with some given probability.

3. **Complexity:** Finally, we consider the assumed computational complexity of the adversary. As above, there are two categories here:

a. **Polynomial time:** The adversary is allowed to run in (probabilistic) polynomial time (and sometimes, expected polynomial time). The specific computational model used differs, depending on whether the adversary is uniform (in which case, it is a probabilistic polynomial-time Turing machine) or non-uniform (in which case, it is modeled by a polynomial-size family of circuits).

b. **Computationally unbounded:** In this model, the adversary has no computational limits whatsoever.

The above distinction regarding the complexity of the adversary yields two very different models for secure computation: the *information-theoretic* model [9, 15] and the *computational* model [77, 35]. In the information-theoretic setting, the adversary is not bound to any complexity class (and in particular, is not assumed to run in polynomial time). Therefore, results in this model hold unconditionally and do not rely on any complexity or cryptographic assumptions. The only assumption used is that parties are connected via ideally *private* channels (i.e., it is assumed that the adversary cannot eavesdrop on or interfere with the communication between honest parties).

In contrast, in the computational setting the adversary is assumed to be polynomial time. Results in this model typically assume cryptographic assumptions like the existence of trapdoor permutations. We note that it is not necessary here to assume that the parties have access to ideally private channels, because such channels can be implemented using public-key encryption. However, it is assumed that the communication channels between parties are authenticated; that is, if two honest parties communicate, then the adversary can eavesdrop but cannot modify any message that is sent. Such authentication can be achieved using digital signatures [38] and a public-key infrastructure.

It is only possible to achieve information-theoretic security in the case of an honest majority [9]. Thus, it is not relevant to the case of two-party computation, which is the focus of this book. We will therefore consider the computational setting only.

We remark that all possible combinations of the above types of adversaries have been considered in the literature.

Stand-alone computation versus composition. All of the above relates to the stand-alone model, where only a single protocol execution takes place (or many take place but only one is "under attack"). A far more realistic model is that of *concurrent general composition* where many secure (and possibly insecure) protocols are executed together [12]. This is a strictly harder problem to solve [14] and has been the focus of much work in the past decade. See [54] for a study of this topic.

Feasibility of secure multiparty computation. The above-described definition of security seems to be very restrictive in that no adversarial success is tolerated, irrespective of its strategy. Thus, one may wonder whether it is even possible to obtain secure protocols under this definition, and if yes, for which distributed computing tasks. Perhaps surprisingly, powerful feasibility results have been established, demonstrating that in fact, *any* distributed computing task can be securely computed. We now briefly state the most central of these results for the case of *malicious adversaries* and *static corruptions* in the *stand-alone model*. Let m denote the number of participating parties and let t denote a bound on the number of parties that may be corrupted:

1. For $t < m/3$ (i.e., when less than a third of the parties can be corrupted), secure multiparty protocols with guaranteed output delivery can be achieved for any function in a point-to-point network, without any setup assumptions. This can be achieved both in the computational set-

ting [35] (assuming the existence of trapdoor permutations) and in the
information-theoretic (private channel) setting [9, 15].

2. For $t < m/2$ (i.e., in the case of a guaranteed honest majority), secure
   multiparty protocols with fairness and guaranteed output delivery can
   be achieved for any function assuming that the parties have access to a
   broadcast channel. This can be achieved in the computational setting [35]
   (under the same assumptions as above), and in the information-theoretic
   setting [73, 4].

3. For $t \geq m/2$ (i.e., when the number of corrupted parties is not limited), secure multiparty protocols (without fairness or guaranteed output delivery)
   can be achieved for any function assuming that the parties have access to
   a broadcast channel and in addition assuming the existence of enhanced
   trapdoor permutations [77, 35, 32]. These feasibility results hold only in
   the computational setting; analogous results for the information-theoretic
   setting cannot be obtained when $t \geq m/2$ [9].

In summary, secure multiparty protocols exist for any distributed computing
task. In the computational model, this holds for all possible numbers of corrupted parties, with the qualification that when no honest majority exists,
then fairness and guaranteed output delivery are not obtained. We note that
the above results all hold with respect to malicious, static adversaries in the
stand-alone model.

**This book – two-parties, static adversaries and the stand-alone
model.** As we have mentioned, adaptive corruption captures a real-world
threat and as such protocols that are secure in the presence of such adversaries provide a strong security guarantee. In addition, the stand-alone model
of computation is not the realistic model in which protocols are executed
today. Nevertheless, the problem of constructing *highly efficient two-party
protocols* that are secure in the presence of *static adversaries in the stand-
alone model* serves as an important stepping stone for constructing protocols
in more complex settings. As we will see, it is already difficult to construct
efficient protocols for static adversaries in the stand-alone model, and we
strongly believe that a broad understanding of the problems that arise in
more restricted settings is needed before progressing to more complex settings. Our experience also shows us that the techniques developed for solving
the problems of secure computation in the stand-alone model with static
adversaries are often useful also in the more complex setting of concurrent
composition (with static or adaptive adversaries). For these reasons, we have
chosen to focus solely on the stand-alone model and static adversaries in this
book.

1.2 The GMW Protocol for Secure Computation

As we have mentioned above, it has been shown that any probabilistic polynomial-time two-party functionality can be securely computed in the presence of malicious adversaries (without fairness or guaranteed output delivery), assuming the existence of enhanced trapdoor permutations [35, 32]. This powerful feasibility result – known as the GMW construction – is obtained in two stages. First, it is shown how to securely compute any functionality in the presence of semi-honest adversaries. Then, a protocol compiler is presented that takes any protocol that is secure in the presence of semi-honest adversaries and outputs a protocol that is secure in the presence of malicious adversaries.

Security for semi-honest adversaries. A secure protocol is constructed based on a Boolean circuit that computes the functionality in question. The basic idea behind the construction is for the parties to iteratively compute the gates in the circuit in an "oblivious manner". This is achieved by having the parties first share their input bits; that is, for every input wire to the circuit, the parties hold random bits α and β so that $\alpha \oplus \beta$ equals the actual input bit associated with that wire. Then, for every (AND/OR/NOT) gate, the parties run a mini-protocol to compute random shares of the output of the gate, based on their given random shares of the inputs to the gate. At the end of the protocol, the parties hold random shares of the output wires which they can send to each other in order to reconstruct the actual output. The security of the protocol is derived from the fact that each party sees only random values (shares) throughout the protocol. Therefore, it learns nothing beyond the output, as required.

Compilation to security for malicious adversaries. The basic idea that stands behind the GMW construction is to have the parties run a suitable protocol that is secure in the presence of semi-honest adversaries, while *forcing* the potentially malicious participants to behave in a semi-honest manner. The GMW compiler therefore takes for input a protocol that is secure against semi-honest adversaries; from here on we refer to this as the "basic protocol". Recall that this protocol is secure in the case where each party follows the protocol specification exactly, using its input and uniformly chosen random tape. We must therefore force a malicious adversary to behave in this way. First and foremost, this involves forcing the parties to follow the prescribed protocol. However, this only makes sense relative to a *given* input and random tape. Furthermore, a malicious party must be forced into using a *uniformly chosen* random tape. This is because the security of the basic protocol may depend on the fact that the party has no freedom in setting its own randomness.[1]

---

[1] A good example of this is the semi-honest 1-out-of-2 oblivious transfer protocol of [25]. The oblivious transfer functionality is defined by $((x_0, x_1), \sigma) \mapsto (\lambda, x_\sigma)$. In the protocol

In light of the above discussion, the GMW protocol compiler begins by having each party commit to its input. Next, the parties run a coin-tossing protocol in order to fix their random tapes (clearly, this protocol must be secure against malicious adversaries). A regular coin-tossing protocol in which both parties receive the same uniformly distributed string is not sufficient here. This is because the parties' random tapes must remain secret. This is solved by augmenting the coin-tossing protocol so that one party receives a uniformly distributed string (to be used as its random tape) and the other party receives a commitment to that string. Now, following these two steps, each party holds its own uniformly distributed random tape and a commitment to the other party's input and random tape. Therefore, each party can be "forced" into working consistently with the committed input and random tape.

We now describe how this behavior is enforced. A protocol specification is a deterministic function of a party's view consisting of its input, random tape and messages received so far. As we have seen, each party holds a commitment to the input and random tape of the other party. Furthermore, the messages sent so far are public. Therefore, the assertion that a new message is computed according to the protocol is of the $\mathcal{NP}$ type (and the party sending the message knows an adequate $\mathcal{NP}$-witness to it). Thus, the parties can use zero-knowledge proofs to show that their steps are indeed according to the protocol specification. As the proofs used are zero-knowledge, they reveal nothing. Furthermore, due to the soundness of the proofs, even a malicious adversary cannot deviate from the protocol specification without being detected. We thus obtain a reduction of the security in the malicious case to the given security of the basic protocol against semi-honest adversaries.

Efficiency and the GMW construction. The complexity of the GMW protocol for semi-honest adversaries is related to the size of the circuit needed to compute the functionality. Specifically, the parties need to run an oblivious transfer (involving asymmetric computations) for every circuit of the gate. Although this results in a significant computational overhead, it is reasonable for functionalities with circuits that are not too large. We remark that Yao's protocol for secure two-party computation is typically more efficient than the GMW construction. This is due to the fact that only symmetric operations are needed for computing every gate of the circuit, and oblivious transfers are only used for the input bits. This means that Yao's protocol scales better

---

of [25], the receiver gives the sender two images of an enhanced trapdoor permutation, where the receiver knows only one of the preimages. The protocol works so that the receiver obtains x_i if it knows the ith preimage (and otherwise it learns nothing of the value of x_i). Thus, were the receiver to know both preimages, it would learn both x_0 and x_1, in contradiction to the security of the protocol. Now, if the receiver can "alter" its random tape, then it can influence the choice of the images of the permutation so that it knows both preimages. Thus, the fact that the receiver uses a truly random tape is crucial to the security.

to large circuits than GMW. In Chapter 3 we present Yao's protocol for semi-honest adversaries in detail.

As we have mentioned, the cost of achieving security in the presence of semi-honest adversaries is not insignificant. However, it is orders of magnitude less than the cost of achieving security in the presence of malicious adversaries. This is due to the fact that general zero-knowledge protocols for $\mathcal{NP}$ require a Karp reduction from a complex computational statement to a language like 3-colorability or Hamiltonicity. Thus, even when the underlying protocol for semi-honest adversaries is highly efficient, the compiled protocol for malicious adversaries is typically not. We conclude that the GMW construction, and in particular the compilation of GMW from security in the presence of semi-honest adversaries to security in the presence of malicious adversaries, is to be viewed as a fundamental feasibility result, and not as a methodology for obtaining protocols in practice.[2]

Despite what we have stated above, the GMW compilation paradigm *has* had considerable influence over the construction of *efficient protocols*. Indeed, one way to efficiently achieve security in the presence of malicious adversaries is to design a protocol that is secure in the presence of semi-honest adversaries (or a different notion that is weaker than security in the presence of malicious adversaries) in a particular way so that one can efficiently prove "correct behavior" in zero-knowledge. One example of a protocol that uses this paradigm can be found in Section 7.4.

In conclusion, the GMW construction proves that any efficient functionality can be securely computed, even in the presence of a powerful malicious adversary. The next step, given this feasibility result, is to construct more efficient protocols for this task with the final aim of obtaining protocols that can be used in practice. This research goal is the focus of this book.

1.3 A Roadmap to the Book

This book is divided into three distinct parts. We now describe in detail the contents of each part and the chapters therein.

1.3.1 Part I – Introduction and Definitions

In this chapter, we have provided a brief overview of the basic notions, concepts and results of secure computation. The aim of this overview is to place

---

[2] We stress that this should in no way be interpreted as a criticism of GMW; the GMW construction is a beautiful proof of the feasibility of achieving secure computation and is one of most fundamental results of theoretical cryptography.

the material covered in this book in its general context. In Chapter 2 we present a number of different definitions of secure two-party computation. We begin by presenting the classic definitions of security in the presence of semi-honest and malicious adversaries. As we have discussed above, on the one hand, the security guarantee provided when considering semi-honest adversaries is often insufficient. On the other hand, although protocols that are secure in the presence of malicious adversaries provide a very strong security guarantee, they are often highly inefficient. This motivates the search for alternative definitions that provide satisfactory security guarantees, and that are more amenable to constructing highly efficient protocols. We consider three such relaxations:

1. *Covert adversaries (Section 2.4):* Physical security in the real world is achieved via deterrence. It is well known that an expert thief can break into almost anybody's house and can steal most cars. If this is the case, then why aren't there more expert thieves and why are most of our houses and cars safe? The answer to this is simply deterrence: most people do not want to go to jail and so choose professions that are within the law. (Of course, there are also many people who do not steal because it is immoral, but this is not relevant to our discussion here.) The notion of security in the presence of covert adversaries utilizes the concept of deterrence in secure computation. Specifically, a protocol that achieves security under this notion does not provide a foolproof guarantee that an adversary cannot cheat. Rather, it guarantees that if an adversary does attempt to cheat, then the honest parties will detect this with some given probability (say 0.5 or 0.9). Now, if such a protocol is run in a context where cheating can be penalized, then this level of security can suffice. For example, if a secure protocol is used by a consortium of cellphone companies who wish to carry out a statistical analysis of the usage behaviors of cellphone users, then a cheating company (who tries to steal customer data from its competitors) will be penalized by removing them from the consortium.

   We remark that security in the presence of malicious adversaries is the analogue of an armed security guard outside your house 24 hours a day. It is much safer to protect your house in this way. However, the costs involved are often not worth the gain.

2. *Non-simulation based definitions (Section 2.6):* The definitions of security for semi-honest, malicious and covert adversaries all follow the ideal/real-model simulation-based paradigm. We consider two relaxations that do not follow this paradigm.

   a. *Privacy only (Section 2.6.1):* As we have discussed, the simulation-based method of defining security (via the ideal/real-model paradigm) guarantees privacy, correctness, independence of inputs and more. However, in some cases, it may suffice to guarantee privacy without the other properties. For example, if a user wishes to search a database so that her search queries are kept private, then privacy alone may suffice.

b. *One-sided simulation (Section 2.6.2):* In many cases, it is very difficult to formalize a definition of security that guarantees privacy only. This is due to the fact that when a party receives output it learns something and we must try to state that it should learn nothing more. However, the output depends on the parties' inputs and if these are not explicit then it is unclear what the output should be. In contrast, it is very easy to define privacy when nothing should be learned; in such a case, privacy can be formalized via indistinguishability in the same way as encryption. The notion of one-sided simulation helps to define security for protocol problems in which only one party is supposed to receive output. In such a case, we require simulation (via the ideal/real-model paradigm) for the party that receives input, and privacy only (via indistinguishability) for the party that does not receive output. Observe that correctness and independence of inputs are not guaranteed when the party who does not receive output is corrupted. However, as in the example for privacy only above, this is sometimes sufficient. The advantage of "one-sided simulation" over "privacy only" is that a general definition can be given for any functionality in which only one party receives output.

We stress that the "right definition" depends very much on the application being considered. In some cases, it is crucial that security in the presence of malicious adversaries be achieved; take for example computation over highly confidential data that can cause significant damage if revealed. However, in many other cases, weaker notions of security can suffice, especially if the alternative is to not use a secure protocol at all (e.g., as may be the case if the best protocols known for a task that provide security for malicious adversaries are not efficient enough for use).

In addition to presenting the above definitions of security, Chapter 2 contains the following additional material. In Section 2.3.3 we discuss the surprising fact that due to a quirk in the definitions, security in the presence of malicious adversaries does not always imply security in the presence of semi-honest adversaries. In Section 2.5 we show that in many cases it suffices to consider restricted types of functionalities, enabling a simpler presentation. Finally, in Section 2.7 we state modular sequential composition theorems that are very useful when proving the security of protocols.

1.3.2 Part II – General Constructions

A general construction is a protocol that can be used for securely computing any functionality. These constructions are typically based on a circuit for computing the functionality, and as such do not utilize any special properties of the functionality being computed. Thus, they cannot be used for complex computations applied to very large inputs. Despite this, it is important to study these constructions for the following reasons. First, many useful techniques and methodologies can be learned from them. Second, in many cases, a larger protocol uses a smaller subprotocol that is obtained via a general construction (an example of this is given in Chapter 8). Finally, as the efficiency of general constructions improves, we are able to use them for more and more real problems [22, 71].

Semi-honest adversaries. In Chapter 3 we present Yao's protocol for achieving secure two-party computation in the presence of semi-honest adversaries [77]. This protocol works by having one party prepare an encrypted or garbled version of the circuit that can be decrypted to yield only one value, the output of the computation. When the circuit being computed is not too large, this protocol is very efficient. Specifically, the parties need $O(1)$ asymmetric computations per input bit, and $O(1)$ symmetric computations per gate of the circuit. In practice, symmetric computations are far more efficient than asymmetric computations. Thus, a circuit with hundreds of thousands of gates can be easily computed.

Malicious adversaries. In Chapter 4 we present a protocol that achieves security in the presence of malicious adversaries [55]. This protocol is based on Yao's protocol for the semi-honest case, and includes significant machinery for preventing the parties from cheating. The basic technique for achieving this is called cut-and-choose. Specifically, one of the main problems that arises when running Yao's protocol with malicious adversaries is that the party who constructs the garbled circuit can construct it incorrectly (since it is encrypted, this cannot be detected). In order to prevent such behavior, we have the party construct many copies of the circuit and then ask it to open half of them. In this way, we can be sure that most of the remaining unopened circuits are correct. It turns out that this intuitive idea is very hard to implement correctly, and many new problems arise when computing with many circuits. As a result, the construction is much less efficient than in the semi-honest case. However, it can still be run on circuits with tens of thousands of gates, as will be discussed below.

Covert adversaries. In Chapter 5 we present a protocol that is based on the same idea as that in Chapter 4 but provides security only in the presence of covert adversaries. The main idea is that in the context of covert adversaries it suffices to use cut-and-choose on many fewer circuits, and it suffices to compute only one circuit at the end. This results in a protocol that is much more efficient than that required to achieve security in the presence of malicious adversaries. Roughly speaking, when the adversary is guaranteed to be caught with probability ϵ if it attempts to cheat, the cost of the protocol is about $O(1/\epsilon)$ times the cost of Yao's semi-honest protocol. Thus, for $\epsilon = 1/2$ it is possible to compute circuits that contain hundreds of thousands of gates, as in the semi-honest case.

Implementations of general protocols. Recent interest in the field of efficient protocols has led to implementations that are useful for understanding the real efficiency behavior of the above protocols. One work which is of relevance here is an implementation of a protocol for securely computing the AES function [71]. That is, one party holds a secret 128-bit symmetric key k for the AES function and the other party holds a 128-bit input x. The computation is such that the first party learns nothing about x, while the second party learns $AES_k(x)$ and nothing else. Such a functionality has many applications, as we will see in Chapter 9. In [71], the exact protocols of Chapters 3, 4 and 5 were implemented for a circuit computing AES which has approximately 33,000 gates. The protocols were implemented using a number of different optimizations. The best optimizations yielded running times of seven seconds for the semi-honest protocol, 95 seconds for the covert protocol and 1,148 seconds for the malicious protocol. Although these running times are not fast enough for real-time applications, they demonstrate that it is feasible to carry out such computations on circuits that are large (tens of thousands of gates). We expect that further efficiency improvements will not be long coming, and believe that these times will be significantly reduced in the not too distant future (especially for the malicious case).

1.3.3 Part III – Specific Constructions

As we have mentioned, the drawback of considering general constructions is that it is not possible to utilize special properties of the functionality being computed. In the final part of the book, we present protocols for specific problems of interest. This part is also divided into two subparts. First, in Chapters 6 and 7 we present some basic tools that are very useful for designing efficient protocols. Then, in Chapters 8 and 9 we study two specific problems as a demonstration of how higher-level protocols can be constructed.

Sigma protocols and efficient zero-knowledge. In Chapter 6 we show how highly efficient zero-knowledge protocols can be constructed. As we have discussed, security in the presence of malicious adversaries is typically achieved by forcing the parties to behave honestly. The immediate way to do this is to force the parties to prove in zero-knowledge that they are following the protocol specification. Needless to say, a straightforward implementation of this idea is very inefficient. For this reason, many try to stay clear of explicit zero-knowledge proofs for enforcing honest behavior. However, in many cases it *is* possible to construct a protocol for which the zero-knowledge proof that is needed is highly efficient. Many of these efficient zero-knowledge protocols are constructed from a simpler primitive called a Σ-protocol. In Chapter 6 we study Σ-protocols in depth and, among other things, present highly efficient generic transformations from Σ-protocols to zero-knowledge proofs of membership and zero-knowledge proofs of knowledge. These transformations are very useful because it is far easier to construct a protocol and prove that it is a Σ-protocol than to construct a protocol and prove that it is a zero-knowledge proof of knowledge.

Oblivious transfer and applications. In Chapter 7 we construct oblivious transfer protocols that are secure under the definitions of privacy only, one-sided simulation, and full simulation-based security in the presence of malicious adversaries. The protocols that are presented progress in a natural way from privacy only through one-sided simulation to full security. The final protocols obtained have only a constant number of exponentiations and as such are very efficient. In addition, we present optimizations for the case where many oblivious transfers need to be run, which is the case in many secure protocols using oblivious transfer. We then conclude with protocols for pseudorandom function evaluation, which is a primitive that also has many applications.

The kth-ranked element and search problems. In Chapters 8 and 9 we show how to securely compute the kth-ranked element of two lists (with a special case being the median) and how to search databases and documents in a secure manner. Admittedly, the choice of these two problems is arbitrary and is based on our personal preferences. Nevertheless, we believe that they are interesting examples of how specific properties of the functionality in question can be used to solve the problem with high efficiency.

Chapter 2
Definitions

In this chapter we present a number of definitions of security for secure computation. Specifically, in Sections 2.2 to 2.4 we present definitions of security for semi-honest, malicious and covert adversaries; all these definitions are based on the ideal/real-model paradigm for formulating security. We begin with the classic definitions of security in the presence of semi-honest and malicious adversaries, and then proceed to the more recent notion of security in the presence of covert adversaries. In Section 2.5, we show that it often suffices to consider restricted types of functionalities, which enables us to simplify the presentation of the general protocols in Chapters 3 to 5. In Section 2.6 we consider two relaxations of these definitions, for the case of malicious adversaries. Finally, in Section 2.7 we conclude with the issue of sequential composition of secure protocols. We stress that since the focus of this book is secure *two-party* computation, all of the definitions are presented for the case of two parties only.

2.1 Preliminaries

We begin by introducing notation and briefly reviewing some basic notions; see [30] for more details. A function $\mu(\cdot)$ is negligible in n, or just negligible, if for every positive polynomial $p(\cdot)$ and all sufficiently large ns it holds that $\mu(n) < 1/p(n)$. A probability ensemble $X = \{X(a,n)\}_{a \in \{0,1\}^*; n \in \mathbb{N}}$ is an infinite sequence of random variables indexed by a and $n \in \mathbb{N}$. (The value a will represent the parties' inputs and n will represent the security parameter.) Two distribution ensembles $X = \{X(a,n)\}_{a \in \{0,1\}^*; n \in \mathbb{N}}$ and $Y = \{Y(a,n)\}_{a \in \{0,1\}^*; n \in \mathbb{N}}$ are said to be computationally indistinguishable, denoted by $X \stackrel{c}{\equiv} Y$, if for every non-uniform polynomial-time algorithm D there exists a negligible function $\mu(\cdot)$ such that for every $a \in \{0,1\}^*$ and every $n \in \mathbb{N}$,

$$|\Pr[D(X(a,n)) = 1] - \Pr[D(Y(a,n)) = 1]| \leq \mu(n).$$

C. Hazay, Y. Lindell, *Efficient Secure Two-Party Protocols*,
Information Security and Cryptography, DOI 10.1007/978-3-642-14303-8_2,
© Springer-Verlag Berlin Heidelberg 2010

All parties are assumed to run in time that is polynomial in the security parameter. (Formally, each party has a security parameter tape upon which that value 1^n is written. Then the party is polynomial in the input on this tape. We note that this means that a party may not even be able to read its entire input, as would occur in the case where its input is longer than its overall running time.) We sometimes use PPT as shorthand for *probabilistic polynomial time*.

For a set X, we denote by $x \leftarrow_R X$ the process of choosing an element x of X under the uniform distribution.

2.2 Security in the Presence of Semi-honest Adversaries

The model that we consider here is that of two-party computation in the presence of *static semi-honest* adversaries. Such an adversary controls one of the parties (statically, and so at the onset of the computation) and follows the protocol specification exactly. However, it may try to learn more information than allowed by looking at the transcript of messages that it received and its internal state. Since we only consider static semi-honest adversaries here, we will sometimes omit the qualification that security is with respect to such adversaries only. The definitions presented here are according to Goldreich in [32].

Two-party computation. A two-party protocol problem is cast by specifying a random process that maps pairs of inputs to pairs of outputs (one for each party). We refer to such a process as a functionality and denote it $f : \{0,1\}^* \times \{0,1\}^* \rightarrow \{0,1\}^* \times \{0,1\}^*$, where $f = (f_1, f_2)$. That is, for every pair of inputs $x, y \in \{0,1\}^n$, the output-pair is a random variable $(f_1(x,y), f_2(x,y))$ ranging over pairs of strings. The first party (with input x) wishes to obtain $f_1(x,y)$ and the second party (with input y) wishes to obtain $f_2(x,y)$. We often denote such a functionality by $(x,y) \mapsto (f_1(x,y), f_2(x,y))$. Thus, for example, the oblivious transfer functionality [72] is specified by $((z_0, z_1), \sigma) \mapsto (\lambda, z_\sigma)$, where λ denotes the empty string. When the functionality f is probabilistic, we sometimes use the notation $f(x,y,r)$, where r is a uniformly chosen random tape used for computing f.

Privacy by simulation. Intuitively, a protocol is secure if whatever can be computed by a party participating in the protocol can be computed based on its input and output only. This is formalized according to the simulation paradigm. Loosely speaking, we require that a party's *view* in a protocol execution be simulatable given only its input and output. This then implies that the parties learn nothing from the protocol *execution* itself, as desired.

Definition of security. We begin with the following notation:

- Let $f = (f_1, f_2)$ be a probabilistic polynomial-time functionality and let π be a two-party protocol for computing f.

- The view of the ith party ($i \in \{1,2\}$) during an execution of π on (x,y) and security parameter n is denoted by $\text{view}_i^\pi(x,y,n)$ and equals $(w, r^i, m_1^i, ..., m_t^i)$, where $w \in \{x,y\}$ (its value depending on the value of i), r^i equals the contents of the ith party's *internal* random tape, and m_j^i represents the jth message that it received.
- The output of the ith party during an execution of π on (x,y) and security parameter n is denoted by $\text{output}_i^\pi(x,y,n)$ and can be computed from its own view of the execution. We denote the joint output of both parties by $\text{output}^\pi(x,y,n) = (\text{output}_1^\pi(x,y,n), \text{output}_2^\pi(x,y,n))$.

Definition 2.2.1 (security w.r.t. semi-honest behavior): *Let $f = (f_1, f_2)$ be a functionality. We say that π securely computes f in the presence of static semi-honest adversaries if there exist probabilistic polynomial-time algorithms S_1 and S_2 such that*

$$\{(S_1(1^n, x, f_1(x,y)), f(x,y))\}_{x,y,n} \overset{c}{\equiv} \{(\text{view}_1^\pi(x,y,n), \text{output}^\pi(x,y,n))\}_{x,y,n},$$

$$\{(S_2(1^n, y, f_2(x,y)), f(x,y))\}_{x,y,n} \overset{c}{\equiv} \{(\text{view}_2^\pi(x,y,n), \text{output}^\pi(x,y,n))\}_{x,y,n},$$

$x, y \in \{0,1\}^*$ *such that* $|x| = |y|$, *and* $n \in \mathbb{N}$.

The above states that the view of a party can be simulated by a probabilistic polynomial-time algorithm given access to the party's *input and output only*. We emphasize that the adversary here is semi-honest and therefore its view in the execution of π is exactly as in the case where both parties follow the protocol specification. We note that it is not enough for the simulator S_i to generate a string indistinguishable from $\text{view}_i^\pi(x,y)$. Rather, the *joint distribution* of the simulator's output and the functionality output $f(x,y)$ must be indistinguishable from $(\text{view}_i^\pi(x,y), \text{output}^\pi(x,y))$. This is necessary for probabilistic functionalities; see [11, 32] for a full discussion.

A simpler formulation for deterministic functionalities. In the case where the functionality f is deterministic, a simpler definition can be used. Specifically, we do not need to consider the joint distribution of the simulator's output with the protocol output. Rather we separately require correctness, meaning that

$$\{\text{output}^\pi(x,y,n))\}_{x,y \in \{0,1\}^*; n \in \mathbb{N}} \overset{c}{\equiv} \{f(x,y)\}_{x,y \in \{0,1\}^*}$$

and, in addition, that there exist PPT S_1 and S_2 such that

$$\{S_1(1^n, x, f_1(x,y))\}_{x,y \in \{0,1\}^*; n \in \mathbb{N}} \overset{c}{\equiv} \{\text{view}_1^\pi(x,y,n)\}_{x,y \in \{0,1\}^*; n \in \mathbb{N}}, \quad (2.1)$$

$$\{S_2(1^n, y, f_2(x,y))\}_{x,y \in \{0,1\}^*; n \in \mathbb{N}} \overset{c}{\equiv} \{\text{view}_2^\pi(x,y,n)\}_{x,y \in \{0,1\}^*; n \in \mathbb{N}} \quad (2.2)$$

The reason this suffices is that when f is deterministic, $\text{output}^\pi(x,y,n)$ must equal $f(x,y)$. Furthermore, the distinguisher for the ensembles can compute

$f(x, y)$ by itself (because it is given x and y, the indices of the ensemble). See [32, Section 7.2.2] for more discussion.

For simplicity of notation, we will often let n be the length of x and y. In this case, the simulators S_1 and S_2 do not need to receive 1^n for input, and we omit n from the VIEW and OUTPUT notations.

An equivalent definition. A different definition of security for two-party computation in the presence of semi-honest adversaries compares the output of a real protocol execution to the output of an *ideal* computation involving an incorruptible trusted third party (as described in the Introduction). The trusted party receives the parties' inputs, computes the functionality on these inputs and returns to each its respective output. Loosely speaking, a protocol is secure if any real-model adversary can be converted into an ideal-model adversary such that the output distributions are computationally indistinguishable. We remark that in the case of semi-honest adversaries, this definition is equivalent to the (simpler) simulation-based definition presented here; see [32]. This formulation of security will be used for defining security in the presence of malicious adversaries below.

Augmented semi-honest adversaries. Observe that by the definition above, a semi-honest party *always* inputs its prescribed input value, even if it is corrupted. We argue that it often makes sense to allow a corrupted semi-honest party to modify its input, as long as it does so before the execution begins. This is due to the following reasons. First, on a subjective intuitive level it seems to us that this is in the spirit of semi-honest behavior because choosing a different input is not "improper behavior". Second, when protocols achieving security in the presence of semi-honest adversaries are used as a stepping stone for obtaining security in the presence of malicious adversaries, it is necessary to allow the semi-honest adversary to modify its input. Indeed, Goldreich introduces the notion of an *augmented semi-honest adversary* that may modify its input before the execution begins when showing how to obtain security against malicious adversaries from protocols that are secure only in the presence of semi-honest adversaries [32, Sec. 7.4.4.1]. Finally, as we discuss in Section 2.3.3, it is natural that any protocol that is secure in the presence of malicious adversaries also be secure in the presence of semi-honest adversaries. Although very counterintuitive, it turns out that this only holds when the semi-honest adversary is allowed to change its input; see Section 2.3.3 for a full discussion. We present the definition of semi-honest adversaries above, where a corrupted party cannot change its input, for historical reasons only. However, we strongly prefer the notion of augmented semi-honest adversaries. We remark that a formal definition of this notion is easily obtained via the ideal/real-model paradigm; see Section 2.3 below.

2.3 Security in the Presence of Malicious Adversaries

In this section, we present the definition of security for the case of malicious adversaries who may use any efficient attack strategy and thus may arbitrarily deviate from the protocol specification. In this case, it does not suffice to construct simulators that can generate the view of the corrupted party. First and foremost, the generation of such a view depends on the actual input used by the adversary; indeed this input affects the actual output received. However, in contrast to the case of semi-honest adversaries, the adversary may not use the input that it is provided. Thus, a simulator for the case where P_1 is corrupted cannot just take x and $f(x, y)$ and generate a view (in order to prove that nothing more than the output is learned), because the adversary may not use x at all. Furthermore, beyond the possibility that a corrupted party may learn more than it should, we require correctness (meaning that a corrupted party cannot cause the output to be incorrectly distributed) and independence of inputs (meaning that a corrupted party cannot make its input depend on the other party's input). As discussed in the overview in Section 1.1, in order to capture these threats, and others, the security of a protocol is analyzed by comparing what an adversary can do in the protocol to what it can do in an ideal scenario that is secure by definition. This is formalized by considering an *ideal* computation involving an incorruptible *trusted third party* to whom the parties send their inputs. The trusted party computes the functionality on the inputs and returns to each party its respective output. Loosely speaking, a protocol is secure if any adversary interacting in the real protocol (where no trusted third party exists) can do no more harm than if it were involved in the above-described ideal computation. See [11, 32] for more discussion on the advantages of this specific formulation.

We remark that since we consider the two-party case, there is no honest majority. It is therefore impossible to achieve fairness in general. Therefore, in the ideal setting we allow the adversary to obtain the corrupted party's output, without the honest party necessarily obtaining its output. We also remark that in defining security for two parties it is possible to consider only the setting where one of the parties is corrupted, or to also consider the setting where none of the parties are corrupted, in which case the adversary seeing the transcript between the parties should learn nothing. Since this latter case can easily be achieved by using encryption between the parties we present the simpler formulation security that assumes that exactly one party is always corrupted.

2.3.1 The Definition

Execution in the ideal model. As we have mentioned, in the case of no honest majority, it is in general impossible to achieve guaranteed output delivery and fairness. This "weakness" is therefore incorporated into the ideal model by allowing the adversary in an ideal execution to abort the execution or obtain output without the honest party obtaining its output. Denote the participating parties by P_1 and P_2 and let $i \in \{1, 2\}$ denote the index of the corrupted party, controlled by an adversary $\mathcal{A}$. An ideal execution for a function $f : \{0, 1\}^* \times \{0, 1\}* \to \{0, 1\}^* \times \{0, 1\}*$ proceeds as follows:

Inputs: Let x denote the input of party P_1, and let y denote the input of party P_2. The adversary $\mathcal{A}$ also has an auxiliary input denoted by z.

Send inputs to trusted party: The honest party P_j sends its received input to the trusted party. The corrupted party P_i controlled by $\mathcal{A}$ may either abort (by replacing the input with a special abort$_i$ message), send its received input, or send some other input of the same length to the trusted party. This decision is made by $\mathcal{A}$ and may depend on the input value of P_i and the auxiliary input z. Denote the pair of inputs sent to the trusted party by (x', y') (note that if $i = 2$ then $x' = x$ but y' does not necessarily equal y, and vice versa if $i = 1$).

Early abort option: If the trusted party receives an input of the form abort$_i$ for some $i \in \{1, 2\}$, it sends abort$_i$ to all parties and the ideal execution terminates. Otherwise, the execution proceeds to the next step.

Trusted party sends output to adversary: At this point the trusted party computes $f_1(x', y')$ and $f_2(x', y')$ and sends $f_i(x', y')$ to party P_i (i.e., it sends the corrupted party its output).

Adversary instructs trusted party to continue or halt: $\mathcal{A}$ sends either continue or abort$_i$ to the trusted party. If it sends continue, the trusted party sends $f_j(x', y')$ to party P_j (where P_j is the honest party). Otherwise, if $\mathcal{A}$ sends abort$_i$, the trusted party sends abort$_i$ to party P_j.

Outputs: The honest party always outputs the output value it obtained from the trusted party. The corrupted party outputs nothing. The adversary $\mathcal{A}$ outputs any arbitrary (probabilistic polynomial-time computable) function of the initial input of the corrupted party, the auxiliary input z, and the value $f_i(x', y')$ obtained from the trusted party.

Let $f : \{0, 1\}^* \times \{0, 1\}^* \to \{0, 1\}^* \times \{0, 1\}^*$ be a two-party functionality, where $f = (f_1, f_2)$, let $\mathcal{A}$ be a non-uniform probabilistic polynomial-time machine, and let $i \in \{1, 2\}$ be the index of the corrupted party. Then, the ideal execution of f on inputs (x, y), auxiliary input z to $\mathcal{A}$ and security parameter n, denoted by $\text{IDEAL}_{f, \mathcal{A}(z), i}(x, y, n)$, is defined as the output pair of the honest party and the adversary $\mathcal{A}$ from the above ideal execution.

Execution in the real model. We next consider the real model in which a real two-party protocol π is executed (and there exists no trusted third party).

In this case, the adversary $\mathcal{A}$ sends all messages in place of the corrupted party, and may follow an arbitrary polynomial-time strategy. In contrast, the honest party follows the instructions of π.

Let f be as above and let π be a two-party protocol for computing f. Furthermore, let $\mathcal{A}$ be a non-uniform probabilistic polynomial-time machine and let $i \in \{1,2\}$ be the index of the corrupted party. Then, the real execution of π on inputs (x,y), auxiliary input z to $\mathcal{A}$ and security parameter n, denoted by $\text{REAL}_{\pi,\mathcal{A}(z),i}(x,y,n)$, is defined as the output pair of the honest party and the adversary $\mathcal{A}$ from the real execution of π.

Security as emulation of a real execution in the ideal model. Having defined the ideal and real models, we can now define security of protocols. Loosely speaking, the definition asserts that a secure party protocol (in the real model) emulates the ideal model (in which a trusted party exists). This is formulated by saying that adversaries in the ideal model are able to simulate executions of the real-model protocol.

Definition 2.3.1 (secure two-party computation): *Let f and π be as above. Protocol π is said to* securely compute f with abort *in the presence of malicious adversaries if for every non-uniform probabilistic polynomial-time adversary $\mathcal{A}$ for the real model, there exists a non-uniform probabilistic polynomial-time adversary $\mathcal{S}$ for the ideal model, such that for every $i \in \{1,2\}$,*

$$\left\{ \text{IDEAL}_{f,\mathcal{S}(z),i}(x,y,n) \right\}_{x,y,z,n} \overset{c}{\equiv} \left\{ \text{REAL}_{\pi,\mathcal{A}(z),i}(x,y,n) \right\}_{x,y,z,n}$$

where $x,y \in \{0,1\}^*$ under the constraint that $|x| = |y|$, $z \in \{0,1\}^*$ and $n \in \mathbb{N}$.

The above definition assumes that the parties (and adversary) know the input lengths (this can be seen from the requirement that $|x| = |y|$ is balanced and so all the inputs in the vector of inputs are of the same length). We remark that some restriction on the input lengths is unavoidable because, as in the case of encryption, to some extent such information is always leaked.

2.3.2 Extension to Reactive Functionalities

Until now we have considered the secure computation of simple functionalities that compute a single pair of outputs from a single pair of inputs. However, not all computations are of this type. Rather, many computations have multiple rounds of inputs and outputs. Furthermore, the input of a party in a given round may depend on its output from previous rounds, and the outputs of that round may depend on the inputs provided by the parties in some or all of the previous rounds. A classic example of this is electronic poker. In this game, in the first phase cards are dealt to the players. Based on these cards, bets are made and cards possibly thrown and dealt. The important thing to notice is that in each round human decisions must be made based on the current status. Thus, new *inputs* are provided in each round (e.g., how much to bet and what cards to throw), and these inputs are based on the current *output* (in this case, the output is the player's current hand and the cards previously played). A more cryptographic example of a multi-phase functionality is that of a commitment scheme. Such a scheme has a distinct commitment and decommitment phase. Thus, it cannot be cast as a standard functionality mapping inputs to outputs.

In the context of secure computation, multi-phase computations are typically called reactive functionalities. Such functionalities can be modeled as a series of functions $(f^1, f^2, \ldots)$ where each function receives some state information and two new inputs. That is, the input to function f^j consists of the inputs (x_j, y_j) of the parties in this phase, along with a state input σ_{j-1} output by f^{j-1}. Then, the output of f^j is defined to be a pair of outputs $f_1^j(x_j, y_j, \sigma_{j-1})$ for P_1 and $f_2^j(x_j, y_j, \sigma_{j-1})$ for P_2, and a state string σ_j to be input into f^{j+1}. We stress that the parties receive only their private outputs, and in particular do *not* receive any of the state information; in the ideal model this is stored by the trusted party. Although the above definition is intuitively clear, a simpler formulation is to define a reactive functionality via a multi-phase probabilistic polynomial-time Turing machine that receives inputs and generates outputs (this is simpler because it is not necessary to explicitly define the state at every stage). The trusted party then runs this machine upon each new pair of inputs it receives and sends the generated outputs. In this formulation, the state information is kept internally by the Turing machine, and not explicitly by the trusted party. We remark that the formal ideal model remains the same, except that the trusted party runs a reactive functionality (i.e., reactive Turing machine) instead of a single function. In addition, once the corrupted party sends abort_i, the ideal execution stops, and no additional phases are run.

2.3.3 Malicious Versus Semi-honest Adversaries

At first sight, it seems that any protocol that is secure in the presence of malicious adversaries is also secure in the presence of semi-honest adversaries. This is because a semi-honest adversary is just a "special case" of a malicious adversary who faithfully follows the protocol specification. Although this is what we would expect, it turns out to be *false* [45]. This anomaly is due to the fact that although a real semi-honest adversary is indeed a special case of a real malicious adversary, this is not true of the respective adversaries in the ideal model. Specifically, the adversary in the ideal model for malicious adversaries is allowed to change its input, whereas the adversary in the ideal model for semi-honest adversary is not. Thus, the adversary/simulator for the case of malicious adversaries has more power than the adversary/simulator for the case of semi-honest adversaries. As such, it may be possible to simulate a protocol in the malicious model, but not in the semi-honest model. We now present two examples of protocols where this occurs.

Example 1 – secure AND. Consider the case of two parties computing the *binary AND* function $f(x, y) = x \wedge y$, where only party P_2 receives output. Note first that if party P_2 uses input 1, then by the output received it can fully determine party P_1's input (if the output is 0 then P_1 had input 0, and otherwise it had input 1). In contrast, if party P_2 uses input 0 then it learns nothing about P_1's input, because the output equals 0 irrespective of the value of P_1's input. The result of this observation is that in the ideal model, an adversary corrupting P_2 can always learn P_1's exact input by sending the trusted party the input value 1. Thus, P_1's input is always revealed. In contrast, in the ideal model with a semi-honest adversary, P_1's input is only revealed if the corrupted party has input 1; otherwise, the adversary learns nothing whatsoever about P_1's input. We use the above observations to construct a protocol that securely computes the binary AND function in the presence of malicious adversaries, but is not secure in the presence of semi-honest adversaries; see Protocol 2.3.2.

PROTOCOL 2.3.2 (A Protocol for Binary AND)

- **Input:** P_1 has an input bit x and P_2 has an input bit y.
- **Output:** The binary value $x \wedge y$ for P_2 only.
- **The protocol:**

    1. P_1 sends P_2 its input bit x.
    2. P_2 outputs the bit $x \wedge y$.

We have the following claims:

Claim 2.3.3 *Protocol 2.3.2 securely computes the binary AND function in the presence of malicious adversaries.*

Proof. We separately consider the case where P_1 is corrupted and the case where P_2 is corrupted. If P_1 is corrupted, then the simulator $\mathcal{S}$ receives from $\mathcal{A}$ the bit that it sends to P_2 in the protocol. This bit fully determines the input of P_1 to the function and so $\mathcal{S}$ just sends it to the trusted party, thereby completing the simulation. In the case where P_2 is corrupted, $\mathcal{S}$ sends input 1 to the trusted party and receives back an output bit b. By the observation above, b is the input of the honest P_1 in the ideal model. Thus, the simulator $\mathcal{S}$ just hands $\mathcal{A}$ the bit $x = b$ as the value that $\mathcal{A}$ expects to receive from the honest P_1 in a real execution. It is immediate that the simulation here is perfect. ■

We stress that the above works because P_2 is the only party to receive output. If P_1 also were to receive output, then $\mathcal{S}$'s simulation in the case of a corrupted P_2 would not work. In order to see this, consider an adversary who corrupts P_2, uses input $y = 0$ and outputs its view in the protocol, including the bit x that it receives from P_1. In this case, S cannot send $y = 1$ to the trusted party because P_1's output would not be correctly distributed. Thus, it must send $y = 0$, in which case the view that it generates for $\mathcal{A}$ cannot always be correct because it does not know the input bit x of P_1.

Claim 2.3.4 *Protocol 2.3.2 does not securely compute the binary AND function in the presence of semi-honest adversaries.*

Proof. Consider the simulator S_2 that is guaranteed to exist for the case where P_2 is corrupted; see (2.2) in Section 2.2. Then, S_2 is given y and $x \wedge y$ and must generate the view of P_2 in the computation. However, this view contains the value x that P_1 sends to P_2 in the protocol. Now, if $y = 0$ and x is random, then there is no way that S_2 can guess the value of x with probability greater than $1/2$. We conclude that the protocol is not secure in the presence of semi-honest adversaries. ∎

Example 2 – set union. Another example where this arises is the problem of set union over a large domain where only one party receives output. Specifically, consider the function $f(X, Y) = (\lambda, X \cup Y)$ where $X, Y \subseteq \{0, 1\}^n$ are sets of the same size, and λ denotes the "empty" output. We claim that the protocol where P_1 sends its set X to P_2 is secure in the presence of malicious adversaries. This follows for the exact same reasons as above because a corrupted P_2 in the malicious model can replace its input set Y with a set Y' of the same size, but containing random values. Since the sets contain values of length n, it follows that the probability that $X \cap Y \neq \phi$ is negligible. Thus, the output that P_2 receives completely reveals the input of P_1. In contrast, if a corrupted party cannot change its input, then when $X \cap Y \neq \phi$ the elements that are common to both sets are hidden. Specifically, if five elements are common to both sets, then P_2 knows that there are five common elements, but does not have any idea as to which are common. Thus, for the same reasons as above, the protocol is not secure in the presence of semi-honest adversaries. Once again, we stress that this works when only one party receives output; in the case where both parties receive output, securely computing this functionality is highly non-trivial.

Discussion. It is our opinion that the above phenomenon should not be viewed as an "annoying technicality". Rather it points to a problem in the definitions that needs to be considered. Our position is that it would be better to define semi-honest adversaries as adversaries that *are allowed to change their input* before the computation starts (e.g., by rewriting the value on their input tape), and once the computation begins must behave in a semi-honest fashion as before. Conceptually, this makes sense because parties are allowed to choose their own input and this is not adversarial behavior. In addition, this model better facilitates the "compilation" of protocols that are secure in the semi-honest model into protocols that are secure in the malicious model.

Indeed, in order to prove the security of the protocol of [35], and specifically the compilation of a protocol for the semi-honest model into one that is secure in the presence of malicious adversaries, Goldreich introduces the notion of augmented semi-honest behavior, which is exactly as described above; see Definition 7.4.24 in Section 7.4.4.1 of [30]. We stress that all protocols presented in this book that are secure in the presence of semi-honest adversaries are also secure in the presence of *augmented semi-honest adversaries*. Furthermore, as stated in the following proposition, security in the malicious model implies security in the augmented semi-honest model, as one would expect.

Proposition 2.3.5 *Let π be a protocol that securely computes a functionality f in the presence of malicious adversaries. Then π securely computes f in the presence of* augmented *semi-honest adversaries.*

Proof. Let π be a protocol that securely computes f in the presence of malicious adversaries. Let $\mathcal{A}$ be an augmented semi-honest real adversary and let $\mathcal{S}$ be the simulator for $\mathcal{A}$ that is guaranteed to exist by the security of π (for every malicious $\mathcal{A}$ there exists such an $\mathcal{S}$, and in particular for an augmented semi-honest $\mathcal{A}$). We construct a simulator $\mathcal{S}'$ for the augmented semi-honest setting, by simply having $\mathcal{S}'$ run $\mathcal{S}$. However, in order for this to work, we have to show that $\mathcal{S}'$ can do everything that $\mathcal{S}$ can do. In the malicious ideal model, $\mathcal{S}$ can choose whatever input it wishes for the corrupted party; since $\mathcal{S}'$ is *augmented* semi-honest, it too can modify the input. In addition, $\mathcal{S}$ can cause the honest party to output abort. However, $\mathcal{S}'$ *cannot* do this. Nevertheless, this is not a problem because when $\mathcal{S}$ is the simulator for an augmented semi-honest $\mathcal{A}$ it can cause the honest party to output abort with at most negligible probability. In order to see this, note that when two honest parties run the protocol, neither outputs abort with non-negligible probability. Thus, when an honest party runs together with an augmented semi-honest adversary, it too outputs abort with at most negligible probability. This is due to the fact that the distribution over the messages it receives in both cases is identical (because a semi-honest real adversary follows the protocol instructions just like an honest party). This implies that the simulator for the malicious case, when applied to an augmented semi-honest real adversary, causes an abort with at most negligible probability. Thus, the augmented semi-honest simulator can run the simulator for the malicious case, as required. ∎

Given the above, it is our position that the definition of *augmented semi-honest adversaries* is the "right way" of modeling semi-honest behavior. As such, it would have been more appropriate to use this definition from scratch. However, we chose to remain with the standard definition of semi-honest adversaries for historical reasons.

2.4 Security in the Presence of Covert Adversaries

2.4.1 Motivation

In this chapter, we present a relatively new adversary model that lies between the semi-honest and malicious models. The motivation behind the definition is that in many real-world settings, parties are willing to actively cheat (and as such are not semi-honest), but only if they are not caught (and as such they are not arbitrarily malicious). This, we believe, is the case in many business, financial, political and diplomatic settings, where honest behavior cannot be assumed, but where the companies, institutions and individuals involved cannot afford the embarrassment, loss of reputation, and negative press associated with being *caught* cheating. It is also the case, unfortunately, in many social settings, e.g., elections for a president of the country club. Finally, in remote game playing, players may also be willing to actively cheat, but would try to avoid being caught, or else they may be thrown out of the game. In all, we believe that this type of *covert* adversarial behavior accurately models many real-world situations. Clearly, with such adversaries, it may be the case that the risk of being caught is weighed against the benefits of cheating, and it cannot be assumed that players would avoid being caught at any price and under all circumstances. Accordingly, the definition explicitly models the probability of catching adversarial behavior, a probability that can be tuned to the specific circumstances of the problem. In particular, we do not assume that adversaries are only willing to risk being caught with negligible probability, but rather allow for much higher probabilities.

The definition. The definition of security here is based on the ideal/real simulation paradigm (as in the definition in Section 2.3), and provides the guarantee that if the adversary cheats, then it will be caught by the honest parties (with some probability). In order to understand what we mean by this, we have to explain what we mean by "cheating". Loosely speaking, we say that an adversary successfully cheats if it manages to do something that is impossible in the ideal model. Stated differently, successful cheating is behavior that cannot be simulated in the ideal model. Thus, for example, an adversary who learns more about the honest parties' inputs than what is revealed by the output has cheated. In contrast, an adversary who uses pseudorandom coins instead of random coins (where random coins are what are specified in the protocol) has not cheated.

We are now ready to informally describe the guarantee provided by this notion. Let $0 < \epsilon \leq 1$ be a value (called the *deterrence factor*). Then, any attempt to cheat by a real adversary $\mathcal{A}$ is detected by the honest parties with probability at least ϵ. Thus, provided that ϵ is sufficiently large, an adversary that wishes not to be caught cheating will refrain from *attempting* to cheat, lest it be caught doing so. Clearly, the higher the value of ϵ, the greater the probability adversarial behavior is caught and thus the greater the *deterrent* to cheat. This notion is therefore called security in the presence of covert adversaries with ϵ-deterrent. Note that the security guarantee does not preclude successful cheating. Indeed, if the adversary decides to cheat it may gain access to the other parties' private information or bias the result of the computation. The only guarantee is that if it attempts to cheat, then there is a fair chance that it will be caught doing so. This is in contrast to standard definitions, where absolute privacy and security are guaranteed for the given type of adversary. We remark that by setting $\epsilon = 1$, the definition can be used to capture a requirement that cheating parties are always caught.

Formalizing the notion. The standard definition of security (see Definition 2.3.1) is such that all possible (polynomial-time) adversarial behavior is simulatable. Here, in contrast, we wish to model the situation that parties may *successfully* cheat. However, if they do so, they are likely to be caught. There are a number of ways of defining this notion. In order to motivate this one, we begin with a somewhat naive implementation of the notion, and show its shortcomings:

1. *First attempt:* Define an adversary to be covert if the distribution over the messages that it sends during an execution is computationally indistinguishable from the distribution over the messages that an honest party would send. Then, quantify over all covert adversaries $\mathcal{A}$ for the real world (rather than all adversaries). A number of problems arise with this definition.

   - The fact that the distribution generated by the adversary can be distinguished from the distribution generated by honest parties does not mean that the honest parties can detect this in any specific execution. Consider for example a coin-tossing protocol where the honest distribution gives even probabilities to 0 and 1, while the adversary manages to double the probability of the 1 outcome. Clearly, the distributions differ. However, in any given execution, even an outcome of 1 does not provide the honest players with sufficient evidence of any wrongdoing. Thus, it is not sufficient that the *distributions* differ. Rather, one needs to be able to detect cheating in any given execution.
   - The fact that the distributions differ does not necessarily imply that the honest parties have an efficient distinguisher. Furthermore, in order to guarantee that the honest parties detect the cheating, they would have to analyze all traffic during an execution. However, this analysis *cannot* be part of the protocol because then the distinguishers used by the honest parties would be known (and potentially bypassed).
   - Another problem is that adversaries may be willing to risk being caught with more than negligible probability, say 10^{-6}. With such an adversary, the proposed definition would provide no security guarantee. In particular, the adversary may be able to *always* learn all parties' inputs, and risk being caught in one run in a million.

2. *Second attempt.* To solve the aforementioned problems, we first require that the protocol itself be responsible for detecting cheating. Specifically, in the case where a party P_i attempts to cheat, the protocol may instruct the honest parties to output a message saying that "party P_i has cheated" (we require that this only happen if P_i indeed cheated). This solves the first two problems. To solve the third problem, we explicitly quantify the probability that an adversary is caught cheating. Roughly, given a parameter ϵ, a protocol is said to be secure against covert adversaries with ϵ-deterrent if any adversary that is not "covert" (as defined in the first attempt) will necessarily be caught with probability at least ϵ.

   This definition captures the spirit of what we want, but is still problematic. To illustrate the problem, consider an adversary that plays honestly with probability 0.99, and cheats otherwise. Such an adversary can only ever be caught with probability 0.01 (because otherwise it is honest). However, when $\epsilon = 1/2$ for example, such an adversary must be caught with probability 0.5, which is impossible. We therefore conclude that an *absolute* parameter cannot be used, and the probability of catching the adversary must be related to the probability that it cheats.

3. *Final attempt.* We thus arrive at the following approach. First, as mentioned, we require that the protocol itself be responsible for detecting cheating. That is, if a party P_i successfully cheats, then with good probability (ϵ), the honest parties in the protocol will all receive a message that "P_i cheated". Second, we do not quantify only over adversaries that are covert (i.e., those that are not detected cheating by the protocol). Rather, we allow all possible adversaries, even completely malicious ones. Then, we require either that this malicious behavior can be successfully simulated (as in Definition 2.3.1), or that the honest parties receive a message that cheating has been detected, and this happens with probability at least ϵ times the probability that successful cheating takes place. We stress that when the adversary chooses to cheat, it may actually learn secret information or cause some other damage. However, since it is guaranteed that such a strategy will likely be caught, there is strong motivation to refrain from doing so. As such, we use the terminology covert adversaries to refer to malicious adversaries that do not wish to be caught cheating.

The above intuitive notion can be interpreted in a number of ways. We present the main formulation here. The definition works by modifying the ideal model so that the ideal-model adversary (i.e., simulator) is explicitly given the ability to cheat. Specifically, the ideal model is modified so that a special cheat instruction can be sent by the adversary to the trusted party. Upon receiving such an instruction, the trusted party tosses coins and with probability ϵ announces to the honest parties that cheating has taken place (by sending the message corrupted$_i$ where party P_i is the corrupted party that sent the cheat instruction). In contrast, with probability $1 - \epsilon$, the trusted party sends the honest party's input to the adversary, and in addition lets the adversary fix the output of the honest party. We stress that in this case the trusted party does not announce that cheating has taken place, and so the adversary gets off scot-free. Observe that if the trusted party announces that cheating has taken place, then the adversary learns absolutely nothing. This is a strong guarantee because when the adversary attempts to cheat, it must take the risk of being caught and gaining nothing.

2.4.2 The Actual Definition

We begin by presenting the modified ideal model. In this model, we add new instructions that the adversary can send to the trusted party. Recall that in the standard ideal model, the adversary can send a special abort$_i$ message to the trusted party, in which case the honest party receives abort$_i$ as output. In the ideal model for covert adversaries, the adversary can send the following additional special instructions:

- *Special input* corrupted$_i$: If the ideal-model adversary sends corrupted$_i$ instead of an input, the trusted party sends corrupted$_i$ to the honest party and halts. This enables the simulation of behavior by a real adversary that always results in detected cheating. (It is not essential to have this special input, but it sometimes makes proving security easier.)
- *Special input* cheat$_i$: If the ideal-model adversary sends cheat$_i$ instead of an input, the trusted party tosses coins and with probability ϵ determines that this "cheat strategy" by P_i was detected, and with probability $1 - \epsilon$ determines that it was not detected. If it was detected, the trusted party sends corrupted$_i$ to the honest party. If it was not detected, the trusted party hands the adversary the honest party's input and gives the ideal-model adversary the ability to set the output of the honest party to whatever value it wishes. Thus, a cheat$_i$ input is used to model a protocol execution in which the real-model adversary decides to cheat. However, as required, this cheating is guaranteed to be detected with probability at least ϵ. Note that if the cheat attempt is not detected then the adversary is given "full cheat capability", including the ability to determine the honest party's output.

The idea behind the new ideal model is that given the above instructions, the adversary in the ideal model can choose to cheat, with the caveat that its cheating is guaranteed to be detected with probability at least ϵ. We stress that since the capability to cheat is given through an "input" that is provided to the trusted party, the adversary's decision to cheat must be made before the adversary learns anything (and thus independently of the honest party's input and the output).

We are now ready to present the modified ideal model. Let $\epsilon : \mathbb{N} \to [0, 1]$ be a function. Then, the ideal execution for a function $f : \{0,1\}^* \times \{0,1\}^* \to \{0,1\}^* \times \{0,1\}^*$ with parameter ϵ proceeds as follows:

Inputs: Let x denote the input of party P_1, and let y denote the input of party P_2. The adversary $\mathcal{A}$ also has an auxiliary input z.

Send inputs to trusted party: The honest party P_j sends its received input to the trusted party. The corrupted party P_i, controlled by $\mathcal{A}$, may abort (by replacing the input with a special abort$_i$ or corrupted$_i$ message), send its received input, or send some other input of the same length to the trusted party. This decision is made by $\mathcal{A}$ and may depend on the input value of P_i and the auxiliary input z. Denote the pair of inputs sent to the trusted party by (x', y').

Abort options: If a corrupted party sends abort$_i$ to the trusted party as its input, then the trusted party sends abort$_i$ to the honest party and halts. If a corrupted party sends corrupted$_i$ to the trusted party as its input, then the trusted party sends corrupted$_i$ to the honest party and halts.

Attempted cheat option: If a corrupted party sends cheat$_i$ to the trusted party as its input, then the trusted party works as follows:

1. With probability ϵ, the trusted party sends corrupted$_i$ to the adversary and the honest party.
2. With probability $1 - \epsilon$, the trusted party sends undetected to the adversary along with the honest party's input. Following this, the adversary sends the trusted party an output value τ of its choice for the honest party. The trusted party then sends τ to P_j as its output (where P_j is the honest party).

If the adversary sent cheat$_i$, then the ideal execution ends at this point. Otherwise, the ideal execution continues below.

Trusted party sends output to adversary: At this point the trusted party computes $f_1(x', y')$ and $f_2(x', y')$ and sends $f_i(x', y')$ to P_i (i.e., it sends the corrupted party its output).

Adversary instructs trusted party to continue or halt: After receiving its output, the adversary sends either continue or abort$_i$ to the trusted party. If the trusted party receives continue then it sends $f_j(x', y')$ to the honest party P_j. Otherwise, if it receives abort$_i$, it sends abort$_i$ to the honest party P_j.

Outputs: The honest party always outputs the output value it obtained from the trusted party. The corrupted party outputs nothing. The adversary $\mathcal{A}$ outputs any arbitrary (probabilistic polynomial-time computable) function of the initial inputs of the corrupted party, the auxiliary input z, and the value $f_i(x', y')$ obtained from the trusted party.

The output of the honest party and the adversary in an execution of the above ideal model is denoted by $\text{IDEALSC}^\epsilon_{f, \mathcal{S}(z), i}(x, y, n)$.

Notice that there are two types of "cheating" here. The first is the classic abort and is used to model "early aborting" due to the impossibility of achieving fairness in general when there is no honest majority. The other type of cheating in this ideal model is more serious for two reasons: first, the ramifications of the cheating are greater (the adversary may learn the honest party's input and may be able to determine its output), and second, the cheating is only guaranteed to be detected with probability ϵ. Nevertheless, if ϵ is high enough, this may serve as a deterrent. We stress that in the ideal model the adversary must decide whether to cheat obliviously of the honest party's input and before it receives any output (and so it cannot use the output to help it decide whether or not it is "worthwhile" cheating). We have the following definition.

Definition 2.4.1 (security – strong explicit cheat formulation [3]): *Let f and π be as in Definition 2.2.1, and let $\epsilon : \mathbb{N} \to [0,1]$ be a function. Protocol π is said to securely compute f in the presence of covert adversaries with ϵ-deterrent if for every non-uniform probabilistic polynomial-time adversary $\mathcal{A}$ for the real model, there exists a non-uniform probabilistic polynomial-time adversary $\mathcal{S}$ for the ideal model such that for every $i \in \{1,2\}$:*

$$\left\{ \mathrm{IDEALSC}^{\epsilon}_{f,\mathcal{S}(z),i}(x,y,n) \right\}_{x,y,z,n} \stackrel{\mathrm{c}}{\equiv} \left\{ \mathrm{REAL}_{\pi,\mathcal{A}(z),i}(x,y,n) \right\}_{x,y,z,n}$$

where $x,y,z \in \{0,1\}^*$ under the constraint that $|x| = |y|$, and $n \in \mathbb{N}$.

2.4.3 Cheating and Aborting

It is important to note that in the above definition, a party that halts midway through the computation may be considered a "cheat" (this is used in an inherent way when constructing protocols later). Arguably, this may be undesirable due to the fact that an honest party's computer may crash (such unfortunate events may not even be that rare). Nevertheless, we argue that as a basic definition it suffices. This is due to the fact that it is possible for all parties to work by storing their input and random tape on disk before they begin the execution. Then, before sending any message, the incoming messages that preceded it are also written to disk. The result of this is that if a party's machine crashes, it can easily reboot and return to its previous state. (In the worst case the party will need to request a retransmit of the last message if the crash occurred before it was written.) We therefore believe that parties cannot truly hide behind the excuse that their machine crashed (it would be highly suspicious that someone's machine crashed in an irreversible way that also destroyed their disk at the critical point of a secure protocol execution).

Despite the above, it is possible to modify the definition so that honest halting is never considered cheating. In order to do this, we introduce the notion of "non-halting detection accuracy" so that if a party halts early, but otherwise does not deviate from the protocol specification, then it is not considered cheating. We formalize this by considering fail-stop adversaries who act semi-honestly except that they may halt early. Formally:

Definition 2.4.2 *Let π be a two-party protocol, let $\mathcal{A}$ be an adversary, and let i be the index of the corrupted party. The honest party P_j is said to* detect cheating *in π if its output in π is* corrupted$_i$*; this event is denoted by* $\mathrm{OUTPUT}_j(\mathrm{REAL}_{\pi,\mathcal{A}(z),i}(x,y,n)) = $ corrupted$_i$*. The protocol π is called* non-halting detection accurate *if for every fail-stop adversary $\mathcal{A}$, the probability that P_j detects cheating in π is negligible.*

Definition 2.4.1 can then be modified by requiring that π be non-halting detection accurate. We remark that although this strengthening is clearly desirable, it may also be prohibitive. Nevertheless, as we will see in Chapter 5, it is possible to efficiently obtain this stronger guarantee for the case of general protocols.

2.4.4 Relations Between Security Models

In order to better understand the definition of security in the presence of covert adversaries, we present two propositions that show the relation between security in the presence of covert adversaries and security in the presence of malicious and semi-honest adversaries.

Proposition 2.4.3 *Let π be a protocol that securely computes some functionality f with abort in the presence of malicious adversaries, as in Definition 2.3.1. Then, π securely computes f in the presence of covert adversaries with ϵ-deterrent, for every $0 \leq \epsilon \leq 1$.*

This proposition follows from the simple observation that according to Definition 2.3.1, there exists a simulator that always succeeds in its simulation. Thus, the same simulator works here (there is simply no need to ever send a cheat input).

Next, we consider the relation between covert and semi-honest adversaries. As we have discussed in Section 2.3.3, security for malicious adversaries only implies security for semi-honest adversaries if the semi-honest adversary is allowed to modify its input before the execution begins. This same argument holds for covert adversaries and we therefore consider *augmented semi-honest* adversaries. We have the following:

Proposition 2.4.4 *Let π be a protocol that securely computes some functionality f in the presence of covert adversaries with ϵ-deterrent, for $\epsilon(n) \geq 1/\mathrm{poly}(n)$. Then, π securely computes f in the presence of augmented semi-honest adversaries.*

Proof. Let π securely compute f in the presence of covert adversaries with ϵ-deterrent, where $\epsilon(n) \geq 1/\mathrm{poly}(n)$. The first observation is that since honest parties cannot send abort, corrupted or cheat instructions in the ideal model, it holds that when both parties are honest in a real execution of π, the values abort and corrupted appear in the output with only negligible probability.

Consider now the case of an augmented semi-honest adversary $\mathcal{A}$ that controls one of the parties, and let $\mathcal{S}$ be the simulator for $\mathcal{A}$. We claim that $\mathcal{S}$ sends abort, corrupted or cheat in the ideal model with at most negligible probability. This is due to the fact that the output of the honest party in an execution with $\mathcal{A}$ is indistinguishable from its output when both parties are honest (because the distribution over the messages received by the honest party in both executions is identical). In particular, the honest party in an execution with $\mathcal{A}$ outputs abort or corrupted with at most negligible probability. Now, in the ideal setting, an honest party outputs abort or corrupted whenever $\mathcal{S}$ sends abort or corrupted (and so it can send these with only negligible probability). Furthermore, an honest party outputs corrupted with probability ϵ whenever $\mathcal{S}$ sends cheat. Since $\epsilon \geq 1/\mathrm{poly}(n)$, it follows that $\mathcal{S}$ can send cheat also with only negligible probability. We therefore have that the ideal model with such an $\mathcal{S}$ is the *standard* ideal model (with no cheating possibility), and the augmented semi-honest simulator can just run $\mathcal{S}$. We stress that this only holds for the augmented semi-honest case, because $\mathcal{S}$ may change the corrupted party's inputs (we have no control over $\mathcal{S}$) and so the semi-honest simulator can only run $\mathcal{S}$ if it too can change the corrupted party's inputs. ∎

We stress that if $\epsilon = 0$ (or is negligible) then the definition of covert adversaries requires nothing, and so the proposition does not hold for this case.

We conclude that, as one may expect, security in the presence of covert adversaries with ϵ-deterrent lies in between security in the presence of malicious adversaries and security in the presence of semi-honest adversaries. If $1/\mathrm{poly}(n) \leq \epsilon(n) \leq 1 - 1/\mathrm{poly}(n)$ then it can be shown that Definition 2.4.1 is strictly different to both the semi-honest and malicious models (this is not difficult to see and so details are omitted). However, as we show below, when $\epsilon(n) = 1 - \mu(n)$, Definition 2.4.1 is equivalent to security in the presence of malicious adversaries (Definition 2.3.1).

Stated differently, the following proposition shows that the definition of security for covert adversaries "converges" to the malicious model as ϵ approaches 1. In order to make this claim technically, we need to deal with the fact that in the malicious model an honest party never outputs corrupted$_i$, whereas this can occur in the setting of covert adversaries even with $\epsilon = 1$.

We therefore define a transformation of any protocol π to π' where the only difference is that if an honest party should output corrupted$_i$ in π, then it outputs abort$_i$ instead in π'. We have the following:

Proposition 2.4.5 *Let π be a protocol and μ a negligible function. Then π securely computes some functionality f in the presence of covert adversaries with $\epsilon(n) = 1 - \mu(n)$ under Definition 2.4.1 if and only if π' securely computes f with abort in the presence of malicious adversaries.*

Proof. The fact that security in the presence of malicious adversaries implies security in the presence of covert adversaries has already been proven in Proposition 2.4.3 (observe that Proposition 2.4.3 holds for all ϵ, including ϵ that is negligibly close to 1). We now prove that security in the presence of covert adversaries under Definition 2.4.1 with ϵ that is negligibly close to 1 implies security in the presence of malicious adversaries. This holds because if the ideal adversary does not send cheat$_i$ then the ideal execution is the same as in the regular ideal model. Furthermore, if it does send cheat$_i$, it is caught cheating with probability that is negligibly close to 1 and so the protocol is aborted. Recall that by Definition 2.4.1, when the adversary is caught cheating it learns nothing and so the effect is the same as an abort in the regular ideal model (technically, the honest party has to change its output from corrupted$_i$ to abort$_i$ as discussed above, but this makes no difference). We conclude that when ϵ is negligibly close to 1, sending cheat$_i$ is the same as sending abort$_i$ and so the security is the same as in the presence of malicious adversaries.   ■

2.5 Restricted Versus General Functionalities

In this section, we show that it often suffices to construct a secure protocol for a restricted type of functionality, and the result to general functionalities can be automatically derived. This is most relevant for *general constructions* that are based on a circuit that computes the functionality in question. As we will see, in these cases the cost of considering the restricted types of functionalities considered here is inconsequential. For the sake of clarity, our general constructions will therefore all be for restricted functionalities of the types defined below.

The claims in this section are all quite straightforward. We therefore present the material somewhat informally, and leave formal claims and proofs as an exercise to the reader.

2.5.1 Deterministic Functionalities

The general definition considers probabilistic functionalities where the output $f(x, y)$ is a random variable. A classic example of a probabilistic functionality is that of coin-tossing. For example, one could define $f(1^n, 1^n)$ to be a uniformly distributed string of length n.

We show that it suffices to consider deterministic functionalities when constructing general protocols for secure computation. Specifically, we show that given a protocol for securely computing any deterministic functionality, it is possible to construct a secure protocol for computing any probabilistic functionality. Let $f = (f_1, f_2)$ be a two-party probabilistic functionality. We denote by $f(x, y; w)$ the output of f upon inputs x and y, and random tape w (the fact that f is probabilistic means that it has a uniformly distributed random tape). Next, define a deterministic functionality $g((x, r), (y, s)) = f(x, y; r \oplus s)$, where (x, r) is P_1's input and (y, s) is P_2's input, and assume that we have a secure protocol π' for computing f'. We now present a secure protocol π for computing f that uses π' for computing f'. Upon respective inputs $x, y \in \{0, 1\}^n$, parties P_1 and P_2 choose uniformly distributed strings $r \leftarrow_R \{0, 1\}^{q(n)}$ and $s \leftarrow_R \{0, 1\}^{q(n)}$, respectively, where $q(n)$ is an upper bound on the number of random bits used to compute f. They then invoke the protocol π' for securely computing f' in order to both obtain $f'((x, r), (y, s)) = f(x, y; r \oplus s)$. The fact that this yields a secure protocol for computing f follows from the fact that as long as either r or s is uniformly distributed, the resulting $w = r \oplus s$ is also uniformly distributed. This reduction holds for the case of semi-honest, malicious and covert adversaries.

Observe that in the case of general protocols that can be used for securely computing any functionality, the complexity of the protocol for computing f' is typically the same as for computing f. This is due to the fact that the complexity of these protocols is related to the size of the circuit computing the functionality, and the size of the circuit computing f' is of the same order as the size of the circuit computing f. The only difference is that the circuit for f' has $q(n)$ additional exclusive-or gates, where $q(n)$ is the length of f's random tape.

2.5.2 Single-Output Functionalities

In the general definition of secure two-party computation, both parties receive output and these outputs may be *different*. However, it is often far simpler to assume that only party P_2 receives output; we call such a functionality single-output. We will show now that this suffices for the general case. That is, we claim that any protocol that can be used to securely compute *any* efficient functionality $f(x, y)$ where only P_2 receives output can be used to securely compute *any* efficient functionality $f = (f_1, f_2)$ where party P_1 receives $f_1(x, y)$ and party P_2 receives $f_2(x, y)$. For simplicity, we will assume that the length of the output of $f_1(x, y)$ is at most n, where n is the security parameter. This can be achieved by simply taking n to be larger in case it is necessary. We show this reduction separately for semi-honest and malicious adversaries, as the semi-honest reduction is more efficient than the malicious one.

Semi-honest adversaries. Let $f = (f_1, f_2)$ be an arbitrary probabilistic polynomial-time computable functionality and define the single-output functionality f' as follows: $f'((x, r), (y, s)) = (f_1(x, y) \oplus r \parallel f_2(x, y) \oplus s)$ where $a \parallel b$ denotes the concatenation of a with b. Now, given a secure protocol π' for computing the single-output functionality f' where P_2 only receives the output, it is possible to securely compute the functionality $f = (f_1, f_2)$ as follows. Upon respective inputs $x, y \in \{0, 1\}^n$, parties P_1 and P_2 choose uniformly distributed strings $r \leftarrow_R \{0, 1\}^{q(n)}$ and $s \leftarrow_R \{0, 1\}^{q(n)}$, respectively, where $q(n)$ is an upper bound on the output length of f on inputs of length n. They then invoke the protocol π' for securely computing f' in order for P_2 to obtain $f'((x, r), (y, s))$; denote the first half of this output by v and the second half by w. Upon receiving (v, w), party P_2 sends v to P_1, which then computes $v \oplus r$ and obtains $f_1(x, y)$. In addition, party P_2 computes $w \oplus s$ and obtains $f_2(x, y)$. It is easy to see that the resulting protocol securely computes f. This is due to the fact that r completely obscures $f_1(x, y)$ from P_2. Thus, neither party learns more than its own input. (In fact, the strings $f_1(x, y) \oplus r$ and $f_2(x, y) \oplus s$ are uniformly distributed and so are easily simulated.)

As in the case of probabilistic versus deterministic functionalities, the size of the circuit computing f' is of the same order as the size of the circuit computing f. The only difference is that f' has one additional exclusive-or gate for every circuit-output wire.

Malicious adversaries. Let $f = (f_1, f_2)$ be as above; we construct a protocol in which P_1 receives $f_1(x, y)$ and P_2 receives $f_2(x, y)$ that is secure in the presence of malicious adversaries. As a building block we use a protocol for computing any efficient functionality, with security for malicious adversaries, with the limitation that only P_2 receives output. As in the semi-honest case, P_2 will also receive P_1's output in encrypted format, and will then hand it to P_1 after the protocol concludes. However, a problem arises in that P_2 can modify the output that P_1 receives (recall that the adversary may be malicious here). In order to prevent this, we add message authentication to the encrypted output.

Let $r, a, b \leftarrow_R \{0, 1\}^n$ be randomly chosen strings. Then, in addition to x, party P_1's input includes the elements r, a and b. Furthermore, define a functionality g (that has only a single output) as follows:

$$g((r, a, b, x), y) = (\alpha, \beta, f_2(x, y))$$

where $\alpha = r + f_1(x, y)$, $\beta = a \cdot \alpha + b$, and the arithmetic operations are defined over $GF[2^n]$. Note that α is a one-time pad encryption of P_1's output $f_1(x, y)$, and β is an information-theoretic message authentication tag of α (specifically, $a\alpha + b$ is a pairwise-independent hash of α). Now, the parties compute the functionality g, using a secure protocol in which only P_2 receives output. Following this, P_2 sends the pair (α, β) to P_1. Party P_1 checks whether $\beta = a \cdot \alpha + b$; if yes, it outputs $\alpha - r$, and otherwise it outputs abort_2.

It is easy to see that P_2 learns nothing about P_1's output $f_1(x, y)$, and that it cannot alter the output that P_1 will receive (beyond causing it to abort), except with probability 2^{-n}. We remark that it is also straightforward to construct a simulator for the above protocol. Formally, proving the security of this transformation requires a modular composition theorem; this is discussed in Section 2.7 below.

As is the case for the previous reductions above, the circuit for computing g is only mildly larger than that for computing f. Thus, the modification above has only a mild effect on the complexity of the secure protocol (assuming that the complexity of the original protocol, where only P_2 receives output, is proportional to the size of the circuit computing f as is the case for the protocol below).

2.5.3 Non-reactive Functionalities

As described in Section 2.3.2, a reactive functionality is one where the computation is carried out over multiple phases, and the parties may choose their inputs in later phases based on the outputs that they have already received. Recall that such a reactive functionality can be viewed as a series of functionalities $(f^1, f^2, \ldots)$ such that the input to f^j is the tuple (x_j, y_j, σ_{j-1}) and the output includes the parties' outputs and state information σ_j; see Section 2.3.2 for more details. We denote by f_1^j and f_2^j the corresponding outputs of parties P_1 and P_2 from f^j, and by σ_j the state output from f^j.

In this section, we show that it is possible to securely compute any reactive functionality given a general protocol for computing non-reactive functionalities. The basic idea behind the reduction is the same as for same-output functionalities (for the semi-honest case) and single-output functionalities (for the malicious case). Specifically, the parties receive the same output as usual, but also receive random shares of the state at each stage; i.e., one party receives a random pad and the other receives the state encrypted by this pad. This ensures that neither party learns the internal state of the reactive functionality. Observe that although this suffices for the semi-honest case, it does not suffice for the case of malicious adversaries, which may modify the values that they are supposed to input. Thus, for the malicious case, we also add a message authentication tag to prevent any party from modifying the share of the state received in the previous stage.

Semi-honest adversaries. Let $(f^1, f^2, \ldots)$ be the series of functionalities defining the reactive functionality. First, we define a series of functionalities $(g^1, g^2, \ldots)$ such that

$$g^j \left((x_j, \sigma_{j-1}^1), (y_j, \sigma_{j-1}^2) \right)$$
$$= \left(\left(f_1^j(x_j, y_j, \sigma_{j-1}^1 \oplus \sigma_{j-1}^2), \sigma_j^1 \right), \left(f_2^j(x_j, y_j, \sigma_{j-1}^1 \oplus \sigma_{j-1}^2), \sigma_j^2 \right) \right)$$

where σ_j^1 and σ_j^2 are uniformly distributed strings under the constraint that $\sigma_j^1 \oplus \sigma_j^2 = \sigma_j$ (the state after the jth stage). That is, g^j receives input (x_j, σ_{j-1}^1) from P_1 and input (y_j, σ_{j-1}^2) from P_2 and then computes f^j on inputs (x_j, y_j, σ_{j-1}) where $\sigma_{j-1} = \sigma_{j-1}^1 \oplus \sigma_{j-1}^2$. In words, g^j receives the parties' inputs to the phase, together with a *sharing* of the state from the previous round. Functionality g^j then outputs the phase outputs to each party, and a sharing of the state from this round of computation.

Malicious adversaries. As we have mentioned, the solution for semi-honest adversaries does not suffice when considering malicious adversaries because nothing prevents the adversary from modifying its share of the state. This is solved by also having party P_1 receive a MAC (message authentication code) key k_1 and a MAC tag $t_j^1 = MAC_{k_2}(\sigma_j^1)$, where the keys k_1, k_2 are chosen randomly in the phase computation and σ_j^1 is the share of the current state that P_1 holds. Likewise, P_2 receives k_2 and $t_j^2 = MAC_{k_1}(\sigma_j^2)$. Then, the functionality in the $(j+1)$th phase receives the parties' phase-inputs, shares σ_j^1 and σ_j^2 of σ_j, keys k_1, k_2. and MAC-tags t_j^1, t_j^2. The functionality checks the keys and MACs and if the verification succeeds, it carries out the phase computation with the inputs and given state. By the security of the MAC, a malicious adversary is unable to change the current state, except with negligible probability.

2.6 Non-simulation-Based Definitions

2.6.1 Privacy Only

The definition of security that follows the ideal/real simulation paradigm provides strong security guarantees. In particular, it guarantees privacy, correctness, independence of inputs and more. However, in some settings, it may be sufficient to guarantee privacy only. We warn that this is not so simple and in many cases it is difficult to separate privacy from correctness and independence of inputs. For example, consider a function f with the property that for every y there exists a x_y such that $f(x_y, y) = y$. Now, if party P_1 can somehow make its input x depend on P_2's input (something which is not possible when independence of inputs is guaranteed), then it may be able to always set $x = x_y$ and learn P_2's input in entirety. (We stress that although this sounds far fetched, such attacks are actually sometimes possible.)

Another difficulty that arises when defining privacy is that it typically depends very much on the function being computed. Intuitively, we would like to require that if two different inputs result in the same output, then no adversarial party can tell which of the two inputs the other party used. In other words, we would like to require that for every adversarial P_1 and input x, party P_1 cannot distinguish whether P_2 used y or y' when the output is $f(x, y)$ and it holds that $f(x, y) = f(x, y')$. However, such a formulation suffers from a number of problems. First, if f is 1–1 no privacy guarantees are provided at all, even if it is hard to invert. Second, the formulation suffers from the exact problem described above. Namely, if it is possible for P_1 to implicitly choose $x = x_y$ based on y (say by modifying a commitment to y that it receives from P_2) so that $f(x_y, y)$ reveals more information about y than "the average x", then privacy is also breached. Finally, we remark that (sequential) composition theorems, like those of Section 2.7, are not known for protocols that achieve privacy only. Thus, it is non-trivial to use protocols that achieve privacy only as subprotocols when solving large protocol problems.

Despite the above problems, it is still sometimes possible to provide a workable definition of privacy that provides non-trivial security guarantees and is of interest. Due to the difficulty in providing a general definition, we will present a definition for one specific function in order to demonstrate how such definitions look. For this purpose, we consider the oblivious transfer function. Recall that in this function, there is a sender S with a pair of input strings (x_0, x_1) and a receiver R with an input bit σ. The output of the function is nothing to the sender and the string x_σ for the receiver. Thus, a secure oblivious transfer protocol has the property that the sender learns nothing about σ while the receiver learns at most one of the strings x_0, x_1. Unfortunately, defining privacy here without resorting to the ideal model is very non-trivial. Specifically, it is easy to define privacy in the presence of a malicious sender S^*; we just say that S^* cannot distinguish the case where R has input 0 from the case where it has input 1. However, it is more difficult to define privacy in the presence of a malicious receiver R^* because it does learn something. A naive approach to defining this says that for some bit b it holds that R^* knows nothing about x_b. However, this value of b may depend on the messages sent during the oblivious transfer and so cannot be fixed ahead of time (see the discussion above regarding independence of inputs).

Fortunately, for the case of *two-message* oblivious transfer (where the receiver sends one message and the sender replies with a single message) it is possible to formally define this. The following definition of security for oblivious transfer is based on [42] and states that replacing one of x_0 and x_1 with some other x should go unnoticed by the receiver. The question of which of x_0, x_1 to replace causes a problem which is solved in the case of a two-message protocol by fixing the first message; see below. (In the definition below we use the following notation: for a two-party protocol with parties S and R, we denote by $\text{VIEW}_S(S(1^n, a), R(1^n, b))$ the view of S in an execution where it has input a, R has input b, and the security parameter is n. Likewise, we denote the view of R by $\text{VIEW}_R(S(1^n, a), R(1^n, b))$.

Definition 2.6.1 *A* **two-message** *two-party probabilistic polynomial-time protocol* (S, R) *is said to be a* private oblivious transfer *if the following holds:*

- NON-TRIVIALITY: *If S and R follow the protocol then after an execution in which S has for input any pair of strings $x_0, x_1 \in \{0, 1\}^*$, and R has for input any bit $\sigma \in \{0, 1\}$, the output of R is x_σ.*
- PRIVACY IN THE CASE OF A MALICIOUS S^*: *For every non-uniform probabilistic polynomial-time S^* and every auxiliary input $z \in \{0, 1\}^*$, it holds that*

$$\{\text{VIEW}_{S^*}(S^*(1^n, z), R(1^n, 0))\}_{n \in \mathbb{N}} \overset{c}{\equiv} \{\text{VIEW}_{S^*}(S^*(1^n, z), R(1^n, 1))\}_{n \in \mathbb{N}}.$$

- PRIVACY IN THE CASE OF A MALICIOUS R^*: *For every non-uniform deterministic polynomial-time receiver R^*, every auxiliary input $z \in \{0, 1\}^*$, and every triple of inputs $x_0, x_1, x \in \{0, 1\}^*$ such that $|x_0| = |x_1| = |x|$ it holds that either:*

$$\{\text{VIEW}_{R^*}(S(1^n, (x_0, x_1)); R^*(1^n, z))\}_{n \in \mathbb{N}} \overset{c}{\equiv} \{\text{VIEW}_{R^*}(S(1^n, (x_0, x)); R^*(1^n, z))\}_{n \in \mathbb{N}}$$

or

$$\{\text{VIEW}_{R^*}(S(1^n, (x_0, x_1)); R^*(1^n, z))\}_{n \in \mathbb{N}} \overset{c}{\equiv} \{\text{VIEW}_{R^*}(S(1^n, (x, x_1)); R^*(1^n, z))\}_{n \in \mathbb{N}}.$$

The way to view the above definition of privacy in the case of a malicious R^* is that R^*'s first message, denoted by $R^*(1^n, z)$, fully determines whether it should receive x_0 or x_1. If it determines for example that it should receive x_0, then its view (i.e., the distribution over S's reply) when S's input is (x_0, x_1) is indistinguishable from its view when S's input is (x_0, x). Clearly this implies that R^* cannot learn anything about x_1 when it receives x_0 and vice versa. In addition, note that since R^* sends its message before receiving anything from S, and since this message fully determines R^*'s input, we have that the problem of independence of inputs discussed above does not arise.

Note that when defining the privacy in the case of a malicious R^* we chose to focus on a *deterministic* polynomial-time receiver R^*. This is necessary in order to fully define the message $R^*(z)$ for any given z, which in turn fully defines the string x_b that $R^*(z)$ does *not* learn. By making R^* non-uniform, we have that this does not weaken the adversary (since R^*'s advice tape can hold its "best coins"). We remark that generalizing this definition to protocols that have more than two messages is non-trivial. Specifically, the problem of independence of inputs described above becomes difficult again when more than two messages are sent.

The above example demonstrates that it is possible to define "privacy only" for secure computation. However, it also demonstrates that this task can be *very difficult*. In particular, we *do not know of a satisfactory definition* of privacy for oblivious transfer with more than two rounds. In general, one can say that when a party does not receive output, it is easy to formalize privacy because it learns nothing. However, when a party does receive output, defining privacy without resorting to the ideal model is problematic (and often it is not at all clear how it can be achieved).

We conclude with one important remark regarding "privacy-only" definitions. As we have mentioned, an important property of security definitions is a composition theorem that guarantees certain behavior when the secure protocol is used as a subprotocol in another larger protocol. No such general composition theorems are known for definitions that follow the privacy-only approach. As such, this approach has a significant disadvantage.

2.6.2 One-Sided Simulatability

Another approach to providing weaker, yet meaningful, security guarantees is that of *one-sided simulation*. This notion is helpful when only one party receives output while the other learns nothing. As discussed above in Section 2.6.1, when a party should learn nothing (i.e., when it has no output), it is easy to define privacy via indistinguishability as for encryption. Specifically, it suffices to require that the party learning nothing is not able to distinguish between *any* two inputs of the other party.[1] In contrast, the difficulty of defining privacy appropriately for the party that does receive output is overcome by requiring full simulation in this case. That is, consider a protocol/functionality where P_2 receives output while P_1 learns nothing. Then, in the case where P_1 is corrupted we require that it not be able to learn anything about P_2's input and formalize this via indistinguishability. However, in the case where P_2 is corrupted, we require the existence of a simulator that can fully simulate its view, as in the definition of Section 2.3. This is helpful because it enables us to provide a general definition of security for problems of this type where only one party receives output. Furthermore, it turns out that in many cases, it *is* possible to achieve high efficiency when "one-sided simulation" is sufficient and "full simulation" is not required.

It is important to note that this is a *relaxed level of security* and does not achieve everything we want. For example, a corrupted P_1 may be able to make its input depend on the other party's input, and may also be able to cause the output to be distributed incorrectly. Thus, this notion is not suitable for all protocol problems. Such compromises seem inevitable given the current state of the art, where highly-efficient protocols that provide full simulation-based security in the presence of malicious adversaries seem very

---

[1] Note that this only makes sense when the party receives no output. Otherwise, if it does receive output, then the other party's input has influence over that output and so it is unreasonable to say that it is impossible to distinguish between any two inputs.

hard to construct. We stress that P_2 cannot carry out any attacks, because full simulation is guaranteed in the case where it is corrupted.

The definition. Let f be a function $f : \{0,1\}^* \times \{0,1\}^* \to \{0,1\}^*$ with only a single output which is designated for P_2. Let $\text{REAL}_{\pi,\mathcal{A}(z),i}(x,y,n)$ denote the outputs of the honest party and the adversary $\mathcal{A}$ (controlling party P_i) after a real execution of protocol π, where P_1 has input x, P_2 has input y, $\mathcal{A}$ has auxiliary input z, and the security parameter is n. Let $\text{IDEAL}_{f,\mathcal{S}(z),i}(x,y,n)$ be the analogous distribution in an ideal execution with a trusted party that computes f for the parties and hands the output to P_2 only. Finally, let $\text{VIEW}^{\mathcal{A}}_{\pi,\mathcal{A}(z),i}(x,y,n)$ denote the view of the adversary after a real execution of π as above. Then, we have the following definition:

Definition 2.6.2 *Let f be a functionality where only P_2 receives output. We say that a protocol π securely computes f with one-sided simulation if the following holds:*

1. *For every non-uniform* PPT *adversary $\mathcal{A}$ controlling P_2 in the real model, there exists a non-uniform* PPT *adversary $\mathcal{S}$ for the ideal model, such that*

$$\left\{ \text{REAL}_{\pi,\mathcal{A}(z),2}(x,y,n) \right\}_{x,y,z,n} \stackrel{\text{c}}{\equiv} \left\{ \text{IDEAL}_{f,\mathcal{S}(z),2}(x,y,n) \right\}_{x,y,z,n}$$

   where $n \in \mathbb{N}$, $x,y,z \in \{0,1\}^*$ and $|x| = |y|$.
2. *For every non-uniform* PPT *adversary $\mathcal{A}$ controlling P_1;*

$$\left\{ \text{VIEW}^{\mathcal{A}}_{\pi,\mathcal{A}(z),1}(x,y,n) \right\}_{x,y,y',z,n} \stackrel{\text{c}}{\equiv} \left\{ \text{VIEW}^{\mathcal{A}}_{\pi,\mathcal{A}(z),1}(x,y',n) \right\}_{x,y,y',z,n} \quad (2.3)$$

   where $n \in \mathbb{N}$, $x,y,y',z \in \{0,1\}^*$ and $|x| = |y| = |y'|$.

Note that the ensembles in (2.3) are indexed by two different inputs y and y' for P_2. The requirement is that $\mathcal{A}$ cannot distinguish between the cases where P_2 used the first input y and the second input y'.

2.7 Sequential Composition – Simulation-Based Definitions

A protocol that is secure under *sequential* composition maintains its security when run multiple times, as long as the executions are run sequentially (meaning that each execution concludes before the next execution begins). Sequential composition theorems are theorems that state "if a protocol is secure in the stand-alone model under definition X, then it remains secure under sequential composition". Thus, we are interested in proving protocols secure under Definitions 2.2.1, 2.3.1 and 2.4.1 (for semi-honest, malicious and covert adversaries), and immediately deriving their security under sequential composition. This is important for two reasons. First, sequential composition constitutes a security goal within itself as security is guaranteed even when parties run many executions, albeit sequentially. Second, sequential composition theorems are useful tools that help in writing proofs of security. Specifically, when constructing a protocol that is made up of a number of secure subprotocols, it is possible to analyze the security of the overall protocol in a modular way, because the composition theorems tell us that the subprotocols remain secure in this setting.

We do not present proofs of the sequential composition theorems for the semi-honest and malicious cases as these already appear in [32]; see Sections 7.3.1 and 7.4.2 respectively. However, we do present a formal *statement* of the theorems as we will use them in our proofs of security of the protocols. In addition, we provide a proof of sequential composition for the case of covert adversaries.

Modular sequential composition. The basic idea behind the formulation of the modular sequential composition theorems is to show that it is possible to design a protocol that uses an ideal functionality as a subroutine, and then analyze the security of the protocol when a trusted party computes this functionality. For example, assume that a protocol is constructed using oblivious transfer as a subroutine. Then, first we construct a protocol for oblivious transfer and prove its security. Next, we prove the security of the protocol that uses oblivious transfer as a subroutine, in a model where the parties have access to a trusted party computing the oblivious transfer functionality. The composition theorem then states that when the "ideal calls" to the trusted party for the oblivious transfer functionality are replaced with real executions of a secure protocol computing this functionality, the protocol remains secure. We begin by presenting the "hybrid model" where parties communicate by sending regular messages to each other (as in the real model) but also have access to a trusted party (as in the ideal model).

The hybrid model. We consider a *hybrid model* where parties both interact with each other (as in the real model) and use trusted help (as in the ideal model). Specifically, the parties run a protocol π that contains "ideal calls" to a trusted party computing some functionalities $f_1, \ldots, f_{p(n)}$. These ideal calls are just instructions to send an input to the trusted party. Upon receiving the output back from the trusted party, the protocol π continues. The protocol π is such that f_i is called before f_{i+1} for every i (this just determines the "naming" of the calls as $f_1, \ldots, f_{p(n)}$ in that order). In addition, if a functionality f_i is *reactive* (meaning that it contains multiple stages like a commitment functionality which has a commit and reveal stage), then no messages are sent by the parties directly to each other from the time that the first message is sent to f_i to the time that all stages of f_i have concluded. We stress that the honest party sends its input to the trusted party in the same round and does not send other messages until it receives its output (this is because we consider *sequential composition* here). Of course, the trusted party may be used a number of times throughout the execution if π. However, each use is independent (i.e., the trusted party does not maintain any state between these calls). We call the regular messages of π that are sent amongst the parties **standard messages** and the messages that are sent between parties and the trusted party **ideal messages**. *We stress that in the hybrid model, the trusted party behaves as in the ideal model of the definition being considered.* Thus, the trusted party computing $f_1, \ldots, f_{p(n)}$ behaves as in Section 2.3 when malicious adversaries are being considered, and as in Section 2.4 for covert adversaries.

Sequential composition – malicious adversaries. Let $f_1, \ldots, f_{p(n)}$ be probabilistic polynomial-time functionalities and let π be a two-party protocol that uses ideal calls to a trusted party computing $f_1, \ldots, f_{p(n)}$. Furthermore, let $\mathcal{A}$ be a non-uniform probabilistic polynomial-time machine and let i be the index of the corrupted party. Then, the $f_1, \ldots, f_{p(n)}$-hybrid execution of π on inputs (x, y), auxiliary input z to $\mathcal{A}$ and security parameter n, denoted $\text{HYBRID}_{\pi, \mathcal{A}(z), i}^{f_1, \ldots, f_{p(n)}}(x, y, n)$, is defined as the output vector of the honest parties and the adversary $\mathcal{A}$ from the hybrid execution of π with a trusted party computing $f_1, \ldots, f_{p(n)}$.

Let $\rho_1, \ldots, \rho_{p(n)}$ be a series of protocols (as we will see ρ_i takes the place of f_i in π). We assume that each ρ_i has a fixed number rounds that is the same for all parties. Consider the real protocol $\pi^{\rho_1, \ldots, \rho_{p(n)}}$ that is defined as follows. All standard messages of π are unchanged. When a party is instructed to send an ideal message α to the trusted party to compute f_j, it begins a real execution of ρ_j with input α instead. When this execution of ρ_j concludes with output y, the party continues with π as if y were the output received by the trusted party for f_j (i.e., as if it were running in the hybrid model).

The composition theorem states that if $\rho_1, \ldots, \rho_{p(n)}$ securely compute $f_1, \ldots, f_{p(n)}$ respectively, and π securely computes some functionality g in the $f_1, \ldots, f_{p(n)}$-hybrid model, then $\pi^{\rho_1, \ldots, \rho_{p(n)}}$ securely computes g (in the real model). As discussed above, the hybrid model that we consider here is where the protocols are run sequentially. Thus, the fact that sequential composition only is considered is implicit in the theorem, via the reference to the hybrid model.

Theorem 2.7.1 (modular sequential composition – malicious): *Let $p(n)$ be a polynomial, let $f_1, \ldots, f_{p(n)}$ be two-party probabilistic polynomial-time functionalities and let $\rho_1, \ldots, \rho_{p(n)}$ be protocols such that each ρ_i securely computes f_i in the presence of malicious adversaries. Let g be a two-party functionality and let π be a protocol that securely computes g in the $f_1, \ldots, f_{p(n)}$-hybrid model in the presence of malicious adversaries. Then, $\pi^{\rho_1, \ldots, \rho_{p(n)}}$ securely computes g in the presence of malicious adversaries.*

Sequential composition – covert adversaries. Let $f_1, \ldots, f_{p(n)}$, π and $\rho_1, \ldots, \rho_{p(n)}$ be as above. Furthermore, define $\pi^{\rho_1, \ldots, \rho_{p(n)}}$ exactly as in the malicious model. Note, however, that in the covert model, a party may receive corrupted$_k$ as output from ρ_j. In this case, as with any other output, it behaves as instructed in π (corrupted$_k$ may be received as output both when the real ρ_j is run and when the trusted party is used to compute f_j because we consider the covert ideal model here). The covert ideal model IDEALSC depends on the deterrent factor because this determines the probability with which the trusted party sends corrupted or undetected. Therefore, we refer to the (f, ϵ)-hybrid model as one where the trusted party computes f and uses the given ϵ. When considering protocols $\rho_1, \ldots, \rho_{p(n)}$ we will refer to the $(f_1, \epsilon_1), \ldots, (f_{p(n)}, \epsilon_{p(n)})$-hybrid model, meaning that the trusted party uses ϵ_i when computing f_i (and the ϵ values may all be different). We have the following:

Theorem 2.7.2 *Let $p(n)$ be a polynomial, let $f_1, \ldots, f_{p(n)}$ be two-party probabilistic polynomial-time functionalities and let $\rho_1, \ldots, \rho_{p(n)}$ be protocols such that each ρ_i securely computes f_i in the presence of covert adversaries with deterrent ϵ_i. Let g be a two-party functionality and let π be a protocol that securely computes g in the $(f_1, \epsilon_1), \ldots, (f_{p(n)}, \epsilon_{p(n)})$-hybrid model in the presence of covert adversaries with ϵ-deterrent. Then, $\pi^{\rho_1, \ldots, \rho_{p(n)}}$ securely computes g in the presence of covert adversaries with ϵ-deterrent.*

Proof (sketch). Theorem 2.7.2 can be derived as an almost immediate corollary from the composition theorem of [11, 32] in the following way. First, define a special functionality interface that follows the instructions of the trusted party in Definition 2.4.1. That is, define a *reactive functionality* (see Section 2.3.2) that receives inputs and writes outputs (this functionality is modeled by an interactive Turing machine). The appropriate reactive functionality here acts exactly like the trusted party (e.g., if it receives a cheat$_i$ message when computing f_ℓ, then it tosses coins and with probability ϵ_ℓ outputs corrupted$_i$ to the honest party and with probability $1 - \epsilon_\ell$ gives the adversary the honest party's input and lets it chooses its output). Next, consider the standard ideal model of Definition 2.3.1 with functionalities of the above form. It is easy to see that a protocol securely computes some functionality f under Definition 2.4.1 *if and only if* it is securely computes the appropriately defined reactive functionality under Definition 2.3.1. This suffices because the composition theorem of [11, 32] can be applied to Definition 2.3.1, yielding the result.   ■

Observe that in Theorem 2.7.2 the protocols $\rho_1, \ldots, \rho_{p(n)}$ and π may all have different deterrent values. Thus the proof of π in the hybrid model must take into account the actual deterrent values $\epsilon_1, \ldots, \epsilon_{p(n)}$ of the protocols $\rho_1, \ldots, \rho_{p(n)}$, respectively.

Part II
General Constructions

In the next three chapters, we present efficient constructions for *general* two-party computation, meaning that the protocols can be used to solve any functionality given a circuit that computes that functionality. The efficiency of these constructions depends on the size of the circuit, and thus these constructions are useful for problems that have reasonably small circuits (depending on the level of security required, this can range from some thousands of gates to a million). In addition, the techniques and paradigms used in these general constructions are important also for constructing efficient protocols for specific problems of interest.

Chapter 3
Semi-honest Adversaries

In this chapter, we present Yao's protocol for secure two-party computation in the presence of semi-honest adversaries. The protocol has a constant number of rounds, and works by having the parties evaluate an "encrypted" or "garbled" circuit such that they learn nothing from the evaluation but the output itself. In particular, all intermediate values in the circuit evaluation (which can reveal more information than is allowed) remain hidden from both parties. We present the protocol for the case of a deterministic, non-reactive, single-output functionality. As we have shown in Section 2.5, this suffices for obtaining the secure computation of any probabilistic, reactive two-party functionality at approximately the same cost.

3.1 An Overview of the Protocol

Let f be a polynomial-time functionality (assume for now that it is deterministic), and let x and y be the parties' respective inputs. The first step is to view the function f as a boolean circuit C. In order to describe Yao's protocol, it is helpful to first recall how such a circuit is computed. Let x and y be the parties' inputs. Then, the circuit $C(x, y)$ is computed gate by gate, from the input wires to the output wires. Once the incoming wires to a gate g have obtained values $\alpha, \beta \in \{0, 1\}$, it is possible to give the outgoing wires of the gate the value $g(\alpha, \beta)$. The output of the circuit is given by the values obtained in the output wires of the circuit. Thus, essentially, computing a circuit involves allocating appropriate zero-one values to the *wires* of the circuit. In the description below, we refer to four different types of wires in a circuit: circuit-input wires (that receive the input values x and y), circuit-output wires (that carry the value $C(x, y)$), gate-input wires (that enter some gate g), and gate-output wires (that leave some gate g).

We now present a high-level description of Yao's protocol. The construction is actually a "compiler" that takes any polynomial-time functionality

C. Hazay, Y. Lindell, *Efficient Secure Two-Party Protocols*,
Information Security and Cryptography, DOI 10.1007/978-3-642-14303-8_3,
© Springer-Verlag Berlin Heidelberg 2010

f, or actually a circuit C that computes f, and constructs a protocol for securely computing f in the presence of semi-honest adversaries. In a secure protocol, the only value learned by a party should be its output. Therefore, the values that are allocated to all wires that are not circuit-output should not be learned by either party (these values may reveal information about the other party's input that could not be otherwise learned from the output). The basic idea behind Yao's protocol is to provide a method of computing a circuit so that values obtained on all wires other than circuit-output wires are never revealed. For every wire in the circuit, two random values are specified such that one value represents 0 and the other represents 1. For example, let w be the label of some wire. Then, two values k_w^0 and k_w^1 are chosen, where k_w^σ represents the bit σ. An important observation here is that even if one of the parties knows the value k_w^σ obtained by the wire w, this does not help it to determine whether $\sigma = 0$ or $\sigma = 1$ (because both k_w^0 and k_w^1 are identically distributed). Of course, the difficulty with such an idea is that it seems to make computation of the circuit impossible. That is, let g be a gate with incoming wires w_1 and w_2 and output wire w_3. Then, given two random values k_1^σ and k_2^τ, it does not seem possible to compute the gate because σ and τ are unknown. We therefore need a method of computing the value of the output wire of a gate (also a random value k_3^0 or k_3^1) given the value of the two input wires to that gate.

In short, this method involves providing "garbled computation tables" that map the random input values to random output values. However, this mapping should have the property that given two input values, it is only possible to learn the output value that corresponds to the output of the gate (the other output value must be kept secret). This is accomplished by viewing the four possible inputs to the gate, k_1^0, k_1^1, k_2^0, and k_2^1, as encryption keys. Then, the output values k_3^0 and k_3^1, which are also keys, are encrypted under the appropriate keys from the incoming wires. For example, let g be an OR gate. Then, the key k_3^1 is encrypted under the pairs of keys associated with the values $(1,1)$, $(1,0)$ and $(0,1)$. In contrast, the key k_3^0 is encrypted under the pair of keys associated with $(0,0)$. See Table 3.1 below.

input wire w_1	input wire w_2	output wire w_3	garbled computation table
k_1^0	k_2^0	k_3^0	$E_{k_1^0}(E_{k_2^0}(k_3^0))$
k_1^0	k_2^1	k_3^1	$E_{k_1^0}(E_{k_2^1}(k_3^1))$
k_1^1	k_2^0	k_3^1	$E_{k_1^1}(E_{k_2^0}(k_3^1))$
k_1^1	k_2^1	k_3^1	$E_{k_1^1}(E_{k_2^1}(k_3^1))$

Table 3.1 Garbled OR Gate

Notice that given the input wire keys k_1^α and k_2^β corresponding to α and β, and the four table values (found in the fourth column of Table 3.1), it is possible to decrypt and obtain the output wire key $k_3^{g(\alpha,\beta)}$. Furthermore, as required above, this is the only value that can be obtained (the other keys on the input wires are not known and so only a single table value can be decrypted). In other words, it is possible to compute the output key $k_3^{g(\alpha,\beta)}$ of a gate, and only that key, without learning anything about the real values α, β or $g(\alpha,\beta)$. (We note that the values of the table are randomly ordered so that a key's position does not reveal anything about the value it is associated with. Despite this random ordering, the specific construction is such that given a pair of input wire keys, it is possible to locate the table entry that is encrypted by those keys.)

So far we have described how to construct a single garbled gate. A garbled circuit consists of garbled gates along with "output decryption tables". These tables map the random values on *circuit-output wires* back to their corresponding real values. That is, for a circuit-output wire w, the pairs $(0, k_w^0)$ and $(1, k_w^1)$ are provided. Then, after obtaining the key k_w^γ on a circuit-output wire, it is possible to determine the actual output bit by comparing the key to the values in the output decryption table.[1] Notice that given the keys associated with inputs x and y, it is possible to (obliviously) compute the entire circuit gate by gate. Then, having obtained the keys on the circuit-output wires, these can be "decrypted" providing the result $C(x, y)$.

The above construction can be described metaphorically using "locked boxes". The basic idea, as above, is that every wire is allocated two padlock keys; one key is associated with the bit 0 and the other with the bit 1. Then, for each gate four doubly locked boxes are provided, where each box is associated with a row in the truth table computing the gate (i.e., one box is associated with inputs $(0, 0)$, another with $(0, 1)$, and so on). The four boxes are locked so that each pair of keys (one from each input wire) opens exactly one box. Furthermore, in each box a single key relating to the output wire of the gate is stored. This key is chosen so that it correctly associates the input bits with the output bit of the gate. (For example, if the keys that open the box are associated with 0 and 1 and the gate computes the AND function, then the key inside the box is the key associated with 0 on the output wire.)

The first important observation is that given the set of keys that are associated with the parties' inputs, it is possible to "compute the circuit" by opening the locked boxes one at a time (for each gate, only one box will open). The process concludes at the output-gate boxes, which can contain the actual output rather than a key. The second important observation is that the computation of the circuit reveals absolutely no information beyond the output itself. This is due to the fact that the keys are not labelled and so it is impossible to know whether a given key is associated with 0 or with 1. This all holds under the assumption that the keys associated with the circuit-input wires are obtained in an "oblivious manner" that does not reveal the association with the parties' inputs. Furthermore, we must assume that only

---

[1] Alternatively, in the output gates it is possible to directly encrypt 0 or 1 instead of k_w^0 or k_w^1, respectively.

a single set of keys is provided (and so in each gate only a single box can be opened). Of course, in the actual garbled-circuit construction, double encryption replaces doubly locked boxes and decryption keys replace physical padlock keys.

We now proceed to informally describe Yao's protocol. In this protocol, one of the parties, henceforth the sender, constructs a garbled circuit and sends it to the other party, henceforth the receiver. The sender and receiver then interact so that the receiver obtains the input-wire keys that are associated with the inputs x and y (this interaction is described below). Given these keys, the receiver then computes the circuit as described, obtains the output and concludes the protocol. This description only shows how the receiver obtains its output, while ignoring the output of the sender. However, the receiver's output can include the sender's output in *encrypted form* (where only the sender knows the decryption key). Then, the receiver can just forward the sender its output at the end of the computation. Since the sender's output is encrypted, the receiver learns nothing more than its own output, as required.

It remains for us to describe how the receiver obtains the keys for the circuit-input wires. Here we differentiate between the inputs of the sender and the inputs of the receiver. Regarding the sender, it simply sends the receiver the values that correspond to its input. That is, if its ith input bit is 0 and the wire w_i receives this input, then the sender just hands the receiver the string k_i^0. Notice that since all of the keys are identically distributed, the receiver can learn nothing about the sender's input from these keys. Regarding the receiver, this is more problematic. The sender cannot hand it all of the keys pertaining to its input (i.e., both the 0 and 1 keys on the receiver's input wires), because this would enable the receiver to compute more than just its output. (For a given input x of the sender, this would enable the receiver to compute $C(x, \tilde{y})$ for every $\tilde{y}$. This is much more information than a single value $C(x, y)$.) On the other hand, the receiver cannot openly tell the sender which keys to send it, because then the sender would learn the receiver's input. The solution to this is to use a 1-out-of-2 oblivious transfer protocol [72, 25]. In such a protocol, a sender inputs two values x_0 and x_1 (in this case, k_w^0 and k_w^1 for some circuit-input wire w), and a receiver inputs a bit σ (in this case, corresponding to its appropriate input bit). The outcome of the protocol is that the receiver obtains the value x_σ (in this case, the key k_w^σ). Furthermore, the receiver learns nothing about the other value $x_{1-\sigma}$, and the sender learns nothing about the receiver's input σ. By having the receiver obtain its keys in this way, we obtain that (a) the sender learns nothing of the receiver's input value, and (b) the receiver obtains only a single set of keys and so can compute the circuit on only a single value, as required. This completes our high-level description of Yao's protocol.

3.2 Tools

3.2.1 "Special" Private-Key Encryption

Our construction uses a private-key encryption scheme that has indistinguishable encryptions for multiple messages. Informally speaking, this means that for every two (known) vectors of messages $\overline{x}$ and $\overline{y}$, no polynomial-time adversary can distinguish an encryption of the vector $\overline{x}$ from an encryption of the vector $\overline{y}$. We stress that according to our construction of Yao's garbled circuit, the encryption scheme must be secure for *multiple messages*. Therefore one-time pads cannot be used. In our proof of security, we will actually use an encryption scheme that is secure under chosen-plaintext attacks (strictly speaking this is not necessary, but it does simplify the presentation). We refer the reader to [32, Chapter 5] for formal definitions of secure encryption.

We will require an additional property from the encryption scheme that we use. Loosely speaking, we require that an encryption under one key will fall in the range of an encryption under another key with negligible probability. We also require that given the key k, it is possible to efficiently verify if a given ciphertext is in the range of k. (These two requirements are very easily satisfied, as demonstrated below.) The reason that we require these additional properties is to enable the receiver to *correctly* compute the garbled circuit. Recall that in every gate, the receiver is given two random keys that enable it to decrypt and obtain the random key for the gate-output wire; see Table 3.1. A problem that immediately arises here is how the receiver can know which value is the intended decryption. (Note that it may be the case that all strings can be decrypted.) By imposing the requirement that encryptions under one key will almost never be valid encryptions under another key, and requiring that this also be efficiently verifiable, it will hold that only one of the values will be valid (except with negligible probability). The receiver will then take the (single) correctly decrypted value as the key for the gate-output wire. We now formally define the requirements on the encryption scheme:

Definition 3.2.1 *Let (G, E, D) be a private-key encryption scheme and denote the range of a key in the scheme by $\mathsf{Range}_n(k) \stackrel{\text{def}}{=} \{E_k(x)\}_{x \in \{0,1\}^n}$.*

1. *We say that (G, E, D) has an* elusive range *if for every probabilistic polynomial-time machine A, every polynomial $p(\cdot)$ and all sufficiently large ns*

$$\Pr_{k \leftarrow G(1^n)}[A(1^n) \in \mathsf{Range}_n(k)] < \frac{1}{p(n)}.$$

2. *We say that (G, E, D) has an* efficiently verifiable range *if there exists a probabilistic polynomial-time machine M such that $M(1^n, k, c) = 1$ if and only if $c \in \mathsf{Range}_n(k)$.*

By convention, for every $c \notin \mathsf{Range}_n(k)$, we have that $D_k(c) = \bot$.

Notice that the requirements for an "elusive range" are quite weak. In particular, the machine A is *oblivious* in that it is given no information on k and no examples of ciphertexts within $\mathsf{Range}_n(k)$. Thus, A must "hit" the range with no help whatsoever.

We now show that it is easy to construct encryption schemes with the above properties. For example, let $\mathcal{F} = \{f_k\}$ be a family of pseudorandom functions, where $f_k : \{0,1\}^n \to \{0,1\}^{2n}$ for $k \in \{0,1\}^n$. Then, define

$$E_k(x) = \langle r, f_k(r) \oplus x0^n \rangle$$

where $x \in \{0,1\}^n$, $r \leftarrow_R \{0,1\}^n$ and $x0^n$ denotes the concatenation of x and 0^n. A similar idea can be used to encrypt a message $x \in \{0,1\}^{\ell(n)}$ of length $\ell(n)$ for an arbitrary polynomial $\ell(\cdot)$, by setting the output of f_k to be of length $\ell(n) + n$ and concatenating n zeroes at the end.[2] (This can be achieved efficiently in practice by using an appropriate mode of operation for block ciphers.) The fact that this encryption scheme has indistinguishable encryptions under chosen-plaintext attacks is well known. Regarding our additional requirements:

1. *Elusive range:* Notice that if a truly random function f_{rand} was used instead of f_k, then the probability that a value c output by the machine A is in the range of $\langle r, f_{\mathrm{rand}}(r) \oplus x0^n \rangle$ is negligible. This follows from the fact that obtaining such a c involves finding a value r and then predicting the last n bits of $f_{\mathrm{rand}}(r)$ (notice that these last n bits are fully revealed in $f_{\mathrm{rand}}(r) \oplus x0^n$). Since f_{rand} is random, this prediction can succeed with probability at most 2^{-n}. Now, by the assumption that f_k is *pseudorandom*, it follows that a polynomial-time machine A will also succeed in generating such a c with at most negligible probability. Otherwise, such an A could be used to distinguish f_k from a random function.

2. *Efficiently verifiable range:* Given k and $c = \langle r, s \rangle$, it is possible to compute $f_k(r)$ and verify that the last n bits of $f_k(r)$ equal the last n bits of s. If so, then it follows that $c \in \mathsf{Range}_n(k)$, and if not, then $c \notin \mathsf{Range}_n(k)$.

We stress that there are many possible ways to ensure that P_2 will obtain the correct output of a garbled gate. For example, as described in [63], explicit (and randomly permuted) indices may be used instead.[3]

Double-encryption security. In Yao's protocol, the private-key encryption scheme is used in order to double encrypt values. As we have described, the protocol works by double encrypting four values, where each double encryption uses a different combination of the keys associated with the input wires. Intuitively, given only two keys, it is possible to decrypt only one of the

---

[2] In fact, the string of 0s can have any super-logarithmic length. We set it to be of length n for simplicity.

[3] We chose this method somewhat arbitrarily. We feel some preference due to the fact that the gate description and circuit construction is the simplest this way. As we will see, however, some price is paid in the proof of correctness.

values. However, formally, this must be proven. We define a double encryption experiment here and prove that any encryption scheme that is secure under chosen-plaintext attacks is secure for double encryption here. We remark that the experiment does not look very natural. However, it is exactly what is needed in our proof of security. Let (G, E, D) be a private-key encryption scheme and assume without loss of generality that $G(1^n)$ returns a string of length n (i.e., the length of a key generated with security parameter 1^n is exactly n). We write $\overline{E}(k_0, k_1, m) = E_{k_0}(E_{k_1}(m))$. The experiment definition is as follows:

Expt$_{\mathcal{A}}^{\text{double}}(n, \sigma)$

1. The adversary $\mathcal{A}$ is invoked upon input 1^n and outputs two keys k_0 and k_1 of length n and two triples of messages (x_0, y_0, z_0) and (x_1, y_1, z_1) where all messages are of the same length.
2. Two keys $k_0', k_1' \leftarrow G(1^n)$ are chosen for the encryption scheme.
3. $\mathcal{A}$ is given the challenge ciphertext

$$\langle \overline{E}(k_0, k_1', x_\sigma), \overline{E}(k_0', k_1, y_\sigma), \overline{E}(k_0', k_1', z_\sigma) \rangle$$

   as well as oracle access to $\overline{E}(\cdot, k_1', \cdot)$ and $\overline{E}(k_0', \cdot, \cdot)$.[4]

4. $\mathcal{A}$ outputs a bit b and this is taken as the output of the experiment.

Security under double encryption simply means that the adversary outputs 1 when $\sigma = 0$ with almost the same probability as it outputs 1 when $\sigma = 1$.

Definition 3.2.2 *An encryption scheme (G, E, D) is secure under chosen double encryption if for every non-uniform probabilistic polynomial-time machine $\mathcal{A}$, every polynomial $p(\cdot)$ and all sufficiently large ns,*

$$\left| \Pr\left[\text{Expt}_{\mathcal{A}}^{\text{double}}(n, 1) = 1\right] - \Pr\left[\text{Expt}_{\mathcal{A}}^{\text{double}}(n, 0) = 1\right] \right| < \frac{1}{p(n)}.$$

We now show that any encryption scheme that is secure (i.e., has indistinguishable encryptions) under chosen-plaintext attacks, is secure under chosen double encryption. We remark that all security here is in the non-uniform model (and so we assume security under chosen-plaintext attacks for non-uniform adversaries). It is well known that under chosen-plaintext attacks, security for a single message implies security for multiple messages (see [32, Section 5.4]), and we will thus assume this in our proof. For the sake of completeness, we define the chosen-plaintext experiment for the case of multiple

---

[4] Note that in the ciphertexts that $\mathcal{A}$ receives, at least one of the keys used is unknown to $\mathcal{A}$. In addition, the oracle access here means that $\mathcal{A}$ can provide any k and m to $\overline{E}(\cdot, k_1', \cdot)$ and receive back $\overline{E}(k, k_1', m)$; likewise for $\overline{E}(k_0', \cdot, \cdot)$.

messages. In fact, we consider only the case of two messages, because this suffices for our proof later.

$\text{Expt}_{\mathcal{A}}^{\text{cpa}}(n, \sigma)$

1. A key $k \leftarrow G(1^n)$ is chosen and the adversary $\mathcal{A}$ is invoked with input 1^n and oracle access to $E_k(\cdot)$. The adversary $\mathcal{A}$ outputs two pairs of messages (x_0, y_0) and (x_1, y_1).
2. The challenge ciphertexts $c_1 = E_k(x_\sigma)$ and $c_2 = E_k(y_\sigma)$ are computed.
3. $\mathcal{A}$ is given the pair (c_1, c_2) as well as continued oracle access to $E_k(\cdot)$
4. $\mathcal{A}$ outputs a bit b and this is taken as the output of the experiment.

The definition of security under chosen-plaintext attacks is analogous to Definition 3.2.2 except that $\text{Expt}_{\mathcal{A}}^{\text{double}}$ is replaced with $\text{Expt}_{\mathcal{A}}^{\text{cpa}}$. We are now ready to state the lemma.

Lemma 3.2.3 *Let (G, E, D) be a private-key encryption scheme that has indistinguishable encryptions under chosen-plaintext attacks in the presence of non-uniform adversaries. Then (G, E, D) is secure under chosen double encryption.*

Proof. In order to prove this lemma, we define a modified experiment, denoted by $\text{Expt}_{\mathcal{A}}^{\text{mod}}(n, \sigma)$, which is exactly the same as $\text{Expt}_{\mathcal{A}}^{\text{double}}(\sigma)$ except that the y part of the challenge ciphertext does not depend on σ. That is, the challenge ciphertext equals $\langle \overline{E}(k_0, k_1', x_\sigma), \overline{E}(k_0', k_1, y_0), \overline{E}(k_0', k_1', z_\sigma) \rangle$; note that x_σ and z_σ are encrypted as before, but y_0 is always encrypted (even if $\sigma = 1$). Clearly,

$$\Pr\left[\text{Expt}_{\mathcal{A}}^{\text{double}}(n, 0) = 1\right] = \Pr\left[\text{Expt}_{\mathcal{A}}^{\text{mod}}(n, 0) = 1\right] \tag{3.1}$$

because in both cases, the encrypted values are x_0, y_0 and z_0. We will prove that for every non-uniform probabilistic polynomial-time adversary and for some negligible function $\mu(\cdot)$, the following two equations hold:

$$\left| \Pr\left[\text{Expt}_{\mathcal{A}}^{\text{mod}}(n, 0) = 1\right] - \Pr\left[\text{Expt}_{\mathcal{A}}^{\text{mod}}(n, 1) = 1\right] \right| < \mu(n); \tag{3.2}$$

$$\left| \Pr\left[\text{Expt}_{\mathcal{A}}^{\text{mod}}(n, 1) = 1\right] - \Pr\left[\text{Expt}_{\mathcal{A}}^{\text{double}}(n, 1) = 1\right] \right| < \mu(n). \tag{3.3}$$

Combining Equations (3.1) to (3.3), we obtain that (G, E, D) is secure under chosen double encryption. We will prove (3.2); (3.3) is proven in an analogous way.

We begin by modifying $\text{Expt}_{\mathcal{A}}^{\text{mod}}$ in the following way. First, we claim that indistinguishability holds even if the adversary $\mathcal{A}$ can choose k_0' by itself. (Observe that once the "y" value is always taken to be y_0, the value σ is found only in the encryptions of x_σ and z_σ. Since both of these values are encrypted with k_1', the adversary $\mathcal{A}$ cannot distinguish between them even if it knows all other keys k_0, k_1 and k_0'.) We can therefore let $\mathcal{A}$ choose k_0, k_1 and k_0'. Given that this is the case, we need not generate $\overline{E}(k_0', k_1, y_0)$ as part of the challenge ciphertext (because given k_0' and k_1, $\mathcal{A}$ can compute it by itself). For the same reason, we can remove the oracle $\overline{E}(k_0', \cdot, \cdot)$ from the experiment. We therefore need only to prove that (3.2) holds for the further modified experiment $\mathsf{Expt}_{\mathcal{A}}^{\mathsf{mod}'}$ defined as follows:

$\mathsf{Expt}_{\mathcal{A}}^{\mathsf{mod}'}(n, \sigma)$

1. The adversary $\mathcal{A}$ is invoked upon input 1^n and outputs three keys k_0, k_1 and k_0' of length n and two pairs of messages (x_0, z_0) and (x_1, z_1) where all messages are of the same length.

2. A key $k_1' \leftarrow G(1^n)$ is chosen for the encryption scheme.

3. $\mathcal{A}$ is given $\langle \overline{E}(k_0, k_1', x_\sigma), \overline{E}(k_0', k_1', z_\sigma) \rangle$ as well as oracle access to $\overline{E}(\cdot, k_1', \cdot)$.

4. $\mathcal{A}$ outputs a bit b and this is taken as the output of the experiment.

From what we have stated above, if we prove that the analogue of (3.2) holds for $\mathsf{Expt}_{\mathcal{A}}^{\mathsf{mod}'}$, then (3.2) itself clearly also holds. However, $\mathsf{Expt}_{\mathcal{A}}^{\mathsf{mod}'}$ is now almost identical to $\mathsf{Expt}_{\mathcal{A}}^{\mathsf{cpa}}$. The only differences are:

1. In $\mathsf{Expt}_{\mathcal{A}}^{\mathsf{mod}'}$ the challenge ciphertext is first encrypted with k_1' (the secret key) and then with k_0 or k_0', whereas in $\mathsf{Expt}_{\mathcal{A}}^{\mathsf{cpa}}$ the challenge ciphertext is encrypted with the k_1' only. However, this clearly does not matter because the adversary attacking the CPA scheme with secret key k_1' can know k_0 and k_0' and so can compute this itself.

2. In $\mathsf{Expt}_{\mathcal{A}}^{\mathsf{mod}'}$ the oracle given to the adversary is $\overline{E}(\cdot, k_1', \cdot)$ whereas in $\mathsf{Expt}_{\mathcal{A}}^{\mathsf{cpa}}$ it is $E_k(\cdot)$. However, since k and k_1' play the same role as the secretly chosen key, it is clear that given oracle $E_{k_1'}(\cdot)$ it is possible to efficiently emulate the oracle $E(\cdot, k_1', \cdot)$. Therefore, this also makes no difference.

We conclude that (3.2) follows from the security of (G, E, D) under chosen-plaintext attacks; the formal proof of this, and (3.3), can be derived in a straightforward way from the above discussion and is thus left as an exercise to the reader. We conclude that (G, E, D) is secure under chosen double encryption, concluding the proof. ∎

3.2.2 Oblivious Transfer

As we have mentioned, the 1-out-of-2 oblivious transfer functionality is defined by $((x_0, x_1), \sigma) \mapsto (\lambda, x_\sigma)$ where λ denotes the empty string. For the sake of self-containment, we will briefly describe the oblivious transfer protocol of [25], which is secure in the presence of semi-honest adversaries and assumes any enhanced trapdoor permutation.[5] Our description will be for the case where $x_0, x_1 \in \{0, 1\}$; when considering semi-honest adversaries, the general case can be obtained by running the single-bit protocol many times in parallel.

PROTOCOL 3.2.4 (Oblivious Transfer)

- **Inputs:** P_1 has $x_0, x_1 \in \{0, 1\}$ and P_2 has $\sigma \in \{0, 1\}$.
- **The protocol:**

  1. P_1 randomly chooses a trapdoor permutation (f, t) from a family of enhanced trapdoor permutations. P_1 sends f (but not the trapdoor t) to P_2.
  2. P_2 chooses a random v_σ in the domain of f and computes $w_\sigma = f(v_\sigma)$. In addition, P_2 chooses a random $w_{1-\sigma}$ in the domain of f, using the "enhanced" sampling algorithm (see Footnote 5). P_2 sends (w_0, w_1) to P_1.
  3. P_1 uses the trapdoor t and computes $v_0 = f^{-1}(w_0)$ and $v_1 = f^{-1}(w_1)$. Then, it computes $b_0 = B(v_0) \oplus x_0$ and $b_1 = B(v_1) \oplus x_1$, where B is a hard-core bit of f. Finally, P_1 sends (b_0, b_1) to P_2.
  4. P_1 computes $x_\sigma = B(v_\sigma) \oplus b_\sigma$ and outputs x_σ.

Oblivious transfer – semi-honest model

Theorem 3.2.5 *Assuming that (f, t) are chosen from a family of enhanced trapdoor permutations, Protocol 3.2.4 securely computes the 1-out-of-2 oblivious transfer functionality in the presence of static semi-honest adversaries.*

Recall that since the oblivious transfer functionality is deterministic, it suffices to use the simplified definition of Equations (2.1) and (2.2). Thus, it is guaranteed that there exist simulators, denoted by S_1^{OT} and S_2^{OT}, that generate the appropriate views of parties P_1 and P_2, respectively. We remark that simulator S_1^{OT} receives P_1's input (x_0, x_1) and outputs a full view of P_1 that includes the input (x_0, x_1), a random tape, and the incoming messages that P_1 expects to see in a real execution (of course, this view output by S_1^{OT} is only computationally indistinguishable from a real view). Notice that P_1 has no output in the oblivious transfer functionality, and so S_1^{OT} receives only P_1's input. The simulator S_2^{OT} receives P_2's input σ and output x_σ and outputs a view, as described above. The proof of Theorem 3.2.5 can be found in [32, Section 7.3.2]. In Chapter 7, we present a number of efficient constructions for oblivious transfer with security for more powerful adversaries.

---

[5] Informally speaking, an enhanced trapdoor permutation has the property that it is possible to sample from its range, so that given the coins used for sampling it is still hard to invert the value. See [32, Appendix C.1] for more details.

3.3 The Garbled-Circuit Construction

In this section, we describe the garbled-circuit construction. Let C be a boolean circuit that receives two inputs $x, y \in \{0,1\}^n$ and outputs $C(x, y) \in \{0,1\}^n$ (for simplicity, we assume that the input length, output length and the security parameter are all of the same length n). We also assume that C has the property that if a circuit-output wire comes from a gate g, then gate g has no wires that are input to other gates.[6] (Likewise, if a circuit-input wire is itself also a circuit-output, then it is not input into any gate.)

We begin by describing the construction of a single garbled gate g in C. The circuit C is boolean, and therefore any gate is represented by a function $g : \{0,1\} \times \{0,1\} \to \{0,1\}$. Now, let the two input wires to g be labelled w_1 and w_2, and let the output wire from g be labelled w_3. Furthermore, let $k_1^0, k_1^1, k_2^0, k_2^1, k_3^0, k_3^1$ be six keys obtained by independently invoking the key-generation algorithm $G(1^n)$; for simplicity, assume that these keys are also of length n. Intuitively, we wish to be able to compute $k_3^{g(\alpha,\beta)}$ from k_1^α and k_2^β, without revealing any of the other three values $k_3^{g(1-\alpha,\beta)}, k_3^{g(\alpha,1-\beta)}$, and $k_3^{g(1-\alpha,1-\beta)}$. The gate g is defined by the following four values:

$$c_{0,0} = E_{k_1^0}(E_{k_2^0}(k_3^{g(0,0)})),$$
$$c_{0,1} = E_{k_1^0}(E_{k_2^1}(k_3^{g(0,1)})),$$
$$c_{1,0} = E_{k_1^1}(E_{k_2^0}(k_3^{g(1,0)})),$$
$$c_{1,1} = E_{k_1^1}(E_{k_2^1}(k_3^{g(1,1)})),$$

where E is from a private-key encryption scheme (G, E, D) that has indistinguishable encryptions under chosen-plaintext attacks, and has an elusive efficiently verifiable range; see Section 3.2.1. The actual gate is defined by a *random permutation* of the above values, denoted by c_0, c_1, c_2, c_3; from here on we call them the **garbled table** of gate g. Notice that given k_1^α and k_2^β, and the values c_0, c_1, c_2, c_3, it is possible to compute the output of the gate $k_3^{g(\alpha,\beta)}$ as follows. For every i, compute $D_{k_2^\beta}(D_{k_1^\alpha}(c_i))$. If more than one decryption returns a non-$\perp$ value, then output **abort**. Otherwise, define k_3^γ to be the only non-$\perp$ value that is obtained. (Notice that if only a single non-$\perp$ value is obtained, then this will be $k_3^{g(\alpha,\beta)}$ because it is encrypted under the given keys k_1^α and k_2^β. Later we will show that except with negligible probability, only one non-$\perp$ value is indeed obtained.)

We are now ready to show how to construct the entire garbled circuit. Let m be the number of *wires* in the circuit C, and let $w_1, \ldots, w_m$ be labels of these wires. These labels are all chosen uniquely with the following exception:

---

[6] This requirement is due to our labelling of gates described below; see Footnote 7. We note that this assumption on C increases the number of gates by at most n.

if w_i and w_j are both output wires from the same gate g, then $w_i = w_j$ (this occurs if the fan-out of g is greater than one). Likewise, if a bit of a party's input enters more than one gate, then all circuit-input wires associated with this bit will have the same label.[7] Next, for every label w_i, choose two independent keys $k_i^0, k_i^1 \leftarrow G(1^n)$; we stress that all of these keys are chosen independently of the others. Now, given these keys, the four garbled values of each gate are computed as described above and the results are permuted randomly. Finally, the output or decryption tables of the garbled circuit are computed. These tables simply consist of the values $(0, k_i^0)$ and $(1, k_i^1)$ where w_i is a *circuit-output wire*. (Alternatively, output gates can just compute 0 or 1 directly. That is, in an output gate, one can define $c_{\alpha, \beta} = E_{k_1^\alpha}(E_{k_2^\beta}(g(\alpha, \beta)))$ for every $\alpha, \beta \in \{0, 1\}$.)

The entire garbled circuit of C, denoted by $G(C)$, consists of the garbled table for each gate and the output tables. We note that the structure of C is given, and the garbled version of C is simply defined by specifying the output tables and the garbled table that belongs to each gate. This completes the description of the garbled circuit.

Correctness. We now claim that the above garbled circuit enables correct computation of the function. That is, given the appropriate input strings and the garbled table for each gate, it is possible to obtain the correct output. It is at this point that we use the "special" properties of the encryption scheme described in Section 3.2.1.

Claim 3.3.1 (correctness): *Let $x = x_1 \cdots x_n$ and $y = y_1 \cdots y_n$ be two n-bit inputs for C. Furthermore, let $w_{in_1}, \ldots, w_{in_n}$ be the labels of the circuit-input wires corresponding to x, and let $w_{in_{n+1}}, \ldots, w_{in_{2n}}$ be the labels of the circuit-input wires corresponding to y. Finally, assume that the encryption scheme used to construct $G(C)$ has an elusive and efficiently verifiable range. Then, given the garbled circuit $G(C)$ and the strings $k_{in_1}^{x_1}, \ldots, k_{in_n}^{x_n}, k_{in_{n+1}}^{y_1}, \ldots, k_{in_{2n}}^{y_n}$, it is possible to compute $C(x, y)$, except with negligible probability.*

Proof. We begin by showing that every gate can be "decrypted" correctly. Specifically, let g be a gate with incoming wires w_1, w_2 and outgoing wire w_3. Then, we show that for every $\alpha, \beta \in \{0, 1\}$, given k_1^α and k_2^β and the garbled table of g, it is possible to determine $k_3^{g(\alpha, \beta)}$, except with negligible probability. More formally, let c_0, c_1, c_2, c_3 be the garbled table of gate g. We wish to find c_i such that $c_i = E_{k_1^\alpha}(E_{k_2^\beta}(k_3^{g(\alpha, \beta)}))$. We claim that, except with negligible probability, there exists a *single* i such that $c_i \in \mathsf{Range}_n(k_1^\alpha)$ and

---

[7] This choice of labelling is not essential and it is possible to provide unique labels for all wires. However, in such a case, the table of a gate with fan-out greater than 1 will have to be redefined so that the keys of all of the wires leaving the gate are encrypted. We chose this labelling because it seems to make for a simpler gate definition. We note, however, that due to this choice, we must assume that if a gate g has an output wire exiting from it, then it does not have another wire that exits it and enters another gate. As we have mentioned, this increases the number of gates by at most n.

$D_{k_1^\alpha}(c_i) \in \mathsf{Range}_n(k_2^\beta)$. In other words, at most one of the values decrypts correctly (from here on we use this informal term to mean what is formally described above).

This follows from the fact that the encryption scheme has an elusive range. Specifically, recall that the gate was constructed by first choosing independent values for the gate-input and gate-output wires $k_1^0, k_1^1, k_2^0, k_2^1, k_3^0, k_3^1$. Next, the values c_0, c_1, c_2 and c_3 are computed. Now, assume that there are (at least) two values c_i, c_j such that $c_i \in \mathsf{Range}(k_1^\alpha)$ and $D_{k_1^\alpha}(c_i) \in \mathsf{Range}_n(k_2^\beta)$; likewise for c_j. Without loss of generality, assume also that $c_i = E_{k_1^\alpha}(E_{k_2^\beta}(k_3^{g(\alpha,\beta)}))$; i.e., assume that c_i should be correctly decrypted. There are two cases regarding c_j:

1. $c_j = E_{k_1^\alpha}(E_{k_2^{1-\beta}}(z))$ *for* $z \in \{k_3^0, k_3^1\}$:

   By our assumption regarding c_j, it follows that $c_j \in \mathsf{Range}(k_1^\alpha)$ and $D_{k_1^\alpha}(c_j) \in \mathsf{Range}_n(k_2^\beta)$. This means that $E_{k_2^{1-\beta}}(z) \in \mathsf{Range}_n(k_2^\beta)$. Next, as mentioned above, recall that $k_2^{1-\beta}, k_3^0$, and k_3^1 are all uniform and independent of k_2^β. Therefore, we can define a machine A that chooses two random keys $k', k'' \leftarrow G(1^n)$ and outputs $c = E_{k'}(k'')$. The probability that $c \in \mathsf{Range}(k)$ for $k \leftarrow G(1^n)$ *equals* the probability that $E_{k_2^{1-\beta}}(z) \in \mathsf{Range}_n(k_2^\beta)$ (recall that $z \in \{k_3^0, k_3^1\}$). Since the encryption scheme (G, E, D) has an elusive range, we conclude that the probability that $c \in \mathsf{Range}_n(k)$ is negligible. Therefore, the probability that $E_{k_2^{1-\beta}}(z) \in \mathsf{Range}_n(k_2^\beta)$ is also negligible. This concludes this case.

2. $c_j = E_{k_1^{1-\alpha}}(z)$ *for* $z = E_{k'}(k'')$ *where* $k' \in \{k_2^0, k_2^1\}$ *and* $k'' \in \{k_3^0, k_3^1\}$:

   In this case, we have that $E_{k_1^{1-\alpha}}(z) \in \mathsf{Range}_n(k_1^\alpha)$. Using the same arguments as above, and noticing once again that $k_1^{1-\alpha}, k'$ and k'' are all independent of k_1^α, we have that this case occurs also with at most negligible probability.

Now, given that in every gate at most one c_i decrypts correctly, we prove the claim. In order to do this, we define that the key k is correct for wire w_i if $k = k_i^\gamma$, where $\gamma \in \{0, 1\}$ is the value obtained on wire w_i when computing the ungarbled circuit C on inputs (x, y). By induction on the circuit, starting from the bottom and working up, we show that for every wire, the correct key for the wire is obtained. This holds for the circuit-input wires by the fact that the keys $k_{in_1}^{x_1}, \ldots, k_{in_n}^{x_n}, k_{in_{n+1}}^{y_1}, \ldots, k_{in_{2n}}^{y_n}$ are given, and is the base case of the induction. Assume that it is true for a gate g with gate-input wires w_i and w_j and let k_i^α and k_j^β be the respective keys held for these wires. Then, by the decryption procedure, it follows that the value $k_\ell^{g(\alpha,\beta)} = D_{k_j^\beta}(D_{k_i^\alpha}(c_{\alpha,\beta}))$ is obtained, where w_ℓ is the output wire of the gate.[8] Furthermore, by the

---

[8] This holds if there are no decryption errors (i.e., if for every k and every x, $D_k(E_k(x)) = x$). If there *is* a negligible error in the decryption, then we will inherit a negligible error probability here.

arguments shown above, this is the only value that is decrypted correctly. Therefore, the correct key for the output wire of gate g is also obtained. This concludes the inductive step.

It follows that the correct keys of the output wires of the circuit are obtained, except with negligible probability. That is, the keys obtained for the circuit-output wires all correspond to the output value $C(x, y)$. Therefore, the value obtained after using the output tables is exactly $C(x, y)$, as required.

■

Removing the error probability. The above construction allows for a negligible probability of error. This is due to two possible events: (a) in some gate more than one value decrypts correctly, or (b) in some gate, the correct value does not decrypt correctly. As we have mentioned in Footnote 8, this second event can occur if the encryption scheme has decryption errors. This problem can be removed by using a scheme *without* decryption errors (this is not a limitation because decryption errors can always be removed [23]).

Regarding the first event causing error, this can be overcome in one of two ways. First, when constructing the circuit, it is possible to check that an error does not occur. Then, if an error has occurred, it is possible to reconstruct the garbled circuit again, repeating until no errors occur. (For this to work, we need to assume that the machine that verifies if a value is in the range of a key runs in deterministic polynomial time, as is the case in our construction. Alternatively, it suffices to assume that it has only a one-sided error and never returns 1 when a value is not in the range.) The problem with this approach is that the construction of the circuit now runs in expected, and not strict, polynomial time. Another approach is to use explicit randomly permuted indices, meaning that the decrypted values in the gates reveal exactly which item in the next table is to be opened. This approach was described in [63].

3.4 Yao's Two-Party Protocol

As we have seen above, given the keys that correspond to the correct input, it is possible to obtain the correct output from the garbled circuit. Thus, the protocol proceeds by party P_1 constructing the garbled circuit and giving it to P_2. Furthermore, P_1 hands P_2 the keys that correspond to $x = x_1 \cdots x_n$. In addition, P_2 must obtain the keys that correspond to its input $y = y_1 \cdots y_n$. However, this must be done carefully, ensuring the following:

1. P_1 should not learn anything about P_2's input string y.
2. P_2 should obtain the keys corresponding to y and no others. (Otherwise, P_2 could compute $C(x, y)$ and $C(x, y')$ for $y' \neq y$, in contradiction to the requirement that $C(x, y)$ and nothing else is learned.)

The above two problems are solved by having P_1 and P_2 run 1-out-of-2 oblivious transfer protocols [72, 25]. That is, for every bit of P_2's input, the parties run an oblivious transfer protocol where P_1's input is (k_{n+i}^0, k_{n+i}^1) and P_2's input is y_i. In this way, P_2 obtains the keys $k_{n+1}^{y_1}, \ldots, k_{2n}^{y_n}$, and only these keys. In addition, P_1 learns nothing about y. The complete description appears in Protocol 3.4.1. We remark that the protocol is written for the case of deterministic single-output functionalities; see Section 2.5 for a justification as to why this suffices.

PROTOCOL 3.4.1 (Yao's Two-Party Protocol)

- **Inputs:** P_1 has $x \in \{0,1\}^n$ and P_2 has $y \in \{0,1\}^n$.
- **Auxiliary input:** A boolean circuit C such that for every $x, y \in \{0,1\}^n$ it holds that $C(x,y) = f(x,y)$, where $f: \{0,1\}^n \times \{0,1\}^n \to \{0,1\}^n$. We require that C is such that if a circuit-output wire leaves some gate g, then gate g has no other wires leading from it into other gates (i.e., no circuit-output wire is also a gate-input wire). Likewise, a circuit-input wire that is also a circuit-output wire enters no gates.
- **The protocol:**

  1. P_1 constructs the garbled circuit $G(C)$ as described in Section 3.3, and sends it to P_2.
  2. Let $w_1, \ldots, w_n$ be the circuit-input wires corresponding to x, and let $w_{n+1}, \ldots, w_{2n}$ be the circuit-input wires corresponding to y. Then,
     a. P_1 sends P_2 the strings $k_1^{x_1}, \ldots, k_n^{x_n}$.
     b. For every i, P_1 and P_2 execute a 1-out-of-2 oblivious transfer protocol in which P_1's input equals (k_{n+i}^0, k_{n+i}^1) and P_2's input equals y_i.
     The above oblivious transfers can all be run in parallel.
  3. Following the above, P_2 has obtained the garbled circuit and $2n$ keys corresponding to the $2n$ input wires to C. Party P_2 then computes the circuit, as described in Section 3.3, obtaining $f(x,y)$.

We now provide a formal proof that Protocol 3.4.1 securely computes the functionality f. Our proof could be simplified by relying on a composition theorem as in Section 2.7 (see [32, Section 7.3.1] for a statement and proof of modular sequential composition for the semi-honest case). However, we have chosen to provide a fully self-contained and direct proof of the security of the protocol. The rest of the proofs in this book will all rely on composition instead.

Theorem 3.4.2 *Let $f: \{0,1\}^* \times \{0,1\}^* \to \{0,1\}^*$ be a polynomial-time two-party single-output functionality. Assume that the oblivious transfer protocol is secure in the presence of static semi-honest adversaries, and that the encryption scheme has indistinguishable encryptions under chosen-plaintext attacks, and has an elusive and efficiently verifiable range. Then, Protocol 3.4.1 securely computes f in the presence of static semi-honest adversaries.*

Proof. Intuitively, since the oblivious transfer protocol is secure, party P_2 receives exactly one key per circuit-input wire. Then, by the security of the encryption scheme, it is only able to decrypt one value in each gate. Furthermore, it has no idea if the value obtained in this decryption corresponds to a 0 or a 1. Therefore, it learns nothing from this computation, except for the output itself. We now formally prove this. Recall that since we consider deterministic functionalities, we can use the simpler formulation of security as stated in Equations (2.1) and (2.2); note that the separate correctness requirement has already been proven in Claim 3.3.1. We prove the simulation separately when P_1 is corrupted and when P_2 is corrupted.

A simplifying convention. In the proof below, we will use the simulators S_1^{OT} and S_2^{OT} that exist for the oblivious transfer functionality in order to generate views for the corrupted parties. In general, a view is represented as the party's input followed by its random tape and concluding with the series of incoming messages. In order to simplify the presentation, we will present the view of a party in a different order. Specifically, we will write the view of a party in Protocol 3.4.1 – *excluding the oblivious transfers* – in the usual way. However, the view of the party in the oblivious transfers is written in full where it appears in the protocol transcript. That is, instead of splitting the view in the oblivious transfers into input, random tape and incoming messages, the input and random tape are written together with the incoming messages. This clearly makes no difference and is just to simplify notation (the standard way of writing the view of a party can be received by a trivial transformation of the view that we write below).

Case 1 – P_1 is corrupted

Note that P_1's view in an execution of π consists only of its view in the oblivious transfer protocols. By the security of the oblivious transfer protocol, P_1's view in the oblivious transfer executions can be generated without knowing P_2's input. The formal proof of this follows a rather standard hybrid argument.

We begin by describing the simulator S_1: Upon input $(x, f(x,y))$, simulator S_1 uniformly chooses a random tape r_C for P_1 and generates the garbled circuit that P_1 would generate with randomness r_C. Then, let $k_{n+1}^0, k_{n+1}^1, \ldots, k_{2n}^0, k_{2n}^1$ be the keys that correspond to P_2's input in the constructed garbled circuit, and let S_1^{OT} be the simulator that is guaranteed to exist for party P_1 in the oblivious transfer protocol. For every $i = 1, \ldots, n$, simulator S_1 invokes the simulator S_1^{OT} upon input (k_{n+i}^0, k_{n+i}^1) in order to obtain P_1's view in the ith oblivious transfer (since P_1 has no output from the oblivious transfer, the simulator is invoked with its input only). Recall that the view generated by S_1^{OT} is made up of the input (in this case (k_{n+i}^0, k_{n+i}^1)), a random tape, and a transcript of messages received. As we have mentioned, we will place the entire view of the party in the oblivious transfers together with the message transcript. We have that S_1 outputs

$$\left(x, r_C, S_1^{OT}(k_{n+1}^0, k_{n+1}^1), \ldots, S_1^{OT}(k_{2n}^0, k_{2n}^1)\right). \tag{3.4}$$

This concludes the description of S_1. We now prove that

$$\{S_1(x, f(x, y))\}_{x,y\in\{0,1\}^*} \stackrel{c}{\equiv} \{\text{view}_1^\pi(x, y)\}_{x,y\in\{0,1\}^*}$$

where $S_1(x, f(x, y))$ is as shown in (3.4) and π denotes Protocol 3.4.1. We first prove a hybrid argument over the simulated views for the oblivious transfers. That is, we define a hybrid distribution H_i in which the first i oblivious transfers are simulated and the last $n - i$ are real. Formally, let $H_i(x, y, r_C)$ denote the distribution

$$\{(x, r_C, S_1^{\text{OT}}(k_{n+1}^0, k_{n+1}^1), \ldots, S_1^{\text{OT}}(k_{n+i}^0, k_{n+i}^1),$$
$$R_1^{\text{OT}}((k_{n+i+1}^0, k_{n+i+1}^1), y_{i+1}), \ldots, R_1^{\text{OT}}((k_{2n}^0, k_{2n}^1), y_n))\}$$

where $R_1^{\text{OT}}((k_{n+j}^0, k_{n+j}^1), y_j)$ denotes the incoming messages from the appropriate *real oblivious transfer execution*; these incoming messages are part of $\text{view}_1^{\text{OT}}((k_{n+j}^0, k_{n+j}^1), y_j)$. Notice that the keys k_{n+j}^0, k_{n+j}^1 here are as defined by the garbled circuit when generated with the random tape r_C. Notice also that when r_C is uniformly chosen, $H_n(x, y, r_C)$ equals the distribution that appears in (3.4); i.e., it equals $S_1(x, f(x, y))$. Furthermore, $H_0(x, y, r_C)$ is exactly the same as $\text{view}_1^\pi(x, y)$. For simplicity, from here on we will assume that x, y, r_C are all of the same length, and in particular, are of length n.

We now prove that $\{H_0(x, y, r_C)\} \stackrel{c}{\equiv} \{H_n(x, y, r_C)\}$. By contradiction, assume that there exists a probabilistic polynomial-time distinguisher D and a polynomial $p(\cdot)$ such that for infinitely many ns (and $x, y, r_C \in \{0,1\}^n$),

$$|\Pr[D(H_0(x, y, r_C)) = 1] - \Pr[D(H_n(x, y, r_C)) = 1]| > \frac{1}{p(n)}.$$

It follows that there exists an i such that for infinitely many x, y, r_C,

$$|\Pr[D(H_i(x, y, r_C)) = 1] - \Pr[D(H_{i+1}(x, y, r_C)) = 1]| > \frac{1}{np(n)}.$$

We now use D to contradict the security of the oblivious transfer protocol. First, note that the only difference between $H_i(x, y, r_C)$ and $H_{i+1}(x, y, r_C)$ is that the random tape and transcript of the $(i + 1)$th oblivious transfer are according to $\text{view}_1^{\text{OT}}((k_{n+i+1}^0, k_{n+i+1}^1), y_{i+1})$ in H_i and according to $S_1^{\text{OT}}(k_{n+i+1}^0, k_{n+i+1}^1)$ in H_{i+1}. Furthermore, given x, y, r_C, i and a view v (which is either $\text{view}_1^{\text{OT}}((k_{n+i+1}^0, k_{n+i+1}^1), y_{i+1})$ or $S_1^{\text{OT}}(k_{n+i+1}^0, k_{n+i+1}^1)$) it is possible to construct a distribution H such that if v is from $\text{view}_1^{\text{OT}}$ then $H = H_i(x, y, r_C)$ and if v is from S_1^{OT} then $H = H_{i+1}(x, y, r_C)$. It therefore follows that for infinitely many inputs, it is possible to distinguish the view of P_1 in a real oblivious transfer execution from its simulated view with the same probability that it is possible to distinguish $H_i(x, y, r_C)$ from $H_{i+1}(x, y, r_C)$. However, this contradicts the security of the oblivious transfer protocol. We therefore conclude that $\{H_0(x, y, r_C)\} \stackrel{c}{\equiv} \{H_n(x, y, r_C)\}$. (We remark that the distinguisher that we construct here is *non-uniform* because it needs to have x, y, r_C and i. For this reason, we defined non-uniform indistinguishability; see the beginning of Section 2.1.) We conclude that

$$\{S_1(x, f(x, y))\} \overset{c}{\equiv} \{(x, r_C, R_1^{\mathrm{OT}}((k_n^0, k_n^1), y_1), \ldots, R_1^{\mathrm{OT}}((k_{2n}^0, k_{2n}^1), y_n))\}$$

as required.

Case 2 – P_2 is corrupted

In this case, we construct a simulator S_2 that is given input $(y, f(x, y))$ and generates the view of P_2 in Protocol 3.4.1. Notice that P_2 expects to receive a garbled circuit, and so S_2 must generate such a circuit. Furthermore, this circuit must be such that P_2 would obtain $f(x, y)$ when computing the circuit according to the protocol instructions. Of course, S_2 cannot just honestly generate the circuit, because it does not know x. (Without knowing x, it would not know which of the keys $k_1^0, k_1^1, \ldots, k_n^0, k_n^1$ to hand to P_2.) It therefore generates a "fake" garbled circuit that always evaluates to $f(x, y)$, irrespective of which keys are used. This is achieved by using gate tables in which all four entries encrypt the same key, and therefore the values of the input wires do not affect the value of the output wire. The crux of the proof is in showing that this circuit is indistinguishable from the real garbled circuit that P_2 receives in a real execution.

In order to show this we use a hybrid argument. We first show that P_2's view in a *real* execution of the protocol is indistinguishable from a hybrid distribution $H_{\mathrm{OT}}(x, y)$ in which the real oblivious transfers are replaced with simulated ones. Next, we consider a series of hybrids $H_i(x, y)$ in which one gate at a time is replaced in the real garbled circuit. The hybrid distributions are such that $H_0(x, y)$ contains a real garbled circuit (and therefore equals $H_{\mathrm{OT}}(x, y)$). In contrast, distribution $H_{|C|}(x, y)$ contains the same fake circuit constructed by S_2 (and, as we will see, therefore equals $S_2(y, f(x, y))$). By a standard hybrid argument, it follows that a distinguisher between $H_0(x, y)$ and $H_{|C|}(x, y)$ can be used to distinguish between two successive hybrids. However, the security of the encryption scheme that is used for generating the gate tables ensures that neighboring hybrids are computationally indistinguishable. We conclude that $H_0(x, y)$ is indistinguishable from $H_{|C|}(x, y)$, and so $\{S_2(y, f(x, y))\} \overset{c}{\equiv} \{\mathrm{view}_2^\pi(x, y)\}$.

We now formally describe $\tilde{S}_2$. Simulator S_2 begins by constructing a fake garbled circuit, denoted by $\tilde{G}(C)$. This is accomplished as follows. For every wire w_i in the circuit C, simulator S_2 chooses two random keys k_i and k_i'. Next, the gates are computed: let g be a gate with input wires w_i, w_j and output wire w_ℓ. Then, g contains encryptions of the single key k_ℓ under *all four combinations* of the keys k_i, k_i', k_j, k_j' that are associated with the input wires to g (in contrast, the key k_ℓ' is not encrypted at all). That is, S_2

computes the values

$$c_{0,0} = E_{k_i}(E_{k_j}(k_\ell)), \qquad c_{0,1} = E_{k_i}(E_{k'_j}(k_\ell)),$$
$$c_{1,0} = E_{k'_i}(E_{k_j}(k_\ell)), \qquad c_{1,1} = E_{k'_i}(E_{k'_j}(k_\ell)),$$

and writes them in random order. This is carried out for all of the gates of the circuit. It remains to describe how the output decryption tables are constructed. Denote the n-bit output $f(x, y)$ by $z_1 \cdots z_n$ (recall that this is part of S_2's input), and denote the circuit-output wires by $w_{m-n+1}, \ldots, w_m$. In addition, for every $i = 1, \ldots, n$, let k_{m-n+i} be the (single) key encrypted in the gate from which wire w_{m-n+i} left, and let k'_{m-n+i} be the other key (as described above). Then, the output decryption table for wire w_{m-n+i} is given by $[(0, k_{m-n+i}), (1, k'_{m-n+i})]$ if $z_i = 0$, and $[(0, k'_{m-n+i}), (1, k_{m-n+i})]$ if $z_i = 1$. This completes the description of the construction of the fake garbled circuit $\tilde{G}(C)$. (Notice that the keys $k_{m-n+1}, \ldots, k_m$ decrypt to $z_1 \cdots z_n = f(x, y)$ exactly.)

Next, S_2 generates the view of P_2 in the phase where it obtains the keys. First, in the simulated view, it sets the keys that P_2 receives from P_1 in Step 2a of Protocol 3.4.1 to be $k_1, \ldots, k_n$. (Recall that $w_1, \ldots, w_n$ are the circuit-input wires associated with P_1's input x and that the keys for these wires are $k_1, k'_1, \ldots, k_n, k'_n$. Here, S_2 takes the keys $k_1, \ldots, k_n$. However, it could have taken $k'_1, \ldots, k'_n$ or any other combination and this would make no difference.) Next, let S_2^{OT} be the simulator that is guaranteed to exist for the oblivious transfer protocol. Then, for every $i = 1, \ldots, n$, simulator S_2 invokes the simulator S_2^{OT} upon input (y_i, k_{n+i}) in order to obtain P_2's view in the ith oblivious transfer. (Here y_i and k_{n+i} are P_2's respective input and output in the ith oblivious transfer. As above, we use the keys $k_{n+1}, \ldots, k_{2n}$ associated with the input wires for y. However, this choice is arbitrary and we could have used $k'_{n+1}, \ldots, k'_{2n}$ or any other combination.) Recall that the view generated by S_2^{OT} is made up of the input (in this case y_i), a random tape, and a transcript of messages received. Recall also that by convention, we place the entire view in the oblivious transfer (including the random tape) together. We therefore have that S_2 outputs

$$\left(y, \tilde{G}(C), k_1, \ldots, k_n, S_2^{\mathrm{OT}}(y_1, k_{n+1}), \ldots, S_2^{\mathrm{OT}}(y_n, k_{2n})\right).$$

This concludes the description of S_2. We now prove that

$$\{S_2(y, f(x, y))\}_{x, y \in \{0,1\}^*} \stackrel{\mathrm{c}}{\equiv} \{\mathrm{view}_2^\pi(x, y)\}_{x, y \in \{0,1\}^*}.$$

First, observe that

$$\{\mathrm{view}_2^\pi(x, y)\}$$
$$\equiv \left\{(y, G(C), k_1^{x_1}, \ldots, k_n^{x_n}, R_2^{\mathrm{OT}}((k_{n+1}^0, k_{n+1}^1), y_1), \ldots, R_2^{\mathrm{OT}}((k_{2n}^0, k_{2n}^1), y_n))\right\}$$

where $R_2^{\text{OT}}((k_{n+i}^0, k_{n+i}^1), y_i)$ denotes the *real transcript* obtained from the view $\text{view}_2^{\text{OT}}((k_{n+i}^0, k_{n+i}^1), y_i)$. We also denote the hybrid distribution where the real oblivious transfers are replaced with simulated ones by

$$H_{\text{OT}}(x, y) = (y, G(C), k_1^{x_1}, \ldots, k_n^{x_n}, S_2^{\text{OT}}(y_1, k_{n+1}^{y_1}), \ldots, S_2^{\text{OT}}(y_n, k_{2n}^{y_n})).$$

We stress that in the hybrid distribution H_{OT}, the garbled circuit $G(C)$ that appears is the real one and not the fake one. We first claim that

$$\{H_{\text{OT}}(x, y)\}_{x,y \in \{0,1\}^*} \stackrel{\text{c}}{\equiv} \{\text{view}_2^\pi(x, y)\}_{x,y \in \{0,1\}^*}. \tag{3.5}$$

The only difference between the distributions in (3.5) is due to the fact that simulated views of the oblivious transfers are provided instead of real ones. Indistinguishability therefore follows from the security of the oblivious transfer protocol. The formal proof of this is almost identical to the case where P_1 is corrupted, and is therefore omitted.

Next, we consider a series of hybrid experiments $H_i(x, y)$ in which one gate at a time is replaced in the real garbled circuit $G(C)$ until the result is the fake garbled circuit $\tilde{G}(C)$. Before we do this, we consider an alternative way of constructing the fake garbled circuit $\tilde{G}(C)$. This alternative construction uses knowledge of both inputs x and y, but results in exactly the same fake garbled circuit as that constructed by S_2 that is given only y and $f(x, y)$. (This is therefore just a mental experiment, or a different description of S_2. Nevertheless, it is helpful in describing the proof.)

The alternative construction works by first traversing the circuit from the circuit-input wires to the circuit-output wires, and labelling all keys as active or inactive. Intuitively, a key is active if it is used in order to compute the garbled circuit upon input (x, y); otherwise it is inactive. Formally, a key k_a^α that is associated with wire w_a is active if when computing the non-garbled circuit C on input (x, y), the bit that is obtained on wire w_a equals α. As expected, an inactive key is just any key that is not active. Now, the alternative construction of $\tilde{G}(C)$ works by first constructing the real garbled circuit $G(C)$. Next, using knowledge of both x and y, all keys in $G(C)$ are labelled active or inactive (given x and y, it is possible to compute $C(x, y)$ and obtain the real values on each wire). Finally, $\tilde{G}(C)$ is obtained by replacing each gate g as follows: Let w_a be the wire that exits gate g. Then, recompute g by encrypting the *active key* on wire w_a with all four combinations of the (active and inactive) keys that are on the wires that enter g. This completes the alternative construction.

We claim that the circuit obtained in this alternative construction is identically distributed to the circuit constructed by $S_2(x, f(x, y))$. First, in both constructions, all gates contain encryptions of a single key only. Second, in both constructions, the order of the ciphertexts in each gate is random. Finally, in both constructions, the output decryption tables yield the same result (i.e., exactly $f(x, y)$). This last observation is due to the fact that in the alternative construction, the output decryption table decrypts active keys to $f(x, y)$ and these active keys are the only ones encrypted in the gates from which the circuit-output wires exit. Likewise, in the circuit $\tilde{G}(C)$, the only keys encrypted in the gates from which the circuit-output wires exit are the keys that decrypt to $f(x, y)$.

Before proceeding we order the gates $g_1, \ldots, g_{|C|}$ of the circuit C as follows: if the input wires to a gate g_ℓ come from gates g_i and g_j, then $i < \ell$ and $j < \ell$; this is called a **topological sort** of the circuit. We are now ready to define the hybrid experiment $H_i(x, y)$.

Hybrid experiment $H_i(x, y)$. In this experiment the view of P_2 in the oblivious transfers is generated in exactly the same way as in $H_{OT}(x, y)$. However, the garbled circuit is constructed differently. As in the alternative construction of $\tilde{G}(C)$, the first step is to construct the real garbled circuit $G(C)$ and then use x and y in order to label all keys in $G(C)$ as active or inactive. Next, the first i gates $g_1, \ldots, g_i$ are modified as in the alternative construction. That is, let w_a be the wire that exits gate g_j for $1 \leq j \leq i$. Then, recompute g_j by encrypting the *active key* on wire w_a with all four combinations of the (active and inactive) keys that are on the wires that enter g_j. The remaining gates $g_{i+1}, \ldots, g_{|C|}$ are left unmodified, and are therefore as in the real garbled circuit $G(C)$.

We claim that the distribution $\{H_0(x, y)\}$ equals $\{H_{OT}(x, y)\}$. This follows from the fact that the only difference is that in $H_0(x, y)$ the keys are labelled active or inactive. However, since nothing is done with this labelling, there is no difference in the resulting distribution. Next, notice that in $H_{|C|}(x, y)$, the circuit that appears in the distribution is exactly the fake garbled circuit $\tilde{G}(C)$ as constructed by S_2. This follows immediately from the fact that in $H_{|C|}$ all gates are replaced, and so the circuit obtained is exactly that of the full alternative construction described above.

We wish to show that $\{H_0(x, y)\} \overset{c}{\equiv} \{H_{|C|}(x, y)\}$. Intuitively, this follows from the indistinguishability of encryptions. Specifically, the only difference between H_0 and $H_{|C|}$ is that the circuit in H_0 is made up of gates that contain encryptions of active and inactive keys, whereas the circuit in $H_{|C|}$ is made up of gates that contain encryptions of active keys only. Since only active keys are seen by P_2 during the computation of the garbled circuit, the difference between H_0 and $H_{|C|}$ cannot be detected.

We prove that $\{H_0(x, y)\} \overset{c}{\equiv} \{H_{|C|}(x, y)\}$ using a hybrid argument. That is, assume that there exists a non-uniform probabilistic polynomial-time distinguisher D and a polynomial $p(\cdot)$ such that for infinitely many ns (and values $x, y \in \{0, 1\}^n$),

$$\left| \Pr[D(H_0(x, y)) = 1] - \Pr[D(H_{|C|}(x, y)) = 1] \right| > \frac{1}{p(n)}.$$

Then, it follows that there exists an i such that

$$|\Pr[D(H_{i-1}(x,y)) = 1] - \Pr[D(H_i(x,y)) = 1]| > \frac{1}{|C| \cdot p(n)}.$$

We use D and x, y, i in order to construct a non-uniform probabilistic polynomial-time distinguisher $\mathcal{A}_E$ for the encryption scheme (G, E, D). The high-level idea here is for $\mathcal{A}_E$ to receive some ciphertexts from which it will construct a partially real and partially fake garbled circuit $G'(C)$. However, the construction will be such that if the ciphertexts received were of one "type", then the resulting circuit is according to $H_{i-1}(x,y)$. However, if the ciphertexts received were of another "type", then the resulting circuit is according to $H_i(x,y)$. In this way, the ability to successfully distinguish $H_{i-1}(x,y)$ from $H_i(x,y)$ yields the ability to distinguish ciphertexts, in contradiction to the security of the encryption scheme. We now formally prove the above intuition, using Lemma 3.2.3 that states that (G, E, D) is secure under chosen double encryption.

A concrete case. First, let us consider the concrete case where g_i is an OR gate, and that wires w_a and w_b enter g_i, and wire w_c exits g_i. Furthermore, assume that the wires w_a and w_b enter gate g_i and no other gate. Finally, assume that when the inputs to the circuit are x and y, the wire w_a obtains the bit 0 and the wire w_b obtains the bit 1. Then, it follows that the keys k_a^0 and k_b^1 are active, and the keys $\boldsymbol{k_a^1}$ and $\boldsymbol{k_b^0}$ are inactive (we mark the inactive keys in bold in order to distinguish them from the active ones). Likewise, the key k_c^1 is active (because $g_i(0,1) = 0 \vee 1 = 1$) and the key $\boldsymbol{k_c^0}$ is inactive. The difference between a real garbled gate g_i and a fake garbled gate g_i is with respect to the encrypted values. Specifically, the real garbled OR gate g_i contains the following values:

$$E_{k_a^0}(\boldsymbol{E_{k_b^0}}(\boldsymbol{k_c^0})), \ E_{k_a^0}(E_{k_b^1}(k_c^1)), \ \boldsymbol{E_{k_a^1}}(\boldsymbol{E_{k_b^0}}(k_c^1)), \ \boldsymbol{E_{k_a^1}}(E_{k_b^1}(k_c^1)) \qquad (3.6)$$

In contrast, the fake garbled OR gate g_i contains the following values which are all encryptions of the active value k_c^1 (recall that the input to g_i equals 0 and 1, and so the output is 1):

$$E_{k_a^0}(\boldsymbol{E_{k_b^0}}(k_c^1)), \ E_{k_a^0}(E_{k_b^1}(k_c^1)), \ \boldsymbol{E_{k_a^1}}(\boldsymbol{E_{k_b^0}}(k_c^1)), \ \boldsymbol{E_{k_a^1}}(E_{k_b^1}(k_c^1)) \qquad (3.7)$$

Thus, in this *concrete* case, the indistinguishability between the gates depends on the indistinguishability of a single encryption (of k_c^0 versus k_c^1) under the inactive key $\boldsymbol{k_b^0}$. (In other cases, the indistinguishability may depend on both inactive keys $\boldsymbol{k_a^1}$ and $\boldsymbol{k_b^0}$, and may depend on more than one encryption under a key; see the general case below.) Indistinguishability follows directly from the definition of $\mathsf{Expt}_{\mathcal{A}}^{\mathrm{double}}$ and security under chosen double encryption (see Definition 3.2.2 in Section 3.2.1).

Formally, we construct a non-uniform probabilistic polynomial-time machine $\mathcal{A}_E$ for $\mathsf{Expt}^{\mathrm{double}}_{\mathcal{A}_E}$ such that

$$\left| \Pr\left[\mathsf{Expt}^{\mathrm{double}}_{\mathcal{A}_E}(n,1)=1\right] - \Pr\left[\mathsf{Expt}^{\mathrm{double}}_{\mathcal{A}_E}(n,0)=1\right] \right| \geq \frac{1}{|C|p(n)}.$$

Upon input 1^n, machine $\mathcal{A}_E$ outputs keys $k_a^0, k_b^1 \leftarrow G(1^n)$ and message triples (k_c^0, k_c^1, k_c^1) and (k_c^1, k_c^1, k_c^1). By the experiment $\mathsf{Expt}^{\mathrm{double}}$, two keys are chosen. For the sake of consistency, we denote them by $\boldsymbol{k_a^1}$ and $\boldsymbol{k_b^0}$. Then, machine $\mathcal{A}_E$ receives either the ciphertexts

$$\langle E_{k_a^0}(\boldsymbol{E_{k_b^0}}(k_c^0)), E_{k_a^1}(\boldsymbol{E_{k_b^1}}(k_c^1)), E_{\boldsymbol{k_a^1}}(\boldsymbol{E_{k_b^0}}(k_c^1)) \rangle \qquad (3.8)$$

or the ciphertexts

$$\langle E_{k_a^0}(\boldsymbol{E_{k_b^0}}(k_c^1)), E_{k_a^1}(\boldsymbol{E_{k_b^1}}(k_c^1)), E_{\boldsymbol{k_a^1}}(\boldsymbol{E_{k_b^0}}(k_c^1)) \rangle \qquad (3.9)$$

depending on whether $\sigma = 0$ (the first case) or $\sigma = 1$ (the second case); note that the only difference is that in the first case the first plaintext is k_c^0 and in the second case the first plaintext is k_c^1. Denote the ciphertexts received by $\mathcal{A}_E$ by (c_1, c_2, c_3). Now, $\mathcal{A}_E$ first computes the value $c = E_{k_a^0}(E_{k_b^1}(k_c^1))$; it can do this by itself because it knows both k_a^0 and k_b^1, as well as k_c^1. Next, given c, $\mathcal{A}_E$ generates the tuple $\langle c_1, c, c_3, c_2 \rangle$. The important point to notice here is that if $\mathcal{A}_E$ received the ciphertexts in (3.8) then $\langle c_1, c, c_3, c_2 \rangle$ is identical to the ciphertexts in (3.6). On the other hand, if $\mathcal{A}_E$ received the ciphertexts in (3.9) then $\langle c_1, c, c_3, c_2 \rangle$ is identical to the ciphertexts in (3.7). Therefore, if it is possible to distinguish between the gates in Equations (3.6) and (3.7) with non-negligible probability, then $\mathcal{A}_E$ can succeed in $\mathsf{Expt}^{\mathrm{double}}_{\mathcal{A}_E}$ with non-negligible probability, in contradiction to the security of the encryption scheme.

This does not yet suffice because we must still show how $\mathcal{A}_E$ can generate the rest of the H_{i-1} or H_i distributions. Notice that $\mathcal{A}_E$ knows the active keys that enter g_i (because it chose them itself), but does not know the inactive keys. We therefore show that the distributions can be constructed without knowledge of the inactive keys $\boldsymbol{k_a^1}$ and $\boldsymbol{k_b^0}$. In order to show this, we distinguish between two cases:

1. *Case 1 – w_b is a circuit-input wire:* In this case, the keys associated with wire w_b do not appear in any gates g_j for $j < i$. However, keys that are associated with circuit-input wires do appear in the distributions H_{i-1} and H_i: the keys $k_i^{x_i}$ appear directly and the keys $k_{n+i}^{y_i}$ are used to generate the view of P_2 in the oblivious transfers. Nevertheless, notice that the keys used here are all *active*. Therefore, $\mathcal{A}_E$ can construct the distributions, as required. We note that $\mathcal{A}_E$ uses the keys k_c^0 and k_c^1 from the experiment in order to construct the gates into which wire w_c enters.

2. *Case 2 – w_b is not a circuit-input wire:* In this case, the keys associated
with wire w_b can appear only in the gate g_j from which w_b exits. However,
by our ordering of the gates, $j < i$. Therefore, in both H_{i-1} and H_i, gate
g_j contains encryptions of the *active* key k_b^0 only. It follows that $\mathcal{A}_E$ can
construct the rest of the distribution, as required. (Again, as above, $\mathcal{A}_E$
uses the keys k_c^0 and k_c^1 in this construction.)

Now, as we have shown above, if $\mathcal{A}_E$ participates in $\mathsf{Expt}_{\mathcal{A}_E}^{\mathsf{double}}(n,0)$, then
the gate g_i is constructed as for a real garbled circuit. In contrast, if $\mathcal{A}_E$
participates in $\mathsf{Expt}_{\mathcal{A}_E}^{\mathsf{double}}(n,1)$, then the gate g_i is constructed as for a fake
garbled circuit. The only dependence between the gate g_i and the rest of the
distribution H_{i-1} or H_i is with respect to the keys k_c^0 and k_c^1; however, these
are known to $\mathcal{A}_E$ and used appropriately. We therefore conclude that if $\mathcal{A}_E$
participates in $\mathsf{Expt}_{\mathcal{A}_E}^{\mathsf{double}}(n,0)$, then it generates a distribution H that equals
$H_{i-1}(x,y)$. In contrast, if it participates in $\mathsf{Expt}_{\mathcal{A}_E}^{\mathsf{double}}(n,1)$, then it generates
a distribution H that equals $H_i(x,y)$. Distinguisher $\mathcal{A}_E$ concludes by running
machine D on the distribution H and outputting whatever D does. By the
contradicting assumption, machine D distinguishes $H_{i-1}(x,y)$ from $H_i(x,y)$
with probability $1/|C|p(n)$. That is, we have that for infinitely many ns

$$\left| \Pr[\mathsf{Expt}_{\mathcal{A}_E}^{\mathsf{double}}(n,0) = 1] - \Pr[\mathsf{Expt}_{\mathcal{A}_E}^{\mathsf{double}}(n,1) = 1] \right|$$

$$= \left| \Pr[D(H_{i-1}(x,y)) = 1] - \Pr[D(H_i(x,y)) = 1] \right| > \frac{1}{|C|p(n)}$$

in contradiction to the security of the encryption scheme. It follows that
$\{H_0(x,y)\} \overset{c}{\equiv} \{H_{|C|}(x,y)\}$. Having proven the argument with respect to a
concrete case, we now move to the general case.

The general case. Let g_i be an arbitrary gate, let w_a and w_b be the wires
entering g_i and let w_c be the wire that exits g_i. Furthermore, let α and β be
the respective values obtained on w_a and w_b in $C(x,y)$. Note that this means
that k_a^α and k_b^β are active, and $\boldsymbol{k_a^{1-\alpha}}$ and $\boldsymbol{k_b^{1-\beta}}$ are inactive. Then, the real
garbled gate g_i contains the following values (in a random order):

$$
\begin{array}{ll}
E_{k_a^\alpha}(E_{k_b^\beta}(k_c^{g_i(\alpha,\beta)})), & E_{k_a^\alpha}(E_{k_b^{1-\beta}}(k_c^{g_i(\alpha,1-\beta)})), \\[2mm]
\boldsymbol{E_{k_a^{1-\alpha}}}(E_{k_b^\beta}(k_c^{g_i(1-\alpha,\beta)})), & \boldsymbol{E_{k_a^{1-\alpha}}}(\boldsymbol{E_{k_b^{1-\beta}}}(k_c^{g_i(1-\alpha,1-\beta)})).
\end{array}
\tag{3.10}
$$

In contrast, the fake garbled gate g_i contains the following values which are
all encryptions of the active value $k_c^{g_i(\alpha,\beta)}$:

$$
\begin{array}{ll}
E_{k_a^\alpha}(E_{k_b^\beta}(k_c^{g_i(\alpha,\beta)})), & E_{k_a^\alpha}(\boldsymbol{E_{k_b^{1-\beta}}}(k_c^{g_i(\alpha,\beta)})), \\[2mm]
\boldsymbol{E_{k_a^{1-\alpha}}}(E_{k_b^\beta}(k_c^{g_i(\alpha,\beta)})), & \boldsymbol{E_{k_a^{1-\alpha}}}(\boldsymbol{E_{k_b^{1-\beta}}}(k_c^{g_i(\alpha,\beta)})).
\end{array}
\tag{3.11}
$$

Thus, the indistinguishability between the gates depends on the indistin-
guishability of encryptions under the inactive keys $\boldsymbol{k_a^{1-\alpha}}$ and $\boldsymbol{k_b^{1-\beta}}$. As above,

we use $\mathsf{Expt}^{\mathsf{double}}$ and security under chosen double encryption. The gate is generated in exactly the same way here as in the concrete case. Now, in the restricted case where both wires w_a and w_b enter the gate g_i only, it is possible to proceed in the same way as in the concrete case above. However, in the more general case wires w_a and w_b may enter multiple gates (in particular, one of the wires may enter a gate g_j with $j > i$). In this case, $\mathcal{A}_E$ *cannot* construct the rest of the circuit given only the active keys, because the inactive keys $\boldsymbol{k}_a^{1-\alpha}$ and $\boldsymbol{k}_b^{1-\beta}$ are used in more than one gate. (We stress that in order to prove the indistinguishability of the neighboring hybrid H_{i-1} and H_i, it is crucial that $\mathcal{A}_E$ is not given these inactive keys. Therefore, it cannot construct these other gates itself.) This is solved by using the *special chosen-plaintext attack* of $\mathsf{Expt}^{\mathsf{double}}$. Recall that in $\mathsf{Expt}^{\mathsf{double}}$, the adversary has access to oracles $\overline{E}(\cdot, k_1', \cdot)$ and $\overline{E}(k_0', \cdot, \cdot)$, where $k_0' = \boldsymbol{k}_a^{1-\alpha}$ and $k_1' = \boldsymbol{k}_b^{1-\beta}$. Here, this means that the adversary can ask for encryptions under these inactive keys, as needed for constructing all of the other gates $g_{i_1}^a, \ldots, g_{i_j}^a$ and $g_{i_1}^b, \ldots, g_{i_j}^b$ that use them. Once again, we have that in $\mathsf{Expt}^{\mathsf{double}}_{\mathcal{A}_E}(n, 0)$ the distribution generated by $\mathcal{A}_E$ is exactly that of $H_{i-1}(x, y)$, whereas in $\mathsf{Expt}^{\mathsf{double}}_{\mathcal{A}_E}(n, 1)$ the distribution generated by $\mathcal{A}_E$ is exactly that of $H_i(x, y)$. Therefore, as above, we conclude that $H_{i-1}(x, y)$ is indistinguishable from $H_i(x, y)$ and so $\{H_0(x, y)\} \stackrel{c}{\equiv} \{H_{|C|}(x, y)\}$.

Concluding the proof. Having proven that $\{H_0(x, y)\} \stackrel{c}{\equiv} \{H_{|C|}(x, y)\}$, we obtain that

$$
\begin{aligned}
&\left\{(y, \tilde{G}(C), k_1^{x_1}, \ldots, k_n^{x_n}, S_2^{\mathrm{OT}}(y_1, k_{n+1}^{y_1}), \ldots, S_2^{\mathrm{OT}}(y_n, k_{2n}^{y_n}))\right\} \\
&\stackrel{c}{\equiv} \left\{(y, G(C), k_1^{x_1}, \ldots, k_n^{x_n}, S_2^{\mathrm{OT}}(y_1, k_{n+1}^{y_1}), \ldots, S_2^{\mathrm{OT}}(y_n, k_{2n}^{y_n}))\right\} \\
&\equiv \{H_{\mathrm{OT}}(x, y)\}.
\end{aligned}
\tag{3.12}
$$

Notice that the first distribution in (3.12) looks almost the same as the distribution $\{S_2(y, f(x, y))\}$. The only difference is that in $S_2(y, f(x, y))$ the keys $k_1, \ldots, k_n, k_{n+1}, \ldots, k_{2n}$ are used instead of the keys $k_1^{x_1}, \ldots, k_n^{x_n}$, $k_{n+1}^{y_1}, \ldots, k_{2n}^{y_n}$. That is, the keys that S_2 takes for the circuit-input wires have no correlation to the actual input (unlike in a real execution). However, in the fake garbled circuit $\tilde{G}(C)$, there is no difference between k_i and k_i' because all combinations of keys are used to encrypt the same (active) key. Thus, the distribution over the keys $k_1, \ldots, k_n, k_{n+1}, \ldots, k_{2n}$ and $k_1^{x_1}, \ldots, k_n^{x_n}, k_{n+1}^{y_1}, \ldots, k_{2n}^{y_n}$ are identical in the fake garbled-circuit construction. We therefore obtain that the first distribution in (3.12) is actually identical to the distribution $\{S_2(y, f(x, y))\}$ and so

$$
\{S_2(y, f(x, y))\}_{x, y \in \{0,1\}^*} \stackrel{c}{\equiv} \{H_{\mathrm{OT}}(x, y)\}_{x, y \in \{0,1\}^*}.
$$

Recalling that by (3.5), $\{H_{\mathrm{OT}}(x,y)\} \overset{\mathrm{c}}{\equiv} \{\mathsf{view}_2^\pi(x,y)\}$, we conclude that

$$\{S_2(y, f(x,y))\}_{x,y \in \{0,1\}^*} \overset{\mathrm{c}}{\equiv} \{\mathsf{view}_2^\pi(x,y)\}_{x,y \in \{0,1\}^*}$$

as required.   ■

By Theorem 3.2.5 it is possible to securely compute the oblivious transfer functionality assuming the existence of enhanced trapdoor permutations. Furthermore, secure encryption schemes as required in Theorem 3.4.2 can be constructed from one-way functions, and so also from enhanced trapdoor permutations. Finally, recall that given a secure protocol for deterministic single-output functionalities, it is possible to obtain a secure protocol for arbitrary probabilistic reactive functionalities. Combining these facts with Theorem 3.4.2, we obtain the following corollary:

Corollary 3.4.3 *Let $f = (f_1, f_2)$ be a probabilistic polynomial-time (possibly reactive) functionality. Then, assuming the existence of enhanced trapdoor permutations, there exists a constant-round protocol that securely computes f in the presence of static semi-honest adversaries.*

3.5 Efficiency of the Protocol

In this book, we analyze the efficiency of protocols according to their round complexity, bandwidth and local computation, where the latter is typically divided into the number of symmetric operations (which are generally very fast) and asymmetric operations (which are slower). The reason for counting the number of symmetric and asymmetric operations is that they usually dominate the local computation of the parties in cryptographic protocols. Needless to say, in protocols or schemes where other computation dominates, this must be taken into account.

In light of the above, we now analyze the complexity of Protocol 3.4.1 by counting the number of rounds of communication, the number of asymmetric and symmetric computations, and the bandwidth (or overall length of messages sent):

- **Number of rounds:** The protocol consists of only four rounds of communication (two messages are sent by each party). Note that the garbled circuit itself and the strings $k_1^{x_1}, \ldots, k_n^{x_n}$ can be sent by P_1 together with the first (or last) round of the oblivious transfer protocol. Thus, we have three rounds of oblivious transfer and a final round where P_2 sends P_1 the output of the circuit computation (the protocol for single-output functionalities does not have this round, but we included it for the general case). We remark that by using a two-round oblivious transfer protocol (see Section 7.2) that is secure in the presence of semi-honest adversaries under the DDH assumption, we have three rounds overall.

- **Asymmetric computations:** The parties run a single oblivious transfer protocol for each bit of the input of P_2. Thus, denoting the input of P_2 by y, we have that the protocol requires $O(|y|)$ asymmetric operations. Being more exact, if we use the oblivious transfer protocol presented as Protocol 3.2.4, then P_1 needs to sample a trapdoor permutation and then P_1 carries out one asymmetric computation (to compute $w_\sigma = f(v_\sigma)$) and P_2 carries out two asymmetric computations (to invert v_0 and v_1). We note that the most expensive operation is typically sampling the permutation-trapdoor pair. However, this can be carried out once for all the $|y|$ oblivious transfers. In addition, the protocol of Section 7.2 can be used, which is more efficient.

- **Symmetric computations:** In order to construct the circuit, P_1 needs to prepare the garbled circuit gates; each such gate requires eight encryptions and so P_1 carries out $8 \cdot |C|$ symmetric computations. Then, in order to compute the circuit, P_2 needs to attempt to decrypt all values in a gate until it finds one that decrypts correctly. Thus, in the worse case P_2 also carries out $8 \cdot |C|$ symmetric computations. However, given that the values in each gate are provided in a random order, the *expected* number of symmetric computations in every gate is four (to compute half) and thus we expect P_2 to carry out $4 \cdot |C|$ symmetric computations.

- **Bandwidth:** The bandwidth is comprised of the messages sent for the oblivious transfers and the garbled circuit (the keys $k_1^{x_1}, \ldots, k_n^{x_n}$ sent by P_1 are dominated by the size of the garbled circuit). Now, once again using Protocol 3.2.4 for the oblivious transfers, and reusing the same trapdoor permutation, we have that P_2 sends P_1 two asymmetric values per bit transferred. Thus, overall $2 \cdot |y|$ asymmetric values are sent. In addition, P_1 replies with two strings of the length of the symmetric keys. Thus denoting by n_{asym} the length of asymmetric values and by n_{sym} the length of symmetric values (or keys), we have that the oblivious transfers involve sending $2 \cdot |y| \cdot (n_{asym} + n_{sym})$ bits. In addition, P_1 sends P_2 the garbled circuit itself, which is of size $O(n_{sym} \cdot |C|)$. We remark that the constant hidden inside the O value is significant here. This is because each encryption adds $2n_{sym}$ bits to the length of the ciphertext (this is the case when using the construction of an encryption scheme with an elusive range that appears in Section 3.2.1) and thus a single encryption of a symmetric key yields a ciphertext of size $3n_{sym}$. Since this is encrypted again (for the double encryption of each key), we have that each double encryption is of size $5n_{sym}$. Each gate has four such encryptions and so we have that the overall size of the circuit is $20 \cdot n_{sym} \cdot |C|$. We stress, however, that it is possible to do this much more efficiently in practice; see [58, 71].

In conclusion, when computing a function with a reasonably small circuit on inputs that are not too large, Yao's protocol is very efficient. In some cases this is what is needed. For example, in Chapter 8 it is necessary to use a subprotocol for comparing two numbers; the circuit for such a computation is small. Of course, it only suffices for the case of semi-honest adversaries.

We will deal with the case of malicious adversaries (which is what is really needed in Chapter 8) in the coming chapter.

Chapter 4
Malicious Adversaries

In this chapter, we show how to securely compute any deterministic, single-output non-reactive functionality in the presence of malicious adversaries. As we have shown in Section 2.5, this suffices for obtaining secure computation of any two-party probabilistic reactive functionality, with almost the same complexity.

4.1 An Overview of the Protocol

The protocol that we present in this chapter is based upon Yao's garbled-circuit construction, as described in Chapter 3. There are a number of issues that must be dealt with when attempting to make Yao's protocol secure against malicious adversaries rather than just semi-honest ones (beyond the trivial observation that the oblivious transfer subprotocol must now be secure in the presence of malicious adversaries).

First and foremost, a malicious P_1 must be forced to construct the garbled circuit correctly so that it indeed computes the desired function. The method that is typically referred to for this task is called *cut-and-choose*. According to this methodology, P_1 constructs many independent copies of the garbled circuit and sends them to P_2. Party P_2 then asks P_1 to open half of them (chosen randomly). After P_1 does so, and party P_2 checks that the opened circuits are correct, P_2 is convinced that most of the remaining (unopened) garbled circuits are also constructed correctly. (If there are many incorrectly constructed circuits, then with high probability, one of those circuits will be in the set that P_2 asks to open.) The parties can then evaluate the remaining unopened garbled circuits as in the original protocol for semi-honest adversaries, and take the majority output value.[1]

---

[1] The reason for taking the majority value as the output is that the aforementioned test only reveals a single incorrectly constructed circuit with probability 1/2. Therefore, if P_1 generates a single or constant number of "bad" circuits, there is a reasonable chance that

C. Hazay, Y. Lindell, *Efficient Secure Two-Party Protocols*,
Information Security and Cryptography, DOI 10.1007/978-3-642-14303-8_4,
© Springer-Verlag Berlin Heidelberg 2010

The cut-and-choose technique described above indeed solves the problem of a malicious P_1 constructing incorrect circuits. However, it also generates new problems! The primary problem that arises is that since there are now many circuits being evaluated, we must make sure that both P_1 and P_2 use the same inputs in each circuit; we call these *consistency checks*. Consistency checks are important since if the parties were able to provide different inputs to different copies of the circuit, then they can learn information that is different from the desired output of the function. It is obvious that P_2 can do so, since it observes the outputs of all circuits, but in fact even P_1, who only gets to see the majority output, can learn additional information: Suppose, for example, that the protocol computes n invocations of a circuit computing the inner product of n-bit inputs. A malicious P_2 could provide the inputs $\langle 10 \cdots 0 \rangle$, $\langle 010 \cdots 0 \rangle, \ldots, \langle 0 \cdots 01 \rangle$, and learn all of P_1's input. If, on the other hand, P_1 is malicious, it could also provide the inputs $\langle 10 \cdots 0 \rangle$, $\langle 010 \cdots 0 \rangle, \ldots, \langle 0 \cdots 01 \rangle$. In this case, P_2 sends it the value which is output by the majority of the circuits, and which is equal to the majority value of P_2's input bits.

Another problem that arises when proving security is that the simulator must be able to fool P_2 and give it incorrect circuits (even though P_2 runs a cut-and-choose test). This is solved using rather standard techniques, like choosing the circuits to be opened via a coin-tossing protocol. Yet another problem is that P_1 might provide corrupt inputs to some of P_2's possible choices in the OT protocols (for shorthand, we will sometimes use "OT" instead of "oblivious transfer"). P_1 might then learn P_2's input based on whether or not P_2 aborts the protocol.

We begin by presenting a high-level overview of the protocol. We then proceed to describe the consistency checks, and finally the full protocol.

4.1.1 High-Level Protocol Description

We work with two security parameters. The parameter n is the security parameter for the commitment schemes, encryption, and the oblivious transfer protocol. The parameter s is a statistical security parameter which specifies how many garbled circuits are used. The difference between these parameters

---

it will not be caught. In contrast, there is only an exponentially small probability that the test reveals no corrupt circuit *and* at the same time a majority of the circuits that are not checked are incorrect. Consequently, with overwhelming probability it holds that if the test succeeds and P_2 takes the majority result of the remaining circuits, the result is correct. We remark that the alternative of aborting in case not all the outputs are the same (namely, where cheating is detected) is not secure and actually yields a concrete attack. The attack works as follows. Assume that P_1 is corrupted and that it constructs all of the circuits correctly except for one. The "incorrect circuit" is constructed so that it computes the exclusive-or of the desired function f with the first bit of P_2's input. Now, if P_2 policy is to abort as soon as two outputs are not the same then P_1 knows that P_2 aborts if, and only if, the first bit of its input is 1.

is due to the fact that the value of n depends on computational assumptions, whereas the value of s reflects the possible error probability that is incurred by the cut-and-choose technique and as such is a "statistical" security parameter. Although it is possible to use a single parameter n, it may be possible to take s to be much smaller than n. Recall that for simplicity, and in order to reduce the number of parameters, we denote the length of the input by n as well.

PROTOCOL 4.1.1 (high-level description): *Parties P_1 and P_2 have respective inputs x and y, and wish to compute the output $f(x, y)$ for P_2.*

1. *The parties decide on a circuit computing f. They then change the circuit by replacing each input wire of P_2 with a gate whose input consists of s new input wires of P_2 and whose output is the exclusive-or of these wires (an exclusive-or gate with s bits of input can be implemented using $s-1$ exclusive-or gates with two bits of input). Consequently, the number of input wires of P_2 increases by a factor of s. This is depicted in Figure 4.1.*
2. *P_1 commits to s different garbled circuits computing f, where s is a statistical security parameter. P_1 also generates additional commitments to the garbled values corresponding to the input wires of the circuits. These commitments are constructed in a special way in order to enable consistency checks.*
3. *For every input bit of P_2, parties P_1 and P_2 run a 1-out-of-2 oblivious transfer protocol in which P_2 learns the garbled values of input wires corresponding to its input.*
4. *P_1 sends to P_2 all the commitments of Step 1.*
5. *P_1 and P_2 run a coin-tossing protocol in order to choose a random string that defines which commitments and which garbled circuits will be opened.*
6. *P_1 opens the garbled circuits and committed input values that were chosen in the previous step. P_2 verifies the correctness of the opened circuits and runs consistency checks based on the decommitted input values.*
7. *P_1 sends P_2 the garbled values corresponding to P_1's input wires in the unopened circuits. P_2 runs consistency checks on these values as well.*
8. *Assuming that all of the checks pass, P_2 evaluates the unopened circuits and takes the majority value as its output.*

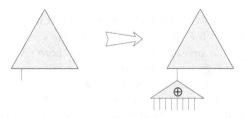

Fig. 4.1 Transforming one of P_2's input wires (Step 1 of Protocol 4.1.1)

4.1.2 Checks for Correctness and Consistency

As can be seen from the above overview, P_1 and P_2 run a number of checks, with the aim of forcing a potentially malicious P_1 to construct the circuits correctly and use the same inputs in (most of) the evaluated circuits. This section describes these checks. Unfortunately, we are unable to present the protocol, nor prove its security, in a modular fashion. Rather, the correctness and consistency checks are closely intertwined with the other parts of the protocol. We will therefore describe the correctness and consistency checks here, and describe the full protocol in Section 4.2. We hope that this improves the readability of the actual protocol.

Encoding P_2's input: As mentioned above, a malicious P_1 may provide corrupt input to one of P_2's possible inputs in an OT protocol. If P_2 chooses to learn this input it will not be able to decode the garbled tables which use this value, and it will therefore have to abort. If P_2 chooses to learn the other input associated with this wire then it will not notice that the first input is corrupt. P_1 can therefore learn P_2's input based on whether or not P_2 aborts. (Note that checking that the circuit is well formed will not help in thwarting this attack, since the attack is based on changing P_1's input to the OT protocol.) The attack is prevented by the parties replacing each input bit of P_2 with s new input bits whose exclusive-or is used instead of the original input (this step was described as Step 0 of Protocol 4.1.1, and is analyzed in Lemma 4.3.3). P_2 therefore has 2^{s-1} ways to encode a 0 input, and 2^{s-1} ways to encode a 1, and given its input it chooses an encoding with uniform probability. The parties then execute the protocol with the new circuit, and P_2 uses oblivious transfer to learn the garbled values of its new inputs. As we will show, if P_1 supplies incorrect values as garbled values that are associated with P_2's input, the probability of P_2 detecting this cheating is *almost independent* (up to a bias of 2^{-s+1}) of P_2's actual input. This is not true if P_2's inputs are not "split" in the way described above. The encoding presented here increases the number of P_2's input bits and, respectively, the number of OTs, from n to ns. In [55] it is shown how to reduce the number of new inputs for P_2 (and thus OTs) to a total of only $O(\max(s, n))$. Due to its complexity, we omit this improvement here.

An unsatisfactory method for proving consistency of P_1's input: Consider the following idea for forcing P_1 to provide the same input to all circuits. Let s be a security parameter and assume that there are s garbled copies of the circuit. Then, P_1 generates two ordered sets of commitments for every one of its input wires to the circuit. Each set contains s commitments: the "0 set" contains commitments to the garbled encodings of 0 for this wire in every circuit, and the "1 set" contains commitments to the garbled encodings of 1 for this wire in every circuit. P_2 receives these commitments from P_1 and then chooses a random subset of the circuits, which will be defined as check circuits. These circuits will never be evaluated and are used only for checking correctness and consistency. Specifically, P_2 asks P_1 to de-garble all of the check circuits and to open the values that correspond to the check circuits in *both* commitment sets. (That is, if circuit i is a check circuit, then P_1 decommits to both the 0 encoding and 1 encoding of all the input wires in circuit i.) Upon receiving the decommitments, P_2 verifies that all opened commitments from the "0 set" correspond to garbled values of 0, and that a similar property holds for commitments from the "1 set".

It now remains for P_2 to evaluate the remaining circuits. In order to do this, P_1 provides (for each of its input wires) the garbled values that are associated with the wire in all of the remaining circuits. Then, P_1 must prove that all of these values come from the same set, without revealing whether the set that they come from is the "0 set" or the "1 set" (otherwise, P_2 will know P_1's input). In this way, on the one hand, P_2 does not learn the input of P_1, and on the other hand, it is guaranteed that all of the values come from the same set, and so P_1 is forced into using the same input in all circuits. This proof can be carried out using, for example, the proofs of partial knowledge of [19]. However, this would require n proofs, each for s values, thereby incurring $O(ns)$ costly asymmetric operations, which we want to avoid.

Proving consistency of P_1's input: P_1 can prove consistency of its inputs without using public-key operations. The proof is based on a cut-and-choose test for the consistency of the commitment sets, which is combined with the cut-and-choose test for the correctness of the circuits. (Note that in the previous proposal, there is only one cut-and-choose test, and it is for the correctness of the circuits.) We start by providing a high-level description of the proof of consistency: The proof is based on P_1 constructing, *for each of its input wires*, s pairs of sets of commitments. One set in every pair contains commitments to the 0 values of this wire in *all* circuits, and the other set is the same with respect to 1. The protocol chooses a random subset of these pairs, and a random subset of the circuits, and checks that these sets provide consistent inputs for these circuits. Then the protocol evaluates the *remaining* circuits, and asks P_1 to open, in each of the *remaining* pairs, and only in one set in every pair, its garbled values for all evaluated circuits. (In this way, P_2 does not learn whether these garbled values correspond to a 0 or to a 1.) In order for the committed sets and circuits to pass P_2's checks, there must be large subsets C and S, of the circuits and commitment sets, respectively, such that every choice of a circuit from C and a commitment set from S results in a circuit and garbled values which compute the desired function f. P_2 accepts the verification stage only if all the circuits and sets it chooses to check are from C and S, respectively. This means that if P_2 does not abort then circuits which are not from C are likely to be a minority of the evaluated circuits, and a similar argument holds for S. Therefore the majority result of the evaluation stage is correct. The exact construction is as follows:

STAGE 1 – COMMITMENTS: P_1 generates s garbled versions of the circuit. Furthermore, it generates commitments to the garbled values of the wires corresponding to P_2's input in each circuit. These commitments are generated in ordered pairs so that the first item in a pair corresponds to the 0 value and the second to the 1 value. The procedure regarding the input bits of P_1 is more complicated (see Figure 4.2 for a diagram explaining this construction). P_1 generates s pairs of sets of committed values for each of its input wires. Specifically, for every input wire i of P_1, it generates s sets of the form $\{W_{i,j}, W'_{i,j}\}_{j=1}^{s}$; we call these commitment sets. Before describing the content of these sets, denote by $k_{i,r}^{b}$ the garbled value that is assigned to the value $b \in \{0, 1\}$ in wire i of circuit r. Then, the sets $W_{i,j}$ and $W'_{i,j}$ both contain $s + 1$ commitments and are defined as follows. Let $b \leftarrow_R \{0, 1\}$ be a random bit, chosen independently for every $\{W_{i,j}, W'_{i,j}\}$ pair. Define $W_{i,j}$ to contain a commitment to b, as well as commitments to the garbled value corresponding to b in wire i in all of the s circuits, and define $W'_{i,j}$ similarly, but with respect to $1 - b$. In other words, $W_{i,j} = \{\mathsf{com}(b), \mathsf{com}(k_{i,1}^{b}), \dots, \mathsf{com}(k_{i,s}^{b})\}$ and $W'_{i,j} = \{\mathsf{com}(1-b), \mathsf{com}(k_{i,1}^{1-b}), \dots, \mathsf{com}(k_{i,s}^{1-b})\}$. The fact that b is chosen randomly means that with probability $1/2$ the set $W_{i,j}$ contains the commitments to values corresponding to 0, and with probability $1/2$ it contains the commitments to values corresponding to 1. We stress that in each of the pairs $(W_{i,1}, W'_{i,1}), \dots, (W_{i,s}, W'_{i,s})$, the values that are committed to are the same because these are all commitments to the garbled values of the ith wire in all s circuits. The only difference is that independent randomness is used in each pair for choosing b and constructing the commitments. We denote the first bit committed to in a commitment set as the indicator bit.

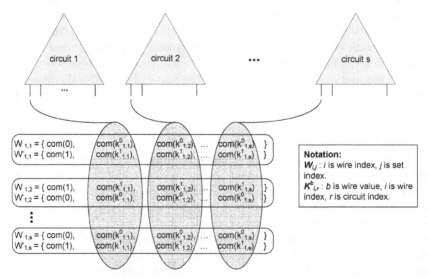

Fig. 4.2 The commitment sets corresponding to P_1's **first** input wire

After constructing these circuits and commitment sets, party P_1 sends to P_2 all of the s garbled circuits (i.e., the garbled gate tables and output tables, but *not* the garbled values corresponding to the input wires), and all the commitment sets. Note that if P_1's input is of length n, then there are sn pairs of commitment *sets*, and a total of $sn(2s + 2) = O(s^2n)$ commitments.

STAGE 2 – CHALLENGE: Two random strings $\rho, \rho' \leftarrow_R \{0, 1\}^s$ are chosen and sent to P_1 (in the actual protocol, these strings are determined via a simple coin-tossing protocol). The string ρ is a challenge indicating which garbled circuits to open, and the string ρ' is a challenge indicating which commitment sets to open. We call the opened circuits check circuits and the unopened ones evaluation circuits. Likewise, we call the opened sets check sets and the unopened ones evaluation sets. A circuit (resp., commitment set) is defined to be a check circuit (resp., check set) if the corresponding bit in ρ (resp., ρ') equals 1; otherwise, it is defined to be an evaluation circuit (resp., evaluation set).

STAGE 3 – OPENING: First, party P_1 opens all the commitments corresponding to P_2's input wires in all of the check circuits. Second, in all of the *check sets* P_1 opens the commitments that correspond to *check circuits*. That is, if circuit r is a check circuit, then P_1 decommits to all of the values $\text{com}(k_{i,r}^0), \text{com}(k_{i,r}^1)$ in check sets, where i is any of P_1's input bits. Finally, for every check set, P_1 opens the commitment to the indicator bit; i.e., the initial value in each of the sets $W_{i,j}, W_{i,j}'$. See Figure 4.3 for a diagram in which the values which are opened are highlighted (the diagram refers to only one of P_1's input wires in the circuit).

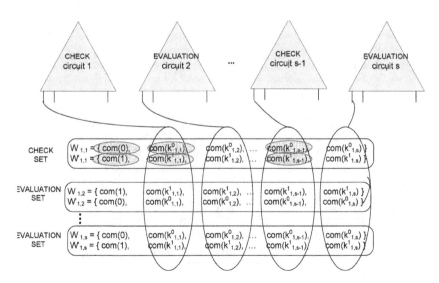

Fig. 4.3 In every check set, the commitment to the indicator bit and the commitments corresponding to check circuits are all opened

STAGE 4 – VERIFICATION: In this step, party P_2 verifies that all of the check circuits were correctly constructed. In addition, it verifies that, with regard to P_1's inputs, all the opened commitments in sets whose first item is a commitment to 0 are to garbled encodings of 0; likewise for 1. These checks are carried out as follows. First, in all of the check circuits, P_2 receives the decommitments to the garbled values corresponding to its own input, and by the order of the commitments P_2 knows which value corresponds to 0 and which value corresponds to 1. Second, for every check circuit, P_2 receives decommitments to the garbled input values of P_1 in all the check sets, along with a bit indicating whether these garbled values correspond to 0 or to 1. It first checks that for every wire, the garbled values of 0 are all equal and the garbled values of 1 are all equal. Then, the above decommitments enable the complete opening of the garbled circuits (i.e., the decryption of all of the garbled tables). Once this has been carried out, it is possible to simply check that the check circuits are all correctly constructed. Namely, that they agree with a specific and agreed-upon circuit computing f.

STAGE 5 – EVALUATION AND VERIFICATION: Party P_1 reveals the garbled values corresponding to its input: If i is a wire that corresponds to a bit of P_1's input and r is an evaluation circuit, then P_1 decommits to the commitments $k_{i,r}^b$ in all of the *evaluation sets*, where b is the value of its input bit. This is depicted in Figure 4.4. Finally, P_2 verifies that (1) for every input wire, all the opened commitments that were opened in evaluation sets contain the same garbled value, and (2) for every i, j P_1 opened commitments of evaluated circuits in exactly one of $W_{i,j}$ and $W'_{i,j}$. If these checks pass, it continues to evaluate the circuit.

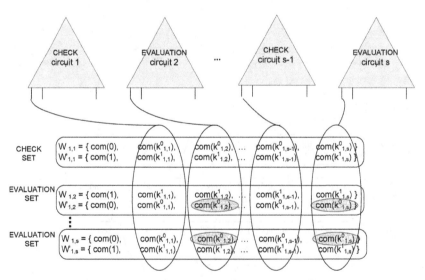

Fig. 4.4 P_1 opens the commitments that correspond to its input in the evaluation sets. In every evaluation set these commitments come from the same item in the pair

Intuition. Having described the mechanism for checking consistency, we now provide some intuition as to why it is correct. A simple cut-and-choose check verifies that most of the evaluated circuits are correctly constructed. The main remaining issue is ensuring that P_1's inputs to most circuits are consistent. If P_1 wants to provide different inputs to a certain wire in two circuits, then *all* the $W_{i,j}$ (or $W'_{i,j}$) sets it opens in *evaluation sets* must contain a commitment to 0 in the first circuit and a commitment to 1 in the other circuit. However, if any of these sets is chosen to be checked, and the circuits are among the checked circuits, then P_2 aborts. This means that if P_1 attempts to provide different inputs to two circuits and they are checked, it is caught. Now, since P_2 outputs the majority output of the evaluated circuits, the effect of P_1 providing different inputs has an influence only if it does this in a majority of the evaluation circuits. However, this means that P_1 provides different inputs in a constant fraction of the circuits (if half of the circuits are evaluation circuits, then it must provide different inputs in at least a quarter of the circuits overall). However, in this case, the probability that P_1 is not caught (which happens whenever it provides different inputs in a circuit that turns out to be a check circuit) is exponentially small in s.

4.2 The Protocol

We now describe the full protocol in detail. We use the notation com to refer to a perfectly binding commitment scheme, and com$_h$ to refer to a perfectly hiding commitment scheme (see [30] for definitions). The protocol below uses oblivious transfer that is secure in the presence of malicious adversaries. Furthermore, it uses "batch" oblivious transfer, which means that many executions are run in parallel. An efficient protocol that achieves this level of security can be found in Chapter 7.

PROTOCOL 4.2.1 (protocol for computing $f(x, y)$):

- **Input:** *P_1 has input $x \in \{0, 1\}^n$ and P_2 has input $y \in \{0, 1\}^n$.*
- **Auxiliary input:** *a statistical security parameter s and the description of a circuit C^0 such that $C^0(x, y) = f(x, y)$.*
- **Specified output:** *party P_2 should receive $f(x, y)$ and party P_1 should receive no output. (Recall that this suffices for the general case where both parties receive possibly different outputs; see Section 2.5.2.)*
- **The protocol:**

  0. CIRCUIT CONSTRUCTION: *The parties replace C^0 with a circuit C which is constructed by replacing each input wire of P_2 by the result of an exclusive-or of s new input wires of P_2, as depicted in Figure 4.1. The number of input wires of P_2 is increased from $|y| = n$ to sn. Let the bit-wise representation of P_2's original input be $y = y_1 \ldots y_n$. Denote its new input as $\hat{y} = \hat{y}_1, \ldots, \hat{y}_{ns}$. P_2 chooses its new input at random subject to the constraint $y_i = \hat{y}_{(i-1)\cdot s+1} \oplus \cdots \oplus \hat{y}_{i\cdot s}$.*

1. COMMITMENT CONSTRUCTION: P_1 constructs the circuits and commits to them, as follows:

   a. P_1 constructs s independent copies of a garbled circuit of C, denoted by $GC_1, \ldots, GC_s$.

   b. P_1 commits to the garbled values of the wires corresponding to P_2's input to each circuit. That is, for every input wire i corresponding to an input bit of P_2, and for every circuit GC_r, P_1 computes the ordered pair $(\mathsf{com}(k_{i,r}^0), \mathsf{com}(k_{i,r}^1))$, where $k_{i,r}^b$ is the garbled value associated with b on input wire i in circuit GC_r.

   c. P_1 computes commitment-sets for the garbled values that correspond to its own inputs to the circuits. That is, for every wire i that corresponds to an input bit of P_1, it generates s pairs of commitment sets $\{W_{i,j}, W'_{i,j}\}_{j=1}^s$, in the following way:

   Denote by $k_{i,r}^b$ the garbled value that was assigned by P_1 to the value $b \in \{0,1\}$ of wire i in GC_r. Then, for every $j = 1, \ldots, s$, party P_1 chooses $b \leftarrow_R \{0,1\}$ (independently for each j) and computes

   $$W_{i,j} = \langle \mathsf{com}(b), \mathsf{com}(k_{i,1}^b), \ldots, com_b(k_{i,s}^b) \rangle, \qquad \text{and}$$
   $$W'_{i,j} = \langle \mathsf{com}(1-b), \mathsf{com}(k_{i,1}^{1-b}), \ldots, \mathsf{com}(k_{i,s}^{1-b}) \rangle.$$

   For each i, j, the sets are constructed using independent randomness, and in particular the value of b is chosen independently for every $j = 1, \ldots, s$. There is a total of ns commitment-sets. We divide them into s supersets, where superset S_j is defined to be the set containing the jth commitment set for all wires. Namely, it is defined as $S_j = \{(W_{1,j}, W'_{1,j}), \ldots, (W_{n,j}, W'_{n,j})\}$.

2. OBLIVIOUS TRANSFERS: For every input bit of P_2, parties P_1 and P_2 run a 1-out-of-2 oblivious transfer protocol in which P_2 receives the garbled values for the wires that correspond to its input bit (in every circuit). That is, let $c_{i,r}^b$ denote the commitment to the garbled value $k_{i,r}^b$ and let $dc_{i,r}^b$ denote the decommitment value for $c_{i,r}^b$. Furthermore, let $i_1, \ldots, i_{ns}$ be the input wires that correspond to P_2's input.

   Then, for every $j = 1, \ldots, ns$, parties P_1 and P_2 run a 1-out-of-2 batch oblivious transfer protocol in which:

   a. P_1's input is the pair of vectors $([dc_{i_j,1}^0, \ldots, dc_{i_j,s}^0], [dc_{i_j,1}^1, \ldots, dc_{i_j,s}^1])$.

   b. P_2's input is its jth input bit $\hat{y}_j$ (and its output should thus be $[dc_{i_j,1}^{\hat{y}_j}, \ldots, dc_{i_j,s}^{\hat{y}_j}]$).

3. SEND CIRCUITS AND COMMITMENTS: P_1 sends to P_2 the garbled circuits (i.e., the gate and output tables), as well as all of the commitments that it prepared above.

4. PREPARE CHALLENGE STRINGS:

   a. P_2 chooses a random string $\rho_2 \leftarrow_R \{0,1\}^s$ and sends $\mathsf{com}_h(\rho_2)$ to P_1.

   b. P_1 chooses a random string $\rho_1 \in \{0,1\}^s$ and sends $\mathsf{com}(\rho_1)$ to P_2.

   c. P_2 decommits, revealing ρ_2.

   d. P_1 decommits, revealing ρ_1.

    e. P_1 and P_2 set $\rho = \rho_1 \oplus \rho_2$.

    The above steps are run a second time, defining an additional string ρ'.[2]

5. DECOMMITMENT PHASE FOR CHECK CIRCUITS: *From here on, we refer to the circuits for which the corresponding bit in ρ is 1 as* check *circuits, and we refer to the other circuits as* evaluation *circuits. Likewise, if the jth bit of ρ' equals 1, then all commitment sets in superset $S_j = \{(W_{i,j}, W'_{i,j})\}_{i=1...n}$ are referred to as* check *sets; otherwise, they are referred to as* evaluation *sets.*

    For every **check circuit** *GC_r, party P_1 operates in the following way:*

    a. For every input wire i corresponding to an input bit of P_2, party P_1 decommits to the pair $(\mathsf{com}(k^0_{i,r}), \mathsf{com}(k^1_{i,r}))$ (namely to both of P_2's inputs).

    b. For every input wire i corresponding to an input bit of P_1, party P_1 decommits to the appropriate values in the check *sets $\{W_{i,j}, W'_{i,j}\}$. Specifically, P_1 decommits to the $\mathsf{com}(k^0_{i,r})$ and $\mathsf{com}(k^1_{i,r})$ values in $(W_{i,j}, W'_{i,j})$ for every check set S_j (see Figure 4.3). In addition, P_1 decommits to the indicator bits of these sets (i.e., to the first committed value in each set).*

    For every pair of **check sets** *$(W_{i,j}, W'_{i,j})$, party P_1 decommits to the first value in each set (i.e., to the value that is supposed to be a commitment to the indicator bit, $\mathsf{com}(0)$ or $\mathsf{com}(1)$).*

6. DECOMMITMENT PHASE FOR P_1'S INPUT IN EVALUATION CIRCUITS: *P_1 decommits to the garbled values that correspond to its inputs in evaluation circuits. Let i be the index of an input wire that corresponds to P_1's input (the following procedure is applied to all such wires). Let b be the binary value that P_1 assigns to input wire i. In every evaluation set $(W_{i,j}, W'_{i,j})$, P_1 chooses the set (out of $(W_{i,j}, W'_{,j})$), which corresponds to the value b. It then opens in this set the commitments that correspond to evaluation circuits, namely, to the values $k^b_{i,r}$, where r is an index of an evaluation circuit (see Figure 4.4).*

7. CORRECTNESS AND CONSISTENCY CHECKS: *P_2 performs the following checks; if any of them fails it aborts.*

    a. Checking correctness of the check circuits: *P_2 verifies that each check circuit GC_i is a garbled version of C. This check is carried out by P_2 first constructing the input tables that associate every garbled value of an input wire to a binary value. The input tables for P_2's inputs are constructed by checking that the decommitments in Step 5a to the*

---

[2] Recall that ρ and ρ' are used to ensure that P_1 constructs the circuits correctly and uses consistent input in each circuit. Thus, it may seem strange that they are generated via a coin-tossing protocol, and not just chosen singlehandedly by P_2. Indeed, in order to prove the security of the protocol when P_1 is corrupted, there is no need for a coin-tossing protocol here. However, having P_2 choose ρ and ρ' singlehandedly creates a problem for the simulation in the case where P_2 is corrupted. We therefore use a coin-tossing protocol instead.

pairs $(\mathsf{com}(k_{i,r}^0), \mathsf{com}(k_{i,r}^1))$ *(where i is a wire index and r is a circuit index) are valid, and then interpreting the first value to be associated with 0 and the second value to be associated with 1.*

Next, P_2 checks the decommitments to P_1's inputs. This check involves first checking that the decommitment values of Step 5b are valid. Then, P_2 verifies that in each pair of check sets, one of $(W_{i,j}, W'_{i,j})$ begins with a commitment to 0 (henceforth the 0 tuple), and the other begins with a commitment to 1 (henceforth the 1 tuple). Then P_2 checks that for every wire, the values that are decommitted to in the 0 tuples in all check sets are all equal, and that a similar property holds for the 1 tuples. P_2 then assigns the logical value of 0 to all of the opened commitments in the 0 tuples, and the logical value of 1 to the opened commitments in the 1 tuples.

Finally, given all the garbled values to the input wires and their associated binary values, P_2 decrypts the circuit and compares it to the circuit C.

b. Verifying P_2's input in the check circuits: *P_2 verifies that P_1's decommitments to the wires corresponding to P_2's input values in the check circuits are correct, and agree with the logical values of these wires (the indicator bits). P_2 also checks that the inputs it learned in the oblivious transfer stage for the check circuits correspond to its actual input. Specifically, it checks that the decommitment values that it received in the oblivious transfer stage open the committed values that correspond to the garbled values of its logical input (namely, that it received the first value in the pair if the input bit is 0 and the second value if it is 1).*[3]

c. Checking P_1's input to evaluation circuits: *Finally, P_2 verifies that for every input wire i of P_1 the following two properties hold:*

  i. *In every evaluation set, P_1 chose one of the two sets and decommitted to all the commitments in it which corresponded to evaluation circuits.*

  ii. *For every evaluation circuit, all of the commitments that P_1 opened in evaluation sets commit to the same garbled value.*

8. CIRCUIT EVALUATION: *If any of the above checks fails, P_2 aborts and outputs $\perp$. Otherwise, P_2 evaluates the evaluation circuits (in the same way as for the semi-honest protocol of Yao). It might be that in certain circuits the garbled values provided for P_1's inputs, or the garbled values learned by P_2 in the OT stage, do not match the tables and so decryption of the circuit fails. In this case P_2 also aborts and outputs $\perp$. Otherwise, P_2 takes the output that appears in most circuits, and outputs it (the proof shows that this value is well defined).*

---

[3] This check is crucial and thus the order of first running the oblivious transfer and then sending the circuits and commitments is not at all arbitrary.

4.3 Proof of Security

The security of Protocol 4.2.1 is stated in the following theorem.

Theorem 4.3.1 *Let $f : \{0,1\}^* \times \{0,1\}^* \to \{0,1\}^*$ be any polynomial-time two-party single-output functionality. Assume that the oblivious transfer protocol is secure in the presence of static malicious adversaries, that com is a perfectly-binding commitment scheme, that com_h is a perfectly-hiding commitment scheme, and that the garbled circuits are constructed as in Chapter 3. Then, Protocol 4.2.1 securely computes f in the presence of static malicious adversaries.*

The theorem is proved in two stages: first for the case where P_1 is corrupted and next for the case where P_2 is corrupted.

4.3.1 Security Against a Malicious P_1

Intuition. The proof constructs an ideal-model adversary/simulator which has access to P_1 and to the trusted party, and can simulate the view of an actual run of the protocol. It uses the fact that the strings ρ, ρ', which choose the circuits and commitment sets that are checked, are uniformly distributed even if P_1 is malicious. The simulator runs the protocol until P_1 opens the commitments of the checked circuits and checked commitment sets, and then rewinds the execution and runs it again with new random ρ, ρ' values. We expect that about one quarter of the circuits will be checked in the first execution *and* evaluated in the second execution. For these circuits, in the first execution the simulator learns the translation between the garbled values of P_1's input wires and the actual values of these wires, and in the second execution it learns the garbled values that are associated with P_1's input (this association is learned from the garbled values that P_1 sends to P_2). Combining the two, it learns P_1's input x, which can then be sent to the trusted party, completing the simulation.

When examining the detailed proof, first note that the strings ρ, ρ' computed in Step 4 of Protocol 4.2.1 are uniformly distributed even in the presence of a malicious P_1. This is due to the perfect hiding of P_2's commitments in the coin-tossing subprotocol. We say that a circuit GC_i and a superset S_j *agree* if the checks in Step 7 of the protocol succeed when considering only the check circuit GC_i and the superset of check sets S_j. In particular, this means that GC_i computes the required function when the garbled values of P_1's input wires are taken from S_j, and that these garbled values agree with the indicator bit of the sets in S_j. This also means that the committed values of the garbled values of P_2's input wires in GC_i are correctly constructed.

(Some circuits might not agree with any set S_j, e.g., if they do not compute f. Other circuits might agree with some supersets and not agree with others.)

We begin by proving two lemmas that will be used in analyzing the simulation (described below). We say that a circuit is ε-bad if more than εs of the supersets disagree with it. The following lemma shows that P_2 aborts (with high probability) if more than εs of the circuits are ε-bad. We can therefore concentrate on the case where εs or fewer of the circuits are bad.

Lemma 4.3.2 *If at least εs of the circuits are ε-bad, then P_2 aborts with probability of at least $1 - 2 \cdot 2^{-\varepsilon s}$.*

Proof. As a warm-up, suppose that there is a single ε-bad circuit. Then the probability of P_2 not aborting is at most $1/2 + 1/2 \cdot 2^{-\varepsilon s}$, i.e., the probability that the bad circuit is not chosen as a check circuit plus the probability that it is a check circuit but none of the disagreeing check sets are chosen (since the circuits and sets are chosen independently, we can just multiply the probabilities in the latter case). Suppose now that there are j different ε-bad circuits. Then the probability of P_2 not aborting is at most $2^{-j} + (1 - 2^{-j})2^{-\varepsilon s} \leq 2^{-j} + 2^{-\varepsilon s}$. Setting $j = \varepsilon s$ yields the lemma. ∎

The following lemma shows that P_2 aborting does not reveal information to P_1 about P_2's input.

Lemma 4.3.3 *For any two different inputs y and y' of P_2 for the function f, the difference between the probability that P_2 aborts Protocol 4.2.1 when its input is y and when its input is y' is at most $n2^{-s+1}$.*

Proof. P_2 may abort Protocol 4.2.1 in Step 7(a) while checking the correctness of the check circuits and the check sets. In this case, the decision to abort is based on P_1's construction of the sets and circuits, and on the random inputs of the parties, and is independent of P_2's input. The same is true of Step 7(c) where P_2 checks P_1's input to the evaluation circuits. In Step 7(b), however, P_2 aborts based on whether the values it learned in the oblivious transfer invocations open P_1's commitments to the garbled values of P_2's input. This case must be examined in detail.

Consider a specific input bit of P_2. In Step 0 of Protocol 4.2.1 the circuit is changed so that this bit is computed as the exclusive-or of s new input bits of P_2. Consider the s new inputs which replace a single input wire of the original circuit. Suppose that P_1 provides in the OT protocol corrupt values to both garbled values of one of P_2's (new) input wires. Then P_2 aborts with probability 1 regardless of its input. If P_1 provides a corrupt OT value to exactly one of the two possible OT choices of $1 \leq j < s$ new wires, then P_2 aborts with probability $1 - 2^{-j}$, again regardless of the actual value of its original input. This holds because the values assigned by P_2 to any *proper subset* of the s bits are independent of P_2's actual input. Assume now that P_1 corrupts one OT value for each of the s new wires (say all '1' values). Then P_2 aborts with probability 1 if its original input had one value ('1' in this example), and aborts with probability $1 - 2^{-s+1}$ if its original input had the other value (in this example, P_2 does not abort if its input is '0' and it chose only '0' inputs in the s OT invocations). Therefore, for any two different inputs y and y' of P_2 of length n bits each, the probability that P_2 aborts the protocol differs by at most $n2^{-s+1}$, as required. ∎

We are now ready to prove the security of the protocol under simulation-based definitions. (For shorthand, from here on by "secure" we mean secure in the presence of malicious adversaries.)

Lemma 4.3.4 *Assume that the oblivious transfer protocol is secure, that the commitment scheme* com_h *is perfectly hiding, and that the commitment scheme* com *is perfectly binding. Then, Protocol 4.2.1 is secure in the case where* P_1 *is corrupted. (We say that a protocol is secure in the case where* P_1 *is corrupted if Definition 2.3.1 holds when the adversary* $\mathcal{A}$ *controls* P_1.)*

Proof. Let $\mathcal{A}$ be an adversary corrupting P_1; we construct an ideal-model adversary/simulator $\mathcal{S}$. Since we assume that the oblivious transfer protocol is secure, we analyze the security of Protocol 4.2.1 in the hybrid model with a trusted party computing the oblivious transfer functionality.

The simulator. The simulator $\mathcal{S}$ chooses a random input y' for P_2 and uses it in all but the last stage of the simulation. $\mathcal{S}$ receives all of the garbled circuits and commitments from $\mathcal{A}$. Simulator $\mathcal{S}$ then runs the coin-tossing phase (for preparing the challenge strings) as P_2 would, and receives all of the required decommitments from $\mathcal{A}$, including the garbled values that supposedly correspond to its input. $\mathcal{S}$ runs all of the checks that P_2 would run. If any of the checks fail, $\mathcal{S}$ sends an abort message to $\mathcal{A}$, sends $\perp$ to the trusted party and halts, outputting whatever $\mathcal{A}$ outputs. Otherwise, $\mathcal{S}$ rewinds $\mathcal{A}$ and returns to the coin-tossing phase. Once again $\mathcal{S}$ runs this phase as P_2 would (but with new randomness) and runs all of the checks that P_2 would run. $\mathcal{S}$ continues this until all of the checks pass for a second time. Let α be the output of $\mathcal{A}$ in this second successful execution (note that an honest P_1 has no output).

Denote by ρ, ρ' the (uniformly distributed) challenge strings from the first execution of $\mathcal{S}$ with $\mathcal{A}$, and denote by $\hat{\rho}, \hat{\rho}'$ the challenge strings from the second execution. Furthermore, denote $\rho = \rho^1 \cdots \rho^s$ and $\hat{\rho} = \hat{\rho}^1 \cdots \hat{\rho}^s$. Now, if there are less than $s/8$ indices i for which $\rho_i = 1$ and $\hat{\rho}_i = 0$, then $\mathcal{S}$ outputs fail_1. Otherwise, let I be a subset of indices of size exactly $s/8$ for which $\rho_i = 1$ and $\hat{\rho}^i = 0$ (it is easier to work with a fixed number of is that have this property, so we choose them here). Then, for every $i \in I$, we have that in the first execution GC_i is a check circuit and in the second execution it is an evaluation circuit. Thus, $\mathcal{S}$ obtains all of the decommitments of GC_i in the first execution (including the association of the garbled values corresponding to P_1's input, i.e., the decommitment to $\mathsf{com}(b)$ in the commitment-sets), and in the second execution it obtains the garbled values corresponding to P_1's input that P_1 sends to P_2. For each such i, it is possible to define P_1's input in circuit GC_i by associating the indicator bit obtained when GC_i was a check circuit with the garbled value sent by P_1 when GC_i was an evaluation circuit. Thus, S obtains $s/8$ possible n-bit input vectors for P_1. If no input value appears more than $s/16$ times, then S outputs fail_2. Otherwise, S sets x to be the value that appears more than $s/16$ times and sends it to the trusted party. S then outputs α (the output of A in the second execution) and halts.

Analysis. We claim that the view of A in the simulation with S is *statistically close* to its view in a hybrid execution of Protocol 4.2.1 with a trusted party computing the oblivious transfer protocol. We first claim that the difference between the probability that P_2 receives "abort" (i.e., $\perp$) in an ideal execution with S and the probability that P_2 outputs "abort" (i.e. $\perp$) in a real execution with A is at most negligible. Observe that in the simulation, S uses a random input for its emulation of P_2 instead of the real y that P_2 holds. This makes a difference when S checks the decommitments for the wires that are associated with P_2's input. (Notice that in the real protocol P_2 also uses its input in oblivious transfer subprotocols. Nevertheless, in the hybrid model that we are analyzing here, A learns nothing about P_2's input in the oblivious transfer because it is ideal.) Nevertheless, by Lemma 4.3.3 we know that the probability of abort is at most negligibly different between the case where P_2 has a random input and the case where it has a specific input y. From here on, we therefore consider the case where P_2 does not abort the protocol (i.e., does not output $\perp$). We now prove that S outputs fail_1 or fail_2 with at most negligible probability.

Bounding fail_1. The proof that fail_1 occurs with negligible probability follows from the Chernoff bound, as follows. Denote an index i as good if $\rho_i = 1$ and $\hat{\rho}_i = 0$. The probability of this event is $1/4$, independently of other indices. Event fail_1 happens if less than $s/8$ of the indices are good. Let $X_i = 1$ if and only if index i is good. Then, $\Pr[X_i = 1] = 1/4$ and the Chernoff bound implies that

$$
\begin{aligned}
\Pr\left[\sum_{i=1}^{s} X_i < \frac{s}{8}\right] &= \Pr\left[\frac{\sum_{i=1}^{s} X_i}{s} < \frac{1}{8}\right] \\
&\leq \Pr\left[\left|\frac{\sum_{i=1}^{s} X_i}{s} - \frac{1}{4}\right| > \frac{1}{8}\right] \\
&< 2 \cdot e^{-\frac{(1/8)^2}{2 \cdot (1/4) \cdot (3/4)} \cdot s} \\
&= 2 \cdot e^{\frac{-s}{24}} \\
&< 2 \cdot 2^{\frac{-s}{17}}.
\end{aligned}
$$

Bounding fail_2. We now show that the event fail_2 occurs with negligible probability. Let $\varepsilon = 1/16$ and denote by many-bad the event that at least εs of the circuits are ε-bad (i.e., the event that $s/16$ of the circuits are $1/16$-*bad*).

Denote by abort the event that S sends $\perp$ to the trusted party. Lemma 4.3.2 shows that $\Pr[\neg\text{abort} \mid \text{many-bad}] \leq 2 \cdot 2^{-s/16}$.

We begin by analyzing the probability that fail_2 occurs given $\neg\text{many-bad}$; i.e., given the event that less than $s/16$ of the circuits are $1/16$-*bad*. Consider the set of $s/8$ circuits GC_i with $i \in I$. The definition of $\neg\text{many-bad}$ implies that a majority of the $s/8$ circuits in I are *not* $1/16$-bad. The circuits which are not $1/16$-bad agree with at least $15s/16$ of the commitment sets. The probability that any of these circuits does not agree with a majority of the evaluation sets is negligible: this event only happens if the number of evaluation sets is less than $s/8$, and the probability of this event happening can be bounded (using the Chernoff bound) by $(s/8) \cdot 2 \cdot e^{-\frac{(3/8)^2}{2 \cdot (1/2)^2} \cdot s} = (s/8) \cdot e^{-\frac{9s}{32}} < 2^{\frac{-s}{2.5}}$. If a circuit agrees with a majority of the evaluation sets then the committed values of these sets open the circuit correctly. In the evaluation step, for each of its input wires P_1 opens the values for all evaluation circuits taken from the same commitment set. P_2 and S check that the values opened for a wire in all sets are equal. For the good circuits in I these values agree with the same logical value (the indicator bit of the set). Therefore in this case a majority of the circuits in I obtain the same logical input, and fail_2 does not occur.

When $\varepsilon = 1/16$, the previous argument shows that

$$\Pr[\text{fail}_2 \mid \neg\text{many-bad}] < 2^{-s/2.5},$$

and Lemma 4.3.2 shows that $\Pr[\neg\text{abort} \mid \text{many-bad}] < 2 \cdot 2^{-s/16}$. We are interested in $\Pr[\text{fail}_2]$, which we bound as follows:

$$\Pr[\text{fail}_2] = \Pr[\text{fail}_2 \wedge \text{abort}] + \Pr[\text{fail}_2 \wedge \neg\text{abort}] = \Pr[\text{fail}_2 \wedge \neg\text{abort}]$$

where the last equality is due to the fact that in the event of fail_2 the simulator S does not send $\perp$ (and so $\neg\text{abort}$ does not occur) and vice versa. Thus, $\Pr[\text{fail}_2 \wedge \text{abort}] = 0$. Now,

$$\Pr[\text{fail}_2 \wedge \neg\text{abort}] = \Pr[\text{fail}_2 \wedge \neg\text{abort} \wedge \text{many-bad}]$$
$$+ \Pr[\text{fail}_2 \wedge \neg\text{abort} \wedge \neg\text{many-bad}]$$
$$\leq \Pr[\neg\text{abort} \wedge \text{many-bad}] + \Pr[\text{fail}_2 \wedge \neg\text{many-bad}].$$

Combining the above and using the fact that for all two events X and Y it holds that $\Pr[X \wedge Y] \leq \Pr[X \mid Y]$ we conclude that

$$\Pr[\text{fail}_2] \leq \Pr[\neg\text{abort} \mid \text{many-bad}] + \Pr[\text{fail}_2 \mid \neg\text{many-bad}]$$
$$< 2 \cdot 2^{-s/16} + 2^{-s/2.5}$$
$$< 3 \cdot 2^{-s/16}.$$

Completing the proof. We now show that conditioned on $\mathcal{S}$ not outputting any fail message, the view of $\mathcal{A}$ in the simulation is *statistically close* to its view in an execution of Protocol 4.2.1. First note that the probability of abort in the real and ideal executions is at most negligibly far apart (this follows from Lemma 4.3.3 and the fact that $\mathcal{S}$ uses a random input instead of the one that the honest P_2 has). Next, consider the case where abort does not occur. Recall that $\mathcal{S}$ just runs the honest P_2's instructions. The only difference is that in the event that all of $\mathcal{S}$'s checks pass in the first execution (which is the event of no abort that we are considering here), it rewinds the second execution until this event occurs again. The final view of $\mathcal{A}$ is then the view that appears in this second execution in which this occurs. Since $\mathcal{S}$ uses independent random coins each time, and follows P_2's instructions each time, the above process results in a distribution that is identical to the view of $\mathcal{A}$ in a real execution with P_2.

We now proceed to show that the *joint distribution* of $\mathcal{S}$'s output (which is just $\mathcal{A}$'s output α) and the honest B_2's output, is computationally indistinguishable from the joint distribution of $\mathcal{A}$ and P_2's output in an execution of Protocol 4.2.1 (where an ideal oblivious transfer is used instead of the OT subprotocol). We will actually show statistical closeness. (This does not mean, however, that the overall protocol gives statistical security because our analysis is in the hybrid model for an oblivious transfer functionality and it depends on the security of the actual oblivious transfer subprotocol used.) In order to prove this, we show that if the real P_2 would have received the set of evaluation circuits and decommitments that $\mathcal{A}$ sent in the second execution with $\mathcal{S}$, and it has input y, then it would compute $f(x, y)$ in a majority of the circuits (where x is the input value that $\mathcal{S}$ sent to the trusted party computing f). This follows from the same argument that was used to show above that fail_2 occurs with negligible probability: with all but negligible probability, most of the evaluation circuits are not ε-bad and they each agree with a majority of the evaluation sets. Denote these circuits as *good* (or ε-good) circuits. In particular, the committed values provided in these sets for P_1's inputs in these circuits correctly decrypt them according to their association with the indicator bit. P_2 also checks that each of P_1's input wires receives the same garbled value in all sets. Therefore, the evaluation step is aborted unless P_1 opens garbled values for the good circuits that agree with the same logical value (the indicator bit of the set). The fact that these circuits are good also implies that P_2 obtains garbled values in the OT stage that agree with its input. As a result, a majority of the evaluation circuits obtain the same logical input (x, y) and compute $f(x, y)$.

It remains to show that $\mathcal{S}$ runs in expected polynomial time. In order to see this, notice that aside from the rewinding, all of $\mathcal{S}$'s work takes a strict polynomial number of steps. Furthermore, each rewinding attempt also takes a strict polynomial number of steps. Now, denote by p the probability that $\mathcal{A}$ responds correctly and so $\mathcal{S}$'s checks all pass. Then, the probability that $\mathcal{S}$ enters the rewinding phase equals p. Furthermore, the expected number of rewinding attempts equals exactly $1/p$ (notice that $\mathcal{S}$ runs exactly the same strategy in each rewinding attempt). Thus, the overall expected running time of $\mathcal{S}$ equals $\text{poly}(n, s) + p \cdot 1/p \cdot \text{poly}(n, s) = \text{poly}(n, s)$. This completes the proof of Lemma 4.3.4 and thus the case where P_1 is corrupted.

(We note one important subtlety in this part of the proof: the sequential composition theorem of Section 2.7 was only proven for the case where the security of the subprotocol is proven via a simulator that runs in strict polynomial time (see [48, 33] for a full discussion of this issue). Thus, the modular sequential composition theorem does not cover the case where the simulator for the oblivious transfer subprotocol runs in expected polynomial time. Despite this, we claim that this is no problem in our specific case. In order to see that $\mathcal{S}$ runs in expected polynomial time even if the oblivious transfer protocol is proven secure using expected polynomial-time simulation, note that we can divide $\mathcal{A}$ into two parts. The first part runs up until the end of the oblivious transfer protocol and outputs state information; the second part takes the state information and continues until the end of the execution. Now, the simulator for the oblivious transfer protocol may result in an expected polynomial-time adversary for the first part of $\mathcal{A}$. However, the second part of $\mathcal{A}$ still runs in strict polynomial time, and $\mathcal{S}$ only rewinds this second part. Therefore, the overall running time of $\mathcal{S}$ – even after replacing the ideal oblivious transfer functionality with a real protocol that may use expected polynomial-time simulation – is expected polynomial time, as required.) ∎

4.3.2 Security Against a Malicious P_2

Intuition. Intuitively, the security in this case is derived from the fact that (a) the oblivious transfer protocol is secure, and so P_2 only learns a single set of keys (corresponding to a single input y) for decrypting the garbled circuits, and (b) the commitment schemes are hiding and so P_2 does not know what input corresponds to the garbled values that P_1 sends it for evaluating the circuit. Of course, in order to formally prove security we construct an ideal-model simulator $\mathcal{S}$ working with an adversary $\mathcal{A}$ that has corrupted P_2. The simulator first extracts $\mathcal{A}$'s input bits from the oblivious transfer protocol, and then sends the input y it obtained to the trusted party and receives back $\tau = f(x, y)$. Given the output, the simulator constructs the garbled circuits. However, rather than constructing them all correctly, for each circuit it tosses a coin and, based on the result, either constructs the circuit correctly, or constructs it to compute the constant function outputting τ (the output is received from the trusted party). In order to make sure that the simulator is not caught cheating, it biases the coin-tossing phase so that all of the correctly constructed garbled circuits are check circuits, and all of the other circuits are evaluation circuits (this is why the protocol uses joint coin-tossing rather than let P_2 alone choose the circuits to be opened). $\mathcal{A}$ then checks the correctly constructed circuits, and is satisfied with the result as if it were interacting with a legitimate P_1. $\mathcal{A}$ therefore continues the execution with the circuits which always output τ. The proof is based on the following lemma:

Lemma 4.3.5 *Assume that the oblivious transfer protocol is secure, that* com_h *is a perfectly-hiding commitment scheme, and that* com *is a perfectly-binding commitment scheme. Then, Protocol 4.2.1 is secure in the case where* P_2 *is corrupted.*

Proof.   As described above, the simulator works by constructing some of the circuits correctly and some of them incorrectly. Before proceeding with the formal proof of the lemma, we show that it is possible to construct such "false circuits", so that $\mathcal{A}$ cannot distinguish between them and correctly constructed circuits.

Lemma 4.3.6 *Given a circuit C and an output value τ (of the same length as the output of C) it is possible to construct a garbled circuit $\widetilde{GC}$ such that:*

1. *The output of $\widetilde{GC}$ is always τ, regardless of the garbled values that are provided for P_1 and P_2's input wires, and*
2. *If $\tau = f(x, y)$, then no non-uniform probabilistic polynomial-time adversary $\mathcal{A}$ can distinguish between the distribution ensemble consisting of $\widetilde{GC}$ and a single arbitrary garbled value for every input wire, and the distribution ensemble consisting of a real garbled version of C, together with garbled values that correspond to x for P_1's input wires, and to y for P_2's input wires.*

Proof (sketch).   The proof of this lemma is taken from the proof of Theorem 3.4.2 (it is not stated in this way there, but is proven). We sketch the construction of $\widetilde{GC}$ here for the sake of completeness, and refer the reader to Section 3.4 for a full description and proof. The first step in the construction of the fake circuit $\widetilde{GC}$ is to choose two random keys k_i and k_i' for every wire w_i in the circuit C. Next, the gate tables of C are computed: let g be a gate with input wires w_i, w_j and output wire w_ℓ. The table of gate g contains encryptions of the single key k_ℓ that is associated with wire w_ℓ, under *all four combinations* of the keys k_i, k_i', k_j, k_j' that are associated with the input wires w_i and w_j to g. (This is in contrast to a real construction of the garbled circuit that involves encrypting both k_ℓ and k_ℓ', depending on the function that the gate in question computes.) That is, the following values are computed:

$$c_{0,0} = E_{k_i}(E_{k_j}(k_\ell)),$$
$$c_{0,1} = E_{k_i}(E_{k_j'}(k_\ell)),$$
$$c_{1,0} = E_{k_i'}(E_{k_j}(k_\ell)),$$
$$c_{1,1} = E_{k_i'}(E_{k_j'}(k_\ell)).$$

The gate table for g is then just a random ordering of the above four values. This process is carried out for all of the gates of the circuit. It remains to describe how the output decryption tables are constructed. Denote the n-bit output τ by $\tau_1 \cdots \tau_n$, and denote the circuit-output wires by $w_{m-n+1}, \ldots, w_m$. In addition, for every $i = 1, \ldots, n$, let k_{m-n+i} be the (single) key encrypted in the gate whose output wire is w_{m-n+i}, and let k'_{m-n+i} be the other key (as described above). Then, the output decryption table for wire w_{m-n+i} is given by $[(0, k_{m-n+i}), (1, k'_{m-n+i})]$ if $\tau_i = 0$, and $[(0, k'_{m-n+i}), (1, k_{m-n+i})]$ if $\tau_i = 1$. This completes the description of the construction of the fake garbled circuit $\widetilde{GC}$.

Notice that by the above construction of the circuit, the output keys (or garbled values) obtained by P_2 for *any* set of input keys (or garbled values), equals $k_{m-n+1}, \ldots, k_m$. Furthermore, by the above construction of the output tables, these keys $k_{m-n+1}, \ldots, k_m$ decrypt to $\tau = \tau_1 \cdots \tau_n = \tau$ exactly. Thus, property (1) of the lemma trivially holds. The proof of property (2) follows from a hybrid argument in which the gate construction is changed one at a time from the real construction to the above fake one (indistinguishability follows from the indistinguishability of encryptions). The construction and proof of this hybrid are described in full in Section 3.3. ∎

We are now ready to begin with the formal proof of Lemma 4.3.5. We denote the number of input wires of P_2 as n' (P_2 had originally n input wires, but in Step 0 of the protocol they are expanded to $n' = ns$ wires, to prevent an attack by P_1). Let $\mathcal{A}$ be an adversary controlling P_2. We construct a simulator $\mathcal{S}$ as follows:

1. $\mathcal{S}$ chooses garbled values for the input wires of P_2 in s garbled circuits. That is, it chooses $n' \cdot s$ pairs of garbled values k_i^0 and k_i^1, and constructs $2n'$ vectors of garbled values of length s. Denote the vectors $v_1^0, v_1^1, \ldots, v_{n'}^0, v_{n'}^1$, where v_i^b contains the garbled values in all circuits that are associated with the bit b for the input wire associated with P_2's ith input bit. Next, $\mathcal{S}$ computes the commitment and decommitment values for these vectors. That is, let c_i^b be a vector of commitments, with the jth element being a commitment to the jth element of v_i^b. Likewise, let dc_i^b be a vector of decommitments, where the jth element of dc_i^b is the decommitment of the jth element of c_i^b.

2. $\mathcal{S}$ invokes $\mathcal{A}$ upon its initial input and obtains the inputs that $\mathcal{A}$ sends to the trusted party computing the oblivious transfer functionality (recall that our analysis is in the hybrid model). Let y_i denote the bit sent by $\mathcal{A}$ that corresponds to the ith oblivious transfer, and let $y = y_1, \ldots, y_n$ (note that y is not necessarily the same as $\mathcal{A}$ and $\mathcal{S}$'s initial input). $\mathcal{S}$ hands $\mathcal{A}$ the vector of decommitments to garbled values $dc_i^{y_i}$ as if they are the output for $\mathcal{A}$ from the trusted party in the ith computation of the oblivious transfer functionality.

3. $\mathcal{S}$ *externally* sends y to the trusted party computing f and receives back $\tau = f(x, y)$.

4. $\mathcal{S}$ chooses a random string $\rho \leftarrow_R \{0,1\}^s$ and constructs s garbled circuits $GC_1, \ldots GC_s$, as follows. Let $\rho = \rho_1, \ldots, \rho_s$. Then, for $i = 1, \ldots, s$, if $\rho_i = 1$ (and so GC_i is a check circuit), simulator $\mathcal{S}$ constructs circuit GC_i correctly (exactly as described in Step 1 of Protocol 4.2.1). Otherwise, if $\rho_i = 0$ (and so GC_i is an evaluation circuit), it constructs circuit $GC_i = \widetilde{GC}$ as described in Lemma 4.3.6. That is, it constructs a garbled circuit whose output is always τ, regardless of the inputs used. The above constructions use the garbled values chosen for the input wires above. That is, the garbled values from v_i^0 and v_i^1 are used to define the input values for the ith wire in all of the s circuits (the jth value in v_i^b defines the value in the jth circuit).

$\mathcal{S}$ constructs the commitments and commitment sets as follows.

- First, for every r such that $\rho_r = 1$ (and so GC_r is a check circuit), the commitment pairs $(\mathsf{com}(k_{i,r}^0), \mathsf{com}(k_{i,r}^1))$ that correspond to P_2's input wires in circuit GC_r are computed correctly (note that $k_{i,r}^b$ is the rth value in v_i^b and $\mathsf{com}(k_{i,r}^b)$ is taken from c_i^b).
- In contrast, for every j for which $\rho_r = 0$ (and so GC_r is an evaluation circuit), these commitment pairs are computed as follows. Assume that P_2's ith input bit is associated with wire i. Then, $\mathcal{S}$ sets $k_{i,r}^{y_i}$ to equal the rth garbled value in the vector $v_i^{y_i}$, and sets $k_{i,r}^{1-y_i}$ to be the string of all zeros. $\mathcal{S}$ then defines the commitment pair to be $(\mathsf{com}(k_{i,r}^0), \mathsf{com}(k_{i,r}^1))$.
- Second, $\mathcal{S}$ chooses a random string $\rho' \leftarrow_R \{0,1\}^s$ and constructs the commitment-sets $W_{i,j}$ and $W_{i,j}'$ (of P_1's inputs), as follows. For every input wire i and for every j such that $\rho_j' = 0$ (i.e., such that the sets $W_{i,j}$ and $W_{i,j}'$ are evaluation sets), $\mathcal{S}$ generates the commitment-set $W_{i,j}$ so that the first commitment is $\mathsf{com}(0)$ and the rest are "correct" (i.e., as instructed in the protocol). It then computes $W_{i,j}'$ *incorrectly*, committing to the exact same values as $W_{i,j}$ (we stress that the commitments are computed using fresh randomness, but they are commitments to the same values).
- Finally, $\mathcal{S}$ constructs the commitment-sets for the values of j such that $\rho_j' = 1$ (i.e., such that $W_{i,j}$ and $W_{i,j}'$ are check sets). Recall that the commitment-set $W_{i,j}$ is made up of an initial indicator commitment (to 0 or 1) followed by s commitments, where the rth commitment corresponds to the rth circuit; denote the rth commitment in $W_{i,j}$ by $W_{i,j}^r$. Now, for every input wire i and every j such that $\rho_j' = 1$:
  - For every r such that $\rho_r = 1$ (corresponding to a check circuit), simulator $\mathcal{S}$ places the correct commitments in $W_{i,j}^r$ and $W_{i,j}'^r$.
  - For every r such that $\rho_r = 0$ (corresponding to an evaluation circuit), simulator $\mathcal{S}$ places commitments to zeros. (These commitments are never opened; see Figures 4.3 and 4.4.)

$\mathcal{S}$ internally hands $\mathcal{A}$ the garbled circuits and commitments that it constructed. (Note that the commitments corresponding to P_1 and P_2's input wires in all of the evaluation circuits contain only a single garbled value from the pair associated with the wire. This will be important later on.)

5. $\mathcal{S}$ simulates the coin-tossing ("prepare challenge strings") phase with $\mathcal{A}$ so that the outcome of $\rho_1 \oplus \rho_2$ equals the string ρ that it chose above. If it fails, it outputs fail and halts. Likewise, the coin-tossing phase for the second challenge string is also simulated so that the outcome is ρ' as chosen above. Again, if it fails, it outputs fail and halts. We describe how this is achieved below.

6. $\mathcal{S}$ opens the commitments for check circuits and check sets for $\mathcal{A}$, exactly as described in Step 5 of Protocol 4.2.1.

7. $\mathcal{S}$ internally hands $\mathcal{A}$ decommitments for the garbled values for each of the input wires corresponding to P_1's input, in each of the evaluation circuits. In order to do this, $\mathcal{S}$ just chooses randomly between $W_{i,j}$ and $W'_{i,j}$ for each evaluation set, and decommits to the garbled values that are associated with the evaluation circuits.

8. $\mathcal{S}$ outputs whatever $\mathcal{A}$ outputs and halts.

If at any time during the simulation, $\mathcal{A}$ aborts (either explicitly or by sending an invalid message that would cause the honest P_1 to abort), $\mathcal{S}$ halts immediately and outputs whatever $\mathcal{A}$ does.

Analysis. We now show that the view of $\mathcal{A}$ in the above simulation by $\mathcal{S}$ is computationally indistinguishable from its view in a hybrid execution with P_1 where the oblivious transfer functionality is computed by a trusted party. (For the sake of clarity, we will refer to $\mathcal{A}$ as a real adversary running a real execution. However, this actually refers to $\mathcal{A}$ running the real protocol but with an ideal oblivious transfer.) We note that since only $\mathcal{A}$ receives output in this protocol, it suffices to consider the view of $\mathcal{A}$ only. Before demonstrating this, we show that the coin-tossing phases can be simulated so that $\mathcal{S}$ outputs fail with at most negligible probability. Intuitively, the simulation of this phase (for ρ) is carried out as follows:

1. $\mathcal{S}$ receives a perfectly-hiding commitment c from $\mathcal{A}$.

2. $\mathcal{S}$ generates a perfectly-binding commitment $\hat{c}$ to a random string $\hat{\rho}$ and internally hands it to $\mathcal{A}$.

3. If $\mathcal{A}$ aborts without decommitting, then $\mathcal{S}$ halts the simulation immediately and outputs whatever $\mathcal{A}$ outputs. Otherwise, let ρ_2 be the value decommitted to by $\mathcal{A}$.

4. $\mathcal{S}$ rewinds $\mathcal{A}$ to after the point that it sends c, and sends it a new commitment $\tilde{c}$ to the string $\rho_1 = \rho \oplus \rho_2$ (where the result of the coin-tossing is supposed to be the string ρ).

5. If $\mathcal{A}$ decommits to ρ_2, then $\mathcal{S}$ has succeeded. Thus, it continues by decommitting to ρ_1, and the result of the coin-tossing is $\rho = \rho_1 \oplus \rho_2$.
   If $\mathcal{A}$ decommits to some $\rho'_2 \neq \rho_2$, then $\mathcal{S}$ outputs ambiguous.
   If $\mathcal{A}$ does not decommit (but rather aborts), then $\mathcal{S}$ continues by sending a new commitment to $\rho_1 = \rho_2 \oplus \rho$. Notice that $\mathcal{S}$ sends a commitment to the

same value ρ_1, but uses fresh randomness in generating the commitments and executes Step 5 of the simulation again.

Unfortunately, as was shown by [34], the above simulation strategy does not necessarily run in expected polynomial time. Rather, it is necessary to first estimate the probability that $\mathcal{A}$ decommits when it receives a commitment $\hat{c}$ to a random value $\hat{\rho}$. Then, the number of rewinding attempts, when $\mathcal{A}$ is given a commitment to $\rho_1 = \rho \oplus \rho_2$, is truncated as some function of the estimate. In [34], it is shown that this strategy yields an expected polynomial-time simulation that fails with only negligible probability (including the probability of outputting ambiguous). Furthermore, the simulation has the property that the view of $\mathcal{A}$ is computationally indistinguishable from its view in a real execution. See Section 6.5.3 for a full detailed analysis of this exact procedure. Of course, the same strategy exactly is used for the simulation of the coin-tossing phase for ρ'.

We now continue with the analysis of the simulation. Intuitively, given that the above coin-tossing simulation succeeds, it follows that all of the check circuits are correctly constructed, as in the protocol (because $\mathcal{S}$ constructs all the circuits for which $\rho_i = 1$ correctly). Thus, the view of $\mathcal{A}$ with respect to these circuits is the same as in a real execution with an honest P_1. Furthermore, the commitments for the evaluation circuits reveal only a single garbled value for each input wire. Thus, Lemma 4.3.6 can be applied.

Formally, we prove indistinguishability in the following way. First, we modify $\mathcal{S}$ into $\mathcal{S}'$, which works in exactly the same way as $\mathcal{S}$ except for how it generates the circuits. Specifically, $\mathcal{S}'$ is given the honest P_1's input value x and constructs all of the circuits correctly. However, it only uses the garbled values corresponding to x in the commitment-sets. That is, if the value $k_{i,\ell}$ is used in all of the commitment sets $W_{i,j}$ and $W'_{i,j}$ with respect to circuit ℓ, then $k_{i,\ell}$ is the garbled value associated with x_i (i.e., the ith bit of x) in circuit ℓ (the other garbled value associated with the wire is associated with $1 - x_i$). Everything else remains the same. In order to see that $\mathcal{A}$'s view in an execution with $\mathcal{S}$ is indistinguishable from its view in an execution with $\mathcal{S}'$, we apply Lemma 4.3.6. In order to apply this claim, recall first that the evaluation circuits with $\mathcal{S}$ are all constructed according to $\widetilde{GC}$, yielding output $\tau = f(x, y)$ where y is the input obtained from $\mathcal{A}$ and x is the honest party's input. In contrast, the evaluation circuits with $\mathcal{S}'$ are all correctly constructed. Note also that the input y obtained by $\mathcal{S}'$ from $\mathcal{A}$ is the same value as that obtained by $\mathcal{S}$, which defines $\tau = f(x, y)$.[4] Finally, note that $\mathcal{S}'$ sends $\mathcal{A}$ the garbled values that correspond to P_1's input x. Thus, by Lemma 4.3.6, $\mathcal{A}$'s view with $\mathcal{S}$ is indistinguishable from its view with $\mathcal{S}'$.

---

[4] Note that there is one oblivious transfer for each input bit. Furthermore, the input of P_1 into these executions is a pair of vectors of s garbled values so that in the ith execution, the first vector contains all of the decommitments for garbled values that correspond to 0 for the ith input wire of P_2, and the second vector contains all of the decommitments for garbled values that correspond to 1 for the ith input wire of P_2. This means that in every circuit, $\mathcal{A}$ receives garbled values that correspond to the same input y.

(The full reduction here works by an adversary obtaining the garbled circuits and values, and then running the simulation of S or S'. Specifically, it generates all the check circuits correctly and uses the garbled values it obtained to generate the evaluation sets and the commitments in the evaluation sets. However, it does not generate the evaluation circuits itself, but uses the ones that it receives. If it receives real circuits, then it will obtain the distribution of S', and if it receives fake garbled circuits, then it will obtain the distribution of S. We therefore conclude by Lemma 4.3.6 that these distributions are indistinguishable. We note that the full proof of this also requires a hybrid argument over the many evaluation circuits, as opposed to the single circuit referred to in Lemma 4.3.6.)

Next, we construct a simulator S'' that works in the same way as S' except that it generates all of the commitments correctly (i.e., as in the protocol specification). Note that this only affects commitments that are never opened. Note also that S'' is given x and so it can do this. The indistinguishability between S' and S'' follows from the hiding property of the commitment scheme com. (The full reduction is straightforward and is therefore omitted.)

Finally, note that the distribution generated by S'' is the same as the one generated by an honest P_1, except for the simulation of the coin-tossing phases. Since, as we have mentioned, the view of A in the simulation of the coin-tossing is indistinguishable from its view in a real execution, we conclude that the view of A in the simulation by S'' is indistinguishable from its view in a real execution with P_1. Combining the above steps, we conclude that A's view in the simulation with S is indistinguishable from its view in a real execution with P_1. Recalling that our analysis actually applies to the hybrid model where the oblivious transfer is ideal, we conclude by applying the sequential composition theorem in order to obtain that A's view in the simulation with S is indistinguishable from its view in a completely real execution with P_1 (where even the oblivious transfer is real). This completes the proof of Lemma 4.3.5 and thus the case where P_2 is corrupted. ∎

Combining the cases. The proof of Theorem 4.3.1 is completed by combining Lemmas 4.3.4 and 4.3.5.

4.4 Efficient Implementation of the Different Primitives

In this section, we describe efficient implementations of the different building blocks of the protocol.

Encryption scheme. As in the semi-honest construction of Chapter 3, the construction uses a symmetric key encryption scheme that has indistinguishable encryptions for multiple messages and an elusive efficiently verifiable range. Informally, this means (1) that for any two (known) messages x and y, no polynomial-time adversary can distinguish between the encryptions of x

and y, and (2) that there is a negligible probability that an encryption under one key falls into the range of encryptions under another key, and given a key k it is easy to verify whether a certain ciphertext is in the range of encryptions with k. See Chapter 3 for a detailed discussion of these properties, and for examples of easy implementations satisfying them. For example, the encryption scheme could be $E_k(m) = \langle r, f_k(r) \oplus m0^n \rangle$, where f_k is a pseudorandom function keyed by k whose output is $|m| + n$ bits long, and r is a randomly chosen value.

Commitment schemes. The protocol uses both unconditionally hiding and unconditionally binding commitments. Our goal should be, of course, to use the most efficient implementations of these primitives, and we therefore concentrate on schemes with $O(1)$ communication rounds (all commitment schemes we describe here have only two rounds). Efficient unconditionally hiding commitment schemes can be based on number-theoretic assumptions, and use $O(1)$ exponentiations (see, e.g., [69]). The most efficient implementation is probably the one due to Damgård, Pedersen, and Pfitzmann, which uses a collision-free hashing function and no other cryptographic primitive [21]; see also [41]. Efficient unconditionally binding commitments can be constructed using the scheme of Naor [60], which has two rounds and is based on using a pseudorandom generator.

Oblivious transfer. The protocol uses an oblivious transfer protocol which is secure in the presence of malicious adversaries (and thus proven secure according to the real/ideal model simulation paradigm). Efficient protocols for this task can be found in Chapter 7.

4.5 Efficiency of the Protocol

The overhead of the protocol depends heavily on the statistical security parameter s. The security proof shows that the adversary's cheating probability is exponentially small in s. In particular, the adversary can cheat with probability at most $2^{-s/17}$. We conjecture that this can be made significantly smaller given a tighter analysis. Nevertheless, we have preferred to present a full and clear proof, rather than to overly optimize the construction at the cost of complicating the proof.

We analyze the complexity of the protocol as above in Section 3.5. We do not count the cost of the oblivious transfers, and refer to Chapter 7 (and in particular Sections 7.4.2 and 7.5 for an exact analysis of this cost):

- **Number of rounds:** The number of rounds of communication in the protocol equals the number of rounds required to run the oblivious transfers plus five.

- **Asymmetric computations:** In Protocol 4.2.1 each input bit of P_2 is replaced with s new input bits and therefore $O(ns)$ oblivious transfers are required. We remark that in [55] it is shown how this can be reduced.
- **Symmetric computations:** Each garbled circuit is of size $|C| = |C^0| + O(ns)$, and requires eight encryptions per gate for constructing the circuit and, as shown in Section 3.5, an expected four encryptions per gate for evaluating it. Since P_2 checks half of the circuits (which involves constructing them from scratch) and evaluates half of the circuits, we conclude that the cost is $8|C| \cdot s + 8|C| \cdot s/2 + 4|C| \cdot s/2 = 14|C|s$ symmetric encryption/decryption operations. In addition, there are $2s(s+1)$ commitments for each of the n inputs of P_1. We conclude that there are approximately $12s|C| + 2s^2n$ symmetric operations.
- **Bandwidth:** The communication overhead of the protocol is dominated by sending s copies of the garbled circuit and the commitment sets. The communication overhead is therefore $O(s|C| + s^2n)$ ciphertexts and commitment values.

4.6 Suggestions for Further Reading

Recently, a number of different approaches have been proposed for constructing efficient general protocols that are secure in the presence of malicious adversaries. In this chapter we have shown just one of these approaches. Other approaches include the following:

- Jarecki and Shmatikov [47] designed a protocol in which the parties efficiently prove, gate by gate, that their behavior is correct. The protocol runs in a constant number of rounds, and is based on the use of a special homomorphic encryption system, which is used to encode the tables of each gate of the circuit (compared to the use of symmetric encryption in Yao's original protocol and in the protocol presented above in this chapter). This protocol works in a completely different way than the protocol presented above because it does not use the cut-and-choose technique. Rather, the circuit constructor proves in zero-knowledge that all gates are correctly constructed; this is carried out efficiently by utilizing properties of the homomorphic encryption scheme. The result is a protocol requiring a constant number of asymmetric operations per gate (and no symmetric operations). Although this approach has the potential to yield an efficient solution for circuits that are not too large, currently the exact number of asymmetric operations required is 720, which is very large. If this constant could be reduced to a small number, this approach would be far more appealing.
- Nielsen and Orlandi [65] use the cut-and-choose technique but in a very different way than in the protocol presented in this chapter. Specifically, the circuit constructor first sends the receiver many gates, and the receiver

checks that they are correctly constructed by asking for some to be opened. After this stage, the parties interact in a way that enables the gates to be securely soldered (like Lego blocks) into a correct circuit. Since it is not guaranteed that all of the gates are correct, but just a vast majority, a fault-tolerant circuit of size $O(s \cdot |C| / \log |C|)$ is constructed, where s is a statistical security parameter. The error as a function of s is 2^{-s} and the constant inside the "O" notation for the number of exponentiations is 32 [66].

- Ishai, Prabhakaran and Sahai [46] present a protocol that works by having the parties simulate a virtual multiparty protocol with an honest majority. The cost of the protocol essentially consists of the cost of running a semi-honest protocol for computing the multiplication of additive shares, for every multiplication carried out by a party in a multiparty protocol with honest majority. Thus, the actual efficiency of the protocol depends heavily on the multiparty protocol to be simulated, and on the semi-honest protocols used for simulating the multiparty protocol. An asymptotic analysis demonstrates that this method may be competitive. However, no concrete analysis has been carried out, and it is currently an open question whether or not it is possible to instantiate this protocol in a way that will be competitive with other known protocols.

- Lindell and Pinkas [57] present a protocol that is built on the same concept as the protocol in this chapter, but is significantly simpler and more efficient. The protocol relies on specific assumptions and tools that we present in the second part of this book. For this reason we chose to present the protocol of [55], and not its improved version [57].

Chapter 5
Covert Adversaries

In this chapter, we present protocols for securely computing any functionality in the model of covert adversaries; see Section 2.4 for motivation and definitions of the model. We begin by presenting a protocol for oblivious transfer that is secure in the covert model. We present this protocol mainly for didactic reasons; it serves as a good warm-up for understanding how to prove security in this model. Nevertheless, we stress that more efficient oblivious transfer protocols are known (both for the covert and malicious settings; see Section 7.4 for an example).

5.1 Oblivious Transfer

In this section we will construct an efficient oblivious transfer protocol that is secure in the presence of covert adversaries with ϵ-deterrent. We will first present the basic scheme that considers a single oblivious transfer and $\epsilon = 1/2$. We will then extend this to enable the simultaneous (batch) execution of many oblivious transfers and also higher values of ϵ. Our constructions all rely on the existence of secure homomorphic encryption schemes.

We remark that it is possible to use the highly efficient oblivious transfer protocol with full simulation-based security that appears in Chapter 7. This is due to the fact that security in the presence of malicious adversaries implies security in the presence of covert adversaries; see Proposition 2.4.3. Nevertheless, we present the protocol here since it is more general and in addition provides a good warm-up to proofs of security in this model.

Homomorphic encryption. Intuitively, a public-key encryption scheme is homomorphic if given two ciphertexts $c_1 = E_{pk}(m_1)$ and $c_2 = E_{pk}(m_2)$ it is possible to efficiently compute $E_{pk}(m_1 + m_2)$ without knowledge of the secret decryption key or the plaintexts. Of course this assumes that the plaintext message space is a group; we actually assume that both the plaintext and ciphertext spaces are groups (with respective group operations $+$ and $\cdot$). A

C. Hazay, Y. Lindell, *Efficient Secure Two-Party Protocols*,
Information Security and Cryptography, DOI 10.1007/978-3-642-14303-8_5,
© Springer-Verlag Berlin Heidelberg 2010

natural way to define this is to require that for all pairs of keys (pk, sk), all $m_1, m_2 \in \mathcal{P}$ and $c_1, c_2 \in \mathcal{C}$ with $m_1 = D_{sk}(c_1)$ and $m_2 = D_{sk}(c_2)$ it hold that $D_{sk}(c_1 \cdot c_2) = m_1 + m_2$. However, we actually need a *stronger property*. Specifically, we require that the result of computing $c_1 \cdot c_2$, when c_2 is a random encryption of m_2, is a random encryption of $m_1 + m_2$ (by a random encryption we mean a ciphertext generated by encrypting the plaintext with uniformly distributed coins). This property ensures that if one party generated c_1 and the other party applied a series of homomorphic operations to c_1 in order to generate c, then the only thing that the first party can learn from c is the underlying plaintext. In particular, it learns nothing about the steps taken to arrive at c (e.g., it cannot know if the second party added m_2 in order to obtain an encryption of $m_1 + m_2$, or if it first added m_3 and then m_4 where $m_2 = m_3 + m_4$). We stress that this holds even if the first party knows the secret key of the encryption scheme. We formalize the above by requiring that the distribution of $\{pk, c_1, c_1 \cdot c_2\}$ is *identical* to the distribution of $\{pk, E_{pk}(m_1), E_{pk}(m_1 + m_2)\}$, where in the latter case the encryptions of m_1 and $m_1 + m_2$ are generated independently of each other, using uniformly distributed random coins. We denote by $E_{pk}(m)$ the random variable generated by encrypting m with public key pk using uniformly distributed random coins. We have the following formal definition.

Definition 5.1.1 *A public-key encryption scheme* (G, E, D) *is* homomorphic *if for all n and all (pk, sk) output by $G(1^n)$, it is possible to define groups* $\mathcal{M}, \mathcal{C}$ *such that:*

- *The plaintext space is $\mathcal{M}$, and all ciphertexts output by E_{pk} are elements of $\mathcal{C}$,[1] and*
- *For every $m_1, m_2 \in \mathcal{M}$ it holds that*

$$\{pk, c_1 = E_{pk}(m_1), c_1 \cdot E_{pk}(m_2)\} \equiv \{pk, E_{pk}(m_1), E_{pk}(m_1 + m_2)\} \quad (5.1)$$

where the group operations are carried out in $\mathcal{C}$ and $\mathcal{M}$, respectively.

Note that in the left distribution in (5.1) the ciphertext c_1 is used to generate an encryption of $m_1 + m_2$ using the homomorphic operation, whereas in the right distribution the encryptions of m_1 and $m_1 + m_2$ are independent. An important observation is that any such scheme supports the multiplication of a ciphertext by a scalar, which can be achieved by computing multiple additions. Such encryption schemes can be constructed under the quadratic residuosity, N-residuosity, decisional Diffie-Hellman (DDH) and other assumptions; see [68, 2] for some references. By convention, no ciphertext is invalid. That is, any value that is not in the ciphertext group $\mathcal{C}$ is interpreted as an encryption of the identity element of the plaintext group $\mathcal{M}$.

---

[1] The plaintext and ciphertext spaces may depend on pk; we leave this implicit.

5.1.1 The Basic Protocol

PROTOCOL 5.1.2 (OT from homomorphic encryption):

- **Inputs:** *Party P_1 has a pair of strings (x_0, x_1) for input; P_2 has a bit σ. Both parties have the security parameter 1^n as auxiliary input. (In order to satisfy the constraints that all inputs are of the same length, it is possible to define $|x_0| = |x_1| = k$ and give P_2 $(\sigma, 1^{2k-1})$.)*
- **Assumption:** *We assume that the group determined by the homomorphic encryption scheme with security parameter n is large enough to contain all strings of length k. Thus, if the homomorphic encryption scheme only works for single bits, we will only consider $k = 1$ (i.e., bit oblivious transfer).*
- **The protocol:**

  1. *Party P_2 chooses two sets of two pairs of keys:*
     a. $(pk_1^0, sk_1^0) \leftarrow G(1^n)$; $(pk_2^0, sk_2^0) \leftarrow G(1^n)$ *using random coins r_G^0, and*
     b. $(pk_1^1, sk_1^1) \leftarrow G(1^n)$; $(pk_2^1, sk_2^1) \leftarrow G(1^n)$ *using random coins r_G^1*
     P_2 sends (pk_1^0, pk_2^0) and (pk_1^1, pk_2^1) to P_1.
  2. Key generation challenge:
     a. *P_1 chooses a random coin $b \leftarrow_R \{0,1\}$ and sends b to P_2.*
     b. *P_2 sends P_1 the random coins r_G^b that it used to generate (pk_1^b, pk_2^b).*
     c. *P_1 checks that the public keys output by the key generation algorithm G when given input 1^n and the appropriate portions of the random tape r_G^b equal pk_1^b and pk_2^b. If this does not hold, or if P_2 did not send any message here, P_1 outputs corrupted$_2$ and halts. Otherwise, it proceeds.*
     Denote $pk_1 = pk_1^{1-b}$ and $pk_2 = pk_2^{1-b}$.
  3. *P_2 chooses two random bits $\alpha, \beta \leftarrow_R \{0,1\}$. Then:*
     a. *P_2 computes*
     $$c_0^1 = E_{pk_1}(\alpha), \quad c_0^2 = E_{pk_2}(1-\alpha),$$
     $$c_1^1 = E_{pk_1}(\beta), \quad c_1^2 = E_{pk_2}(1-\beta),$$
     using random coins r_0^1, r_0^2, r_1^1 and r_1^2, respectively.
     b. *P_2 sends (c_0^1, c_0^2) and (c_1^1, c_1^2) to P_1.*
  4. Encryption generation challenge:
     a. *P_1 chooses a random bit $b' \leftarrow_R \{0,1\}$ and sends b' to P_2.*
     b. *P_2 sends $r_{b'}^1$ and $r_{b'}^2$ to P_1 (i.e., P_2 sends an opening to the ciphertexts $c_{b'}^1$ and $c_{b'}^2$).*
     c. *P_1 checks that one of the ciphertexts $\{c_{b'}^1, c_{b'}^2\}$ is an encryption of 0 and the other is an encryption of 1. If not (including the case where no message is sent by P_2), P_1 outputs corrupted$_2$ and halts. Otherwise, it continues to the next step.*
  5. *P_2 sends a "reordering" of the ciphertexts $\{c_{1-b'}^1, c_{1-b'}^2\}$. Specifically, if $\sigma = 0$ it sets c_0 to be the ciphertext that is an encryption of 1, and sets c_1 to be the ciphertext that is an encryption of 0. Otherwise, if*

$\sigma = 1$, it sets c_0 to be the encryption of 0, and c_1 to be the encryption of 1. (Only the ordering needs to be sent and not the actual ciphertexts. Furthermore, this can be sent together with the openings in Step 4b.)

6. P_1 uses the homomorphic property and c_0, c_1 as follows.
   a. P_1 computes $\tilde{c}_0 = x_0 \cdot_E c_0$ (this operation is relative to the key pk_1 or pk_2 depending on whether c_0 is an encryption under pk_1 or pk_2)
   b. P_1 computes $\tilde{c}_1 = x_1 \cdot_E c_1$ (this operation is relative to the key pk_1 or pk_2 depending on whether c_1 is an encryption under pk_1 or pk_2)
   P_1 sends $\tilde{c}_0$ and $\tilde{c}_1$ to P_2. (Notice that one of the ciphertexts is encrypted with key pk_1 and the other is encrypted with key pk_2.)
7. If $\sigma = 0$, P_2 decrypts $\tilde{c}_0$ and outputs the result (if $\tilde{c}_0$ is encrypted under pk_1 then P_2 outputs $x_0 = D_{sk_1}(\tilde{c}_0)$; otherwise it outputs $x_0 = D_{sk_2}(\tilde{c}_0)$). Otherwise, if $\sigma = 1$, P_2 decrypts $\tilde{c}_1$ and outputs the result.
8. If at any stage during the protocol, P_1 does not receive the next message that it expects to receive from P_2 or the message it receives is invalid and cannot be processed, it outputs abort$_2$ (unless it was already instructed to output corrupted$_2$). Likewise, if P_2 does not receive the next message that it expects to receive from P_1 or it receives an invalid message, it outputs abort$_1$.

We remark that the reordering message of Step 5 can actually be sent by P_2 together with the message in Step 4b. Furthermore, the messages of the key-generation challenge can be piggybacked on later messages, as long as they conclude before the final step. We therefore have that the number of rounds of communication can be exactly *four* (each party sends two messages).

Before proceeding to the proof of security, we present the intuitive argument showing why Protocol 5.1.2 is secure. We begin with the case where P_2 is corrupted. First note that if P_2 follows the instructions of the protocol, it learns only a single value x_0 or x_1. This is because one of c_0 and c_1 is an encryption of 0. If it is c_0, then $\tilde{c}_0 = x_0 \cdot_E c_0 = E_{pk}(0 \cdot x_0) = E_{pk}(0)$ (where $pk \in \{pk_1, pk_2\}$), and so $\tilde{c}_0$ is a random encryption of 0 that is independent of x_0, implying that nothing is learned about x_0; similarly if it is c_1 then $\tilde{c}_1 = E_{pk}(0)$ and so nothing is learned about x_1. However, in general, P_2 may not generate the encryptions $c_0^1, c_1^1, c_0^2, c_1^2$ properly (and so it may be that at least one of the pairs (c_0^1, c_0^2) and (c_1^1, c_1^2) are *both* encryptions of 1, in which case P_2 could learn both x_0 and x_1). This is prevented by the encryption-generation challenge. That is, if P_2 tries to cheat in this way then it is guaranteed to be caught with probability at least $1/2$. The above explains why a malicious P_2 can learn only one of the outputs, unless it is willing to be caught cheating with probability $1/2$. This therefore demonstrates that "privacy" holds. However, we actually need to prove security via simulation, which involves showing how to *extract* P_2's implicit input and how to *simulate* its view. Extraction works by first providing the corrupted P_2 with the encryption-challenge bit $b' = 0$ and then rewinding it and providing it with the challenge $b' = 1$. If the corrupted P_2 replies to both challenges, then the simulator can construct σ from the opened ciphertexts and the reordering provided. Given this input, the simulation can be completed in a straight-forward manner; see the proof below. A crucial point here is that if P_2 does not reply to both challenges then an honest P_1 would output $\mathsf{corrupted}_2$ with probability $1/2$, and so this corresponds to a cheat_2 input in the ideal world.

We now proceed to discuss why the protocol is secure in the presence of a corrupt P_1. In this case, it is easy to see that P_1 cannot learn anything about P_2's input because the encryption scheme is semantically secure (and so a corrupt P_1 cannot determine σ from the unopened ciphertexts). However, as above, we need to show how extraction and simulation work. Extraction here works by providing encryptions so that in one of the pairs (c_0^1, c_0^2) or (c_1^1, c_1^2) both of the encrypted values are 1. If this pair is the one used (and not the one opened), then we have that $\tilde{c}_0$ is an encryption of x_0 and $\tilde{c}_1$ is an encryption of $\tilde{c}_1$. An important point here is that unlike a real P_2, the simulator can do this without being "caught". Specifically, the simulator generates the ciphertexts so that for a random $b' \leftarrow_R \{0, 1\}$ it holds that $c_{1-b'}^1$ and $c_{1-b'}^2$ are both encryptions of 1, whereas $c_{b'}^1$ and $c_{b'}^2$ are generated correctly, one being an encryption of 0 and the other an encryption of 1. Then, the simulator "hopes" that the corrupted P_1 asks it to open the ciphertexts $c_{b'}^1$ and $c_{b'}^2$, which look like they should. In such a case, the simulator proceeds and succeeds in extracting both x_0 and x_1. However, if the corrupted P_1 asks the simulator to open the other ciphertexts (which are clearly invalid), the simulator just rewinds the corrupted P_1 and tries again. Thus, extraction can be achieved. Regarding the simulation of P_1's view, this follows from the fact that the only differences between the above and a real execution are the values encrypted in the ciphertexts $c_0^1, c_0^2, c_1^1, c_1^2$. These distributions are therefore indistinguishable by the semantic security of the encryption scheme.

We now formally prove that Protocol 5.1.2 meets Definition 2.4.1 with $\epsilon = \frac{1}{2}$.

Theorem 5.1.3 *Assuming that (G, E, D) constitutes a semantically secure homomorphic encryption scheme, Protocol 5.1.2 securely computes the oblivious transfer functionality $((x_0, x_1), \sigma) \mapsto (\lambda, x_\sigma)$ in the presence of covert adversaries with ϵ-deterrent for $\epsilon = \frac{1}{2}$.*

Proof. We will separately consider the case where P_2 is corrupted and the case where P_1 is corrupted. We note that although we construct two different simulators (one for each corruption case), a single simulator as required by the definition can be constructed by simply combining them and working appropriately given the identity of the corrupted party.

Corrupted P_2: Let $\mathcal{A}$ be a real adversary that controls P_2. We construct a simulator $\mathcal{S}$ that works as follows:

1. $\mathcal{S}$ receives $(\sigma, 1^{2k-1})$ and z as input and invokes $\mathcal{A}$ on this input.
2. $\mathcal{S}$ plays the honest P_1 with $\mathcal{A}$ as P_2.

3. When S reaches the key generation challenge step, it first sends $b = 0$ and receives back A's response. Then, S rewinds A, sends $b = 1$ and receives back A's response.

   a. If both of the responses from A would cause a corrupted output (meaning a response that would cause P_1 to output corrupted$_2$ in a real execution), S sends corrupted$_2$ to the trusted party, simulates the honest P_1 aborting due to detected cheating, and outputs whatever A outputs.
   b. If A sends back exactly one response that would cause a corrupted output, then S sends cheat$_2$ to the trusted party.
      i. If the trusted party replies with corrupted$_2$, then S rewinds A and hands it the query for which A's response would cause a corrupted output. S then simulates the honest P_1 aborting due to detected cheating, and outputs whatever A outputs.
      ii. If the trusted party replies with undetected and the honest P_1's input pair (x_0, x_1), then S plays the honest P_1 with input (x_0, x_1) in the remainder of the execution with A as P_2. At the conclusion, S outputs whatever A outputs.
   c. If neither of A's responses causes a corrupted output, then S rewinds A, gives it a random b and proceeds as below.

4. S receives ciphertexts $c_0^1, c_0^2, c_1^1, c_1^2$ from A.
5. Next, in the encryption-generation challenge step, S first sends $b' = 0$ and receives back A's response, which includes the reordering of the ciphertexts (recall that the reordered messages are actually sent together with the ciphertext openings). Then, S rewinds A, sends $b' = 1$ and receives back A's response.

   a. If both of the responses from A would cause a corrupted output, S sends corrupted$_2$ to the trusted party, simulates the honest P_1 aborting due to detected cheating, and outputs whatever A outputs.
   b. If A sends back exactly one response that would cause a corrupted output, then S sends cheat$_2$ to the trusted party.
      i. If the trusted party replies with corrupted$_2$, then S rewinds A and hands it the query for which A's response would cause a corrupted output. S then simulates the honest P_1 aborting due to detected cheating, and outputs whatever A outputs.
      ii. If the trusted party replies with undetected and the honest P_1's input pair (x_0, x_1), then S plays the honest P_1 with input (x_0, x_1) and completes the execution with A as P_2. (Note that P_1 has not yet used its input at this stage of the protocol. Thus, S has no problem completing the execution like an honest P_1.) At the conclusion, S outputs whatever A outputs.
   c. If neither of A's responses would cause a corrupted output, then S uses the reorderings to determine the value of σ. Specifically, S chooses a random b' and takes the reordering that relates to $c_{1-b'}^1$ and $c_{1-b'}^2$ (if

$c_{1-b'}^1$ is an encryption of 1, then $\mathcal{S}$ determines $\sigma = 0$ and otherwise it determines $\sigma = 1$). The value b' that is chosen is the one that $\mathcal{S}$ sends to $\mathcal{A}$ and appears in the final transcript.

$\mathcal{S}$ sends σ to the trusted party and receives back $x = x_\sigma$. Simulator $\mathcal{S}$ then completes the execution playing the honest P_1 and using $x_0 = x_1 = x$.

6. If at any point $\mathcal{A}$ sends a message that would cause the honest P_1 to halt and output abort$_2$, simulator $\mathcal{S}$ immediately sends abort$_2$ to the trusted party, halts the simulation and proceeds to the final "output" step.

7. *Output:* At the conclusion, $\mathcal{S}$ outputs whatever $\mathcal{A}$ outputs.

This completes the description of $\mathcal{S}$. Denoting Protocol 5.1.2 as π and noting that I here equals $\{R\}$ (i.e., P_2 is corrupted), we need to prove that for $\epsilon = \frac{1}{2}$,

$$\left\{\text{IDEALSC}_{\text{OT},\mathcal{S}(z),2}^\epsilon(((x_0,x_1),\sigma),n)\right\} \stackrel{c}{\equiv} \left\{\text{REAL}_{\pi,\mathcal{A}(z),2}(((x_0,x_1),\sigma),n)\right\}.$$

It is clear that the simulation is perfect if $\mathcal{S}$ sends corrupted$_2$ or cheat$_2$ at any stage. This is due to the fact that the probability that an honest P_1 outputs corrupted$_2$ in the simulation is identical to the probability in a real execution (probability 1 in the case where $\mathcal{A}$ responds incorrectly to both challenges and probability $1/2$ otherwise). Furthermore, in the case where $\mathcal{S}$ sends cheat$_2$ and receives back undetected, it concludes the execution using the true input of P_1. The simulation until the last step is perfect (it involves merely sending random challenges); therefore the completion using the true P_1's input yields a perfect simulation. The above is clearly true of abort$_2$ as well (because this can only occur before the last step where P_1's input is used).

It remains to analyze the case where $\mathcal{S}$ does not send corrupted$_2$, cheat$_2$ or abort$_2$ to the trusted party. Notice that in this case, $\mathcal{A}$ responded correctly to both the key generation and encryption generation challenges. In particular, this implies that the keys pk_1 and pk_2 are correctly generated, and that $\mathcal{S}$ computes σ based on the encrypted values sent by $\mathcal{A}$ and the reordering.

Now, if $\sigma = 0$, then $\mathcal{S}$ hands $\mathcal{A}$ the ciphertexts $\tilde{c}_0 = E_{pk}(x_0)$ and $\tilde{c}_1 = E_{pk'}(0)$, where $pk, pk' \in \{pk_1, pk_2\}$ and $pk \neq pk'$, and if $\sigma = 1$, it hands $\mathcal{A}$ the ciphertexts $\tilde{c}_0 = E_{pk}(0)$ and $\tilde{c}_1 = E_{pk'}(x_1)$. This follows from the instructions of $\mathcal{S}$ and the honest party ($\mathcal{S}$ plays the honest party with $x_0 = x_1 = x_\sigma$ and so $\tilde{c}_\sigma$ is an encryption of x_σ and $\tilde{c}_{1-\sigma}$ is an encryption of 0). The important point to notice is that these messages are distributed identically to the honest P_1's messages in a real protocol; the fact that $\mathcal{S}$ does not know $x_{1-\sigma}$ makes no difference because for every x' it holds that $x' \cdot E_{pk}(0) = E_{pk}(0)$. We note that this assumes that the homomorphic property of the encryption scheme holds, but this is given by the fact that pk_1 and pk_2 are correctly formed. Regarding the rest of the messages sent by $\mathcal{S}$, these are generated independently of P_1's input and so exactly like an honest P_1.

We conclude that the view of $\mathcal{A}$ as generated by the simulator $\mathcal{S}$ is *identical* to the distribution generated in a real execution. Thus, its output is identically distributed in both cases. (Since P_1 receives no output, we do not need to consider the output distribution of the honest P_1 in the real and ideal executions.) We conclude that

$$\left\{ \text{IDEALSC}^{\epsilon}_{\text{OT},\mathcal{S}(z),2}(((x_0,x_1),\sigma),n) \right\} \equiv \left\{ \text{REAL}_{\pi,\mathcal{A}(z),2}(((x_0,x_1),\sigma),n) \right\},$$

completing this corruption case.

Corrupted P_1: Let $\mathcal{A}$ be a real adversary that controls P_1. We construct a simulator $\mathcal{S}$ that works as follows:

1. $\mathcal{S}$ receives (x_0, x_1) and z and invokes $\mathcal{A}$ on this input.
2. $\mathcal{S}$ interacts with $\mathcal{A}$ and plays the honest P_2 until Step 3 of the protocol.
3. In Step 3 of the protocol, $\mathcal{S}$ works as follows:

   a. $\mathcal{S}$ chooses random bits $b, \alpha \leftarrow_R \{0,1\}$;
   b. $\mathcal{S}$ computes:
   $$c^1_b = E_{pk_1}(\alpha), \qquad c^2_b = E_{pk_2}(1-\alpha),$$
   $$c^1_{1-b} = E_{pk_1}(1), \qquad c^2_{1-b} = E_{pk_2}(1);$$

   c. $\mathcal{S}$ sends $c^1_0, c^2_0, c^1_1, c^2_1$ to $\mathcal{A}$.

4. In the next step (Step 4 of the protocol), $\mathcal{A}$ sends a bit b'. If $b' = b$, then $\mathcal{S}$ opens the ciphertexts c^1_b and c^2_b as the honest P_2 would (note that the ciphertexts are "correctly" constructed). Otherwise, $\mathcal{S}$ returns to Step 3 of the simulation above (i.e., it returns to the beginning of Step 3 of the protocol) and tries again with fresh randomness.[2]
5. $\mathcal{S}$ sends a random reordering of the ciphertexts c^1_{1-b} and c^2_{1-b} (the actual order does not matter because they are both encryptions of 1).
6. The simulator $\mathcal{S}$ receives from $\mathcal{A}$ the ciphertexts $\tilde{c}_0$ and $\tilde{c}_1$. $\mathcal{S}$ computes $x_0 = D_{sk_1}(\tilde{c}_0)$ and $x_1 = D_{sk_2}(\tilde{c}_1)$ (or $x_0 = D_{sk_2}(\tilde{c}_0)$ and $x_1 = D_{sk_1}(\tilde{c}_1)$, depending on which of c_0, c_1 is encrypted with pk_1 and which with pk_2), and sends the pair (x_0, x_1) to the trusted party as P_1's input.
7. If at any stage in the simulation $\mathcal{A}$ does not respond, or responds with an invalid message that cannot be processed, then $\mathcal{S}$ sends abort$_1$ to the trusted party for P_1's inputs. (Such behavior from $\mathcal{A}$ can only occur before the last step and so before any input (x_0, x_1) has already been sent to the trusted party.)
8. $\mathcal{S}$ outputs whatever $\mathcal{A}$ outputs.

---

[2] This yields an expected polynomial-time simulation because these steps are repeated until $b' = b$. A strict polynomial-time simulation can be achieved by just halting after n attempts. The probability that $b' \neq b$ in all of these attempts can be shown to be negligible, based on the hiding property of the encryption scheme.

Notice that $\mathcal{S}$ never sends cheat_1 to the trusted party. Thus we actually prove standard security in this corruption case. That is, we prove that

$$\left\{ \mathrm{IDEAL}_{\mathrm{OT},\mathcal{S}(z),1}((x_0,x_1,\sigma),n) \right\} \overset{\mathrm{c}}{\equiv} \left\{ \mathrm{REAL}_{\pi,\mathcal{A}(z),1}((x_0,x_1,\sigma),n) \right\}. \quad (5.2)$$

By Proposition 2.4.3, this implies security for covert adversaries as well. In order to prove Eq. (5.2), observe that the only difference between the view of the adversary $\mathcal{A}$ in a real execution and in the simulation by $\mathcal{S}$ is due to the fact that $\mathcal{S}$ does not generate c_b^1, c_b^2 correctly. Thus, intuitively, Eq. (5.2) follows from the security of the encryption scheme. That is, we begin by showing that if the view of $\mathcal{A}$ in the real and ideal executions can be distinguished, then it is possible to break the security of the encryption scheme. We begin by showing that the view of $\mathcal{A}$ when interacting with an honest P_1 with input $\sigma = 0$ is indistinguishable from the view of $\mathcal{A}$ when interacting in a simulation with $\mathcal{S}$.

Let $\mathcal{A}'$ be an adversary that attempts to distinguish encryptions under a key pk.[3] Adversary $\mathcal{A}'$ receives a key pk, chooses a random bit $\gamma \leftarrow_R \{0,1\}$ and a random index $\ell \leftarrow_R \{1,2\}$ and sets $pk_\ell^{1-\gamma} = pk$. It then chooses the keys $pk_{3-\ell}^{1-\gamma}$, pk_1^γ and pk_2^γ by itself and sends $\mathcal{A}$ the keys (pk_1^0, pk_2^0) and (pk_1^1, pk_2^1). When $\mathcal{A}$ replies with a bit b, adversary $\mathcal{A}'$ acts as follows. If $b = \gamma$, $\mathcal{A}'$ opens the randomness used in generating (pk_1^b, pk_2^b) as the honest P_2 would ($\mathcal{A}'$ can do this because it chose $(pk_1^\gamma, pk_2^\gamma)$ by itself and $\gamma = b$). If $b \neq \gamma$, $\mathcal{A}'$ cannot open the randomness as an honest P_2 would. Therefore, $\mathcal{A}'$ just halts. If $\mathcal{A}$ continues, then it sets $pk_1 = pk_1^{1-\gamma}$ and $pk_2 = pk_2^{1-\gamma}$ (and so pk_ℓ is the public key pk that $\mathcal{A}'$ is "attacking"). Now, $\mathcal{A}'$ computes the ciphertexts $c_0^1, c_0^2, c_1^1, c_1^2$ in the following way. $\mathcal{A}'$ chooses α and β at random, as the honest P_2 would. Then, for a random ζ adversary $\mathcal{A}'$ computes $c_\zeta^1 = E_{pk_1}(\alpha)$, $c_\zeta^2 = E_{pk_2}(1 - \alpha)$, and $c_{1-\zeta}^{3-\ell} = E_{pk_{3-\ell}}(1)$. However, $\mathcal{A}'$ does not compute $c_{1-\zeta}^\ell = E_{pk_\ell}(1)$. Rather, it outputs a pair of plaintexts $m_0 = 0, m_1 = 1$ for the external encryption game and receives back $c = E_{pk}(m_b) = E_{pk_\ell}(m_b)$ (for $b \leftarrow_R \{0,1\}$). Adversary $\mathcal{A}'$ sets $c_{1-\zeta}^\ell = c$ (i.e., to equal the challenge ciphertext) and continues playing the honest P_2 until the end. In this simulation, $\mathcal{A}'$ sets the reordering so that c_0 equals $c_{1-\zeta}^{3-\ell}$ (that is, it is an encryption of 1).

The key point here is that if $\mathcal{A}'$ does not halt and $b = 0$, then the simulation by $\mathcal{A}'$ is identical to a real execution between $\mathcal{A}$ and an honest P_2 that has input $\sigma = 0$ (because $c_0 = c_{1-\zeta}^{3-\ell}$ is an encryption of 1 and $c_1 = c_{1-\zeta}^\ell$ is an encryption of 0, as required). In contrast, if $\mathcal{A}'$ does not halt and $b = 1$, then the simulation by $\mathcal{A}'$ is identical to the simulation carried out by $\mathcal{S}$ (because in this case they are both encryptions of 1). Finally, note that $\mathcal{A}'$ halts with probability exactly $1/2$ in both cases (this is due to the fact that

---

[3] The game that $\mathcal{A}'$ plays is that it receives a key pk, outputs a pair of plaintexts m_0, m_1, receives back a *challenge ciphertext* $E_{pk}(m_b)$ for some $b \in \{0,1\}$, and outputs a "guess" bit b'. An encryption scheme is indistinguishable if the probability that $\mathcal{A}'$ outputs $b' = b$ is negligibly close to $\frac{1}{2}$.

the distribution of the keys is identical for both choices of γ). Combining the above together, we have that if it is possible to distinguish the view of $\mathcal{A}$ in the simulation by $\mathcal{S}$ from a real execution with P_2 that has input 0, then it is possible to distinguish encryptions. Specifically, $\mathcal{A}'$ can just run the distinguisher that exists for these views and output whatever the distinguisher outputs.

The above shows that the view of $\mathcal{A}$ in the simulation is indistinguishable from its view in a real execution with an honest P_2 with input $\sigma = 0$. However, we actually have to show that when the honest P_2 has input $\sigma = 0$, the *joint distribution* of $\mathcal{A}$ and the honest P_2's outputs in a real execution are indistinguishable from the joint distribution of $\mathcal{S}$ and the honest P_2's outputs in the ideal model. The point to notice here is that the output of the honest P_2 in both the real and ideal models is the value obtained by decrypting $\tilde{c}_0$ using key $pk_{3-\ell}$. (In the real model this is what the protocol instructs the honest party to output and in the ideal model this is the value that $\mathcal{S}$ sends to the trusted party as P_1's input x_0.) However, in this reduction $\mathcal{A}'$ knows the associated secret-key to $pk_{3-\ell}$, because it chose $pk_{3-\ell}$ itself. Thus, $\mathcal{A}'$ can append the decryption of $\tilde{c}_0$ to the view of $\mathcal{A}$, thereby generating a joint distribution. It follows that if $\mathcal{A}'$ received an encryption of $m_0 = 0$, then it generates the joint distribution of the outputs in the real execution, and if it received an encryption of $m_1 = 1$, then it generates the joint distribution of the outputs in the ideal execution. By the indistinguishability of the encryption scheme we have that the real and ideal distributions are indistinguishable, completing the proof of Eq. (5.2) for the case where $\sigma = 0$. The case for $\sigma = 1$ follows from an almost identical argument as above. Combining these two cases, we have the output distribution generated by the simulator in the ideal model is computationally indistinguishable from the output distribution of a real execution.

It remains to show that $\mathcal{S}$ runs in expected polynomial-time. Note that $\mathcal{S}$ rewinds if in the simulation it holds that $b' \neq b$. Now, in the case where the ciphertexts $c_0^1, c_0^2, c_1^1, c_1^2$ are generated as by the honest party (each pair containing an encryption of 0 and an encryption of 1), the probability that $b' \neq b$ is exactly $1/2$ because the value of b' is information-theoretically hidden. In contrast, in the simulation this is not the case because c_b^1, c_b^2 are "correctly" constructed, whereas c_{1-b}^1, c_{1-b}^2 are both encryptions of 1. Nevertheless, if the probability that $b' \neq b$ is non-negligibly far from $1/2$, then this can be used to distinguish an encryption of 0 from an encryption of 1 (the actual reduction can be derived from the reduction already carried out above and is thus omitted). It follows that the expected number of rewindings is at most slightly greater than two, implying that the overall simulation runs in expected polynomial time. As we have mentioned in Footnote 2, the simulation can be made to run in strict polynomial time by aborting if for n consecutive trials it holds that $b' \neq b$. By the argument given above, such an abort can only occur with negligible probability. This concludes the proof of this corruption case, and thus of the theorem. ∎

Discussion. We conclude this section with some important remarks and discussion regarding the construction and proof:

The covert technique. The proof of Protocol 5.1.2 in the case where P_2 is corrupted relies heavily on the fact that the simulator can send cheat and therefore does not need to complete a "standard" simulation. Take for example the case where $\mathcal{A}$ (controlling P_2) only replies with one valid response to the encryption generation challenge. In this case, P_2 can learn both x_0 and x_1 with probability $1/2$. However, the simulator in the ideal model can never learn both x_0 and x_1. Therefore, the simulator cannot generate the correct distribution. However, by allowing the simulator to declare a cheat, it can complete the simulation as required. This demonstrates why it is possible to achieve higher efficiency for this definition of security. We remark that the above protocol is *not* non-halting detection accurate (see Definition 2.4.2). For example, a cheating P_2 can send $c_0^1 = E_{pk_1}(\alpha)$ and $c_0^2 = E_{pk_1}(\alpha)$. Then, if P_1 chooses $b' = 1$ (thus testing c_1^1 and c_1^2), the adversary succeeds in cheating and learning both of P_1's inputs. However, if P_1 chooses $b' = 0$, P_2 can just abort at this point. This means that such an early abort must be considered an attempt to cheat, and so P_1 running with a fail-stop P_2 must also output corrupted$_2$.

Malicious versus covert P_1. We stress that we have actually proven something stronger. Specifically, we have shown that Protocol 5.1.2 is secure in the presence of a covert P_2 with ϵ-deterrent for $e = 1/2$ as stated. However, we have also shown that Protocol 5.1.2 is (fully) secure with abort in the presence of a *malicious* P_1.

Efficiently recognizable public keys. We remark that in the case where it is possible to efficiently recognize that a public key is in the range of the key generator of the public-key encryption scheme, it is possible to skip the key generation challenge step in the protocol (P_1 can verify for itself if the key is valid).

5.1.2 Extensions

String oblivious transfer. In Protocol 5.1.2, x_0 and x_1 are elements in the group over which the homomorphic encryption scheme is defined. If this group is large, then we can carry out string oblivious transfer. This is important because later we will use Protocol 5.1.2 to exchange symmetric encryption keys. However, if the group contains only 0 and 1, then this does not suffice. In order to extend Protocol 5.1.2 to deal with string oblivious transfer, even when the group has only two elements, we only need to change the last two steps of the protocol. Specifically, instead of P_1 computing a single encryption for x_0 and a single encryption for x_1, it computes an encryption for each bit. That is, denote the bits of x_0 by $x_0^1, \ldots, x_0^n$, and likewise for x_1. Then, P_1 computes $\tilde{c}_0 = x_0^1 \cdot_E c_0, \ldots, x_0^n \cdot_E c_0$ and $\tilde{c}_1 = x_1^1 \cdot_E c_1, \ldots, x_1^n \cdot_E c_1$.

Note that P_2 can still only obtain one of the strings because if $\sigma = 0$ then $\tilde{c}_1$ just contains encryptions to zeroes, and vice versa if $\sigma = 1$.

Batch oblivious transfer. We will use Protocol 5.1.2 in Yao's protocol for secure two-party computation. This means that we will run one oblivious transfer for every bit of the input. In principle, these oblivious transfers can be run in parallel, as long as the protocol being used remains secure under parallel composition. The classical notion of parallel composition considers the setting where the honest parties run each execution obliviously of the others (this is often called "stateless composition"). We do not know how to prove that our protocol composes in parallel in this sense. Nevertheless, we can modify Protocol 5.1.2 so that it is possible to simultaneously run many oblivious transfers with a cost that is *less* than that of running Protocol 5.1.2 the same number of times in parallel. We call this *batch oblivious transfer* in order to distinguish it from "parallel oblivious transfer" (which considers stateless parallel composition, as described above). The batch oblivious transfer functionality is defined as follows:

$$((x_1^0, x_1^1), \ldots, (x_n^0, x_n^1), (\sigma_1, \ldots, \sigma_n)) \mapsto (\lambda, (x_1^{\sigma_1}, \ldots, x_n^{\sigma_n})).$$

Thus, we essentially have n oblivious transfers where in the ith such transfer, party P_1 has input (x_i^0, x_i^1) and P_2 has input σ_i.

The extension to Protocol 5.1.2 works as follows. First, the same public-key pair (pk_1, pk_2) can be used in all executions. Therefore, Steps 1 and 2 remain unchanged. Then, Step 3 is carried out *independently* for all n bits $\sigma_1, \ldots, \sigma_n$. That is, for every i, two pairs of ciphertexts encrypting 0 and 1 (in random order) are sent. The important change comes in Step 4. Here, the *same* challenge bit b' is used for every i. P_1 then replies as it should, opening the appropriate $c_{b'}^1$ and $c_{b'}^2$ ciphertexts for every i. The protocol then concludes by P_1 computing the $\tilde{c}_0$ and $\tilde{c}_1$ ciphertexts for every i, and P_2 decrypting.

The proof of the above extension is almost identical to the proof of Theorem 5.1.3. The main point is that since only a single challenge is used for both the key generation challenge and encryption generation challenge, the probability of achieving $b' = b$ (as needed for the simulation) and $b = \gamma$ (as needed for the reduction to the security of the encryption scheme) remains $1/2$. Furthermore, the probability that a corrupted P_2 will succeed in cheating remains the same because if there is any i for which the encryptions are not correctly formed, then P_2 will be caught with probability $1/2$.

Higher values of ϵ. Finally, we show how it is possible to obtain higher values of ϵ with only minor changes to Protocol 5.1.2. The basic idea is to increase the probability of catching a corrupted P_2 in the case where it attempts to generate an invalid key pair or send ciphertexts in Step 3 that do not encrypt the same value. Let $k = \text{poly}(n)$ be an integer. Then, first P_2 generates k pairs of public keys $(pk_1^1, pk_2^1), \ldots, (pk_1^k, pk_2^k)$ instead of just two pairs. P_1 then asks P_2 to reveal the randomness used in generating all the pairs except for one (the unrevealed key pair is the one used in the continuation of the protocol). Note that if a corrupted P_2 generated even one key pair incorrectly, then it is caught with probability $1 - 1/k$. Likewise, in Step 3, P_2 sends k pairs of ciphertexts where in each pair one ciphertext is an encryption of 0 and the other is an encryption of 1. Then, P_1 asks P_2 to open all pairs of encryptions except for one pair. Clearly, P_1 still learns nothing about σ because the reordering is only sent on the ciphertext pair that is not opened. Furthermore, if P_2 generates even one pair of ciphertexts so that the ciphertexts are not correctly formed, then it will be caught with probability $1 - 1/k$. The rest of the protocol remains the same. We conclude that the resulting protocol is secure in the presence of covert adversaries with ϵ-deterrent where $\epsilon = 1 - 1/k$. Note that this works as long as k is polynomial in the security parameter and thus ϵ can be made to be very close to 1, if desired. (Of course, this methodology *cannot* be used to make ϵ negligibly close to 1, because then k has to be super-polynomial.)

Summary. We conclude with the following theorem, derived by combining the extensions above:

Theorem 5.1.4 *Assume that there exist semantically secure homomorphic encryption schemes. Then, for any $k = \mathrm{poly}(n)$ there exists a protocol that securely computes the* batch string oblivious transfer *functionality*

$$((x_1^0, x_1^1), \ldots, (x_n^0, x_1^n), (\sigma_1, \ldots, \sigma_n)) \mapsto (\lambda, (x_1^{\sigma_1}, \ldots, x_n^{\sigma_n}))$$

in the presence of covert adversaries with ϵ-deterrent for $\epsilon = 1 - \frac{1}{k}$.

Efficiency of the batch protocol. The protocol has four rounds of communication, and involves generating $2k$ encryption keys and carrying out $2kn$ encryption operations, $2n$ homomorphic multiplications and n decryptions. Note that the amortized complexity of each oblivious transfer is $2k$ encryptions, two scalar multiplications with the homomorphic encryption scheme and one decryption. (The key generation which is probably the most expensive is run $2k$ times independently of n. Therefore, when many oblivious transfers are run, this becomes insignificant.)

5.2 Secure Two-Party Computation

In this section, we show how to securely compute any two-party functionality in the presence of covert adversaries. We present a protocol for the strong explicit cheat formulation, with parameters that can be set to obtain a wide range of values for the ϵ-deterrent. Our protocol is based on Yao's protocol for semi-honest adversaries; see Chapter 3.

5.2.1 Overview of the Protocol

We begin by providing an intuitive explanation of what is needed to prevent malicious adversaries from cheating in Yao's protocol, without being detected. Much of this discussion appeared in Chapter 4. Nevertheless, we repeat it here because the focus is slightly different, and because we want the material in this chapter to be independent of Chapter 4.

Intuitively, there are two main reasons why the original protocol of Yao is not secure when the parties may be malicious. First, the circuit constructor P_1 may send P_2 a garbled circuit that computes a completely different function. Second, the oblivious transfer protocol that is used when the parties can be malicious must be secure for this case. The latter problem is solved here by using the protocol guaranteed by Theorem 5.1.4. The first problem is solved by having P_1 send P_2 a number of garbled circuits; denote this number by ℓ. Then, P_2 asks P_1 to open all but one of the circuits (chosen at random) in order to check that they are correctly constructed. This opening takes place before P_1 sends the keys corresponding to its input, so nothing is revealed by opening the circuits. The protocol then proceeds similarly to the semi-honest case. The main point here is that if the unopened circuit is correct, then this will constitute a secure execution that can be simulated. However, if it is not correct, then with probability $1 - 1/\ell$ party P_1 will have been caught cheating and so P_2 will output corrupted$_1$ (recall, ℓ denotes the number of circuits sent). While the above intuition forms the basis for our protocol, the actual construction of the appropriate simulator is somewhat delicate, and requires a careful construction of the protocol. We note some of these subtleties hereunder.

First, it is crucial that the oblivious transfers are run before the garbled circuits are sent by P_1 to P_2. This is due to the fact that the simulator sends a corrupted P_2 a fake garbled circuit that evaluates to the exact output received from the trusted party (and only this output), as described in Lemma 4.3.6. However, in order for the simulator to receive the output from the trusted party, it must first send it the input used by the corrupted P_2. This is achieved by first running the oblivious transfers, from which the simulator is able to extract the corrupted P_2's input.

The second subtlety relates to an issue we believe may be a problem for many other implementations of Yao that use cut-and-choose. The problem is that the adversary can construct (at least in theory) a single garbled circuit with two sets of keys to all the input wires, where one set of keys decrypts the circuit to the specified one (computing the function f) and another set of keys decrypts the circuit to an incorrect one (computing a different function f'). This is a problem because the adversary can supply "correct keys" to the circuits that are opened and checked (which thus appear as circuits computing f), and "incorrect keys" to the circuit that is computed (and thus the function computed is f' and not f). Such a strategy cannot be carried out without risk of detection for the keys that are associated with P_2's input because these keys are obtained by P_2 in the oblivious transfers *before* the garbled circuits are even sent (thus if incorrect keys are sent for one of the circuits, P_2 will detect this if that circuit is opened). However, it is possible for a corrupt P_1 to carry out this strategy for the input wires associated with its own input. We prevent this by having P_1 commit to these keys and send the commitments together with the garbled circuits. Then, instead of P_1 just sending the keys associated with its input, it sends the appropriate decommitments.

A third subtlety that arises is due to the following potential attack. Consider a corrupted P_1 that behaves exactly like an honest P_1 except that in the oblivious transfers, it inputs an invalid key in the place of the key associated with 0 as the first bit of P_2. The result is that if the first bit of P_2's input is 1, then the protocol succeeds and no problem arises. However, if the first bit of P_2's input is 0, then the protocol will always fail and P_2 will always detect cheating. Thus, P_1's decision to cheat may depend on P_2's private input, something that is impossible in the ideal model. In order to solve this problem, we use a circuit that computes the function $g(x, \tilde{y}^1, \ldots, \tilde{y}^m) = f(x, \oplus_{i=1}^m \tilde{y}^i)$, instead of a circuit that directly computes f. Then, upon input y, party P_2 chooses random $\tilde{y}^1, \ldots, \tilde{y}^{m-1}$ and sets $\tilde{y}^m = (\oplus_{i=1}^{m-1} \tilde{y}^i) \oplus y$. This makes no difference to the result because $\oplus_{i=1}^m \tilde{y}^i = y$ and so $g(x, \tilde{y}^1, \ldots, \tilde{y}^m) = f(x, y)$. However, this modification makes every bit of P_2's input uniform when considering any proper subset of $\tilde{y}^1, \ldots, \tilde{y}^m$. This helps because as long as P_1 does not provide invalid keys for all m shares of y, the probability of failure is independent of P_2's actual input (because any set of $m-1$ shares is independent of y). Since $m-1$ invalid shares are detected with probability $1 - 2^{-m+1}$ we have that P_2 detects the cheating by P_1 with this probability, independently of its input value. This method is used in Chapter 4 for the same purpose. However, there we must set m to equal the security parameter.

Intuitively, an adversary can cheat by providing an incorrect circuit or by providing invalid keys for shares. However, it is then detected with the probabilities described above. Below, we show that when using ℓ circuits and splitting P_2's input into m shares, we obtain $\epsilon = (1 - 1/\ell)(1 - 2^{-m+1})$. This enables us to play around with the values of m and ℓ in order to optimize efficiency versus ϵ-deterrent. For example, if we wish to obtain $\epsilon = 1/2$ we can use the following parameters:

1. *Set $\ell = 2$ and $m = n$:* This yields $\epsilon = (1 - 1/2)(1 - 2^{-n+1})$ which is negligibly close to 1/2. However, since in Yao's protocol we need to run an oblivious transfer for every one of P_2's input bits, this incurs a blowup of the number of oblivious transfers (and thus exponentiations) by n. Thus, this setting of parameters results in a considerable computational blowup.
2. *Set $\ell = 3$ and $m = 3$:* This yields $\epsilon = (1 - 1/3)(1 - 1/4) = 1/2$. The computational cost incurred here is much less than before because we only need three oblivious transfers for each of P_2's input bits. Furthermore, the cost of sending three circuits is not much greater than two, and so the overall complexity is much better.

Before proceeding to the protocol, we provide one more example of parameters. In order to achieve $\epsilon = 9/10$ it is possible to set $\ell = 25$ and $m = 5$ (setting $\ell = m = 10$ gives 0.898, which is very close). This gives a significantly higher value of ϵ. We remark that such a setting of ϵ also assumes a value of $\epsilon = 9/10$ for the oblivious transfer protocol. As we have seen, this involves a blowup of five times more computation than for oblivious transfer with $\epsilon = 1/2$.

5.2.2 The Protocol for Two-Party Computation

We are now ready to describe the actual protocol. As in Chapter 4, we present a protocol where only P_2 receives output. The modifications required for the general case are described in Section 2.5.

PROTOCOL 5.2.1 (two-party computation of a function f):

- **Inputs:** *Party P_1 has input x and party P_2 has input y, where $|x| = |y|$. In addition, both parties have parameters ℓ and m, and a security parameter n. For simplicity, we will assume that the lengths of the inputs are n.*
- **Auxiliary input:** *Both parties have the description of a circuit C for inputs of length n that computes the function f. The input wires associated with x are $w_1, \ldots, w_n$ and the input wires associated with y are $w_{n+1}, \ldots, w_{2n}$.*
- **The protocol:**

  1. *Parties P_1 and P_2 define a new circuit C' that receives $m + 1$ inputs $x, \tilde{y}^1, \ldots, \tilde{y}^m$ each of length n, and computes the function $f(x, \oplus_{i=1}^m \tilde{y}^i)$. Note that C' has $n + mn$ input wires. Denote the input wires associated with x by $w_1, \ldots, w_n$, and the input wires associated with $\tilde{y}^i$ by $w_{n+(i-1)m+1}, \ldots, w_{n+im}$, for $i = 1, \ldots, n$.*

  2. *Party P_2 chooses $m - 1$ random strings $\tilde{y}^1, \ldots, \tilde{y}^{m-1} \leftarrow_R \{0,1\}^n$ and defines $\tilde{y}^m = (\oplus_{i=1}^{m-1} \tilde{y}^i) \oplus y$, where y is P_2's original input (note that $\oplus_{i=1}^m \tilde{y}^i = y$). The value $\tilde{y} \stackrel{\text{def}}{=} \tilde{y}^1, \ldots, \tilde{y}^m$ serves as P_2's new input of length mn to C'. (The input wires associated with P_2's new input are $w_{n+1}, \ldots, w_{n+mn}$.)*

  3. *For each $i = 1, \ldots, mn$ and $\beta = 0, 1$, party P_1 chooses ℓ encryption keys by running $G(1^n)$, the key generator for the encryption scheme, ℓ times. The jth key associated with a given i and β is denoted by $k^j_{w_{n+i}, \beta}$; note that this is the key associated with the bit β for the input wire w_{n+i} in the jth circuit. The result is an ℓ-tuple, denoted by*

$$[k^1_{w_{n+i}, \beta}, \ldots, k^\ell_{w_{n+i}, \beta}].$$

(This tuple constitutes the keys that are associated with the bit β for the input wire w_{n+i} in all ℓ circuits.)

4. P_1 and P_2 run mn executions of an oblivious transfer protocol, as follows. In the ith execution, party P_1 inputs the pair

$$\left([k^1_{w_{n+i},0}, \ldots, k^\ell_{w_{n+i},0}], [k^1_{w_{n+i},1}, \ldots, k^\ell_{w_{n+i},1}]\right)$$

and party P_2 inputs the bit $\tilde{y}^i$ (P_2 receives keys $[k^1_{w_{n+i},\tilde{y}^i}, \ldots, k^\ell_{w_{n+i},\tilde{y}^i}]$ as output). The executions are run using a batch oblivious transfer functionality, as in Theorem 5.1.4. If a party receives a corrupted$_i$ or abort$_i$ message as output from the oblivious transfer, it outputs it and halts.

5. Party P_1 constructs ℓ garbled circuits $GC_1, \ldots, GC_\ell$ using independent randomness (the circuits are garbled versions of C' described above). The keys for the input wires $w_{n+1}, \ldots, w_{n+mn}$ in the garbled circuits are taken from above (i.e., in GC_j the keys associated with w_{n+i} are $k^j_{w_{n+i},0}$ and $k^j_{w_{n+i},1}$). The keys for the inputs wires $w_1, \ldots, w_n$ are chosen randomly, and are denoted in the same way.

   P_1 sends the ℓ garbled circuits to P_2.

6. P_1 commits to the keys associated with its inputs. That is, for every $i = 1, \ldots, n$, $\beta = 0, 1$ and $j = 1, \ldots, \ell$, party P_1 computes

$$c^j_{w_i,\beta} = \mathsf{com}(k^j_{w_i,\beta}; r^j_{i,\beta})$$

where com is a perfectly-binding commitment scheme, $\mathsf{com}(x; r)$ denotes a commitment to x using randomness r, and $r^j_{i,\beta}$ is a random string of sufficient length to commit to a key of length n.

   P_1 sends all of the above commitments. The commitments are sent as ℓ vectors of pairs (one vector for each circuit); in the jth vector the ith pair is $\{c^j_{w_i,0}, c^j_{w_i,1}\}$ in a random order (the order is randomly chosen independently for each pair).

7. Party P_2 chooses a random index $\gamma \leftarrow_R \{1, \ldots, \ell\}$ and sends γ to P_1.

8. P_1 sends P_2 all of the keys for the input wires in all garbled circuits except for GC_γ (this enables a complete decryption of the garbled circuit), together with the associated mappings and the decommitment values. (I.e., for every $i = 1, \ldots, n + mn$ and $j \neq \gamma$, party P_1 sends the keys and mappings $(k^j_{w_i,0}, 0), (k^j_{w_i,1}, 1)$. In addition, for every $i = 1, \ldots, n$ and $j \neq \gamma$ it sends the decommitments $r^j_{i,0}, r^j_{i,1}$.)

9. P_2 checks that everything that it received is in order. That is, it checks
   - that the keys it received for all input wires in circuits GC_j ($j \neq \gamma$) indeed decrypt the circuits (when using the received mappings), and the decrypted circuits are all C';
   - that the decommitment values correctly open all the commitments $c^j_{w_i,\beta}$ that were received, and these decommitments reveal the keys $k^j_{w_i,\beta}$ that were sent for P_1's wires;
   - that the keys received in the oblivious transfers earlier match the appropriate keys that it received in the opening (i.e., if it received

$[k_i^1, \ldots, k_i^\ell]$ in the ith oblivious transfer, then it checks that k_i^j from
the oblivious transfer equals $k_{w_{n+i}, \tilde{y}^i}^j$ from the opening).

If all the checks pass, it proceeds to the next step. If not, it outputs
corrupted$_1$ and halts. In addition, if P_2 does not receive this message at
all, it outputs corrupted$_1$.

10. P_1 sends decommitments to the input keys associated with its input for
the unopened circuit GC_γ. That is, for $i = 1, \ldots, n$, party P_1 sends P_2
the key k_{w_i, x_i}^γ and decommitment r_{i, x_i}^γ, where x_i is the ith bit of P_1's
input.

11. P_2 checks that the values received are valid decommitments to the commitments received above. If not, it outputs abort$_1$. If so, it uses the keys
to compute $C'(x, \tilde{y}) = C'(x, \tilde{y}^1, \ldots, \tilde{y}^m) = C(x, y)$, and outputs the result. If the keys are not correct (and so it is not possible to compute the
circuit), or if P_2 does not receive this message at all, it outputs abort$_1$.

Note that steps 8–10 are actually a single step of P_1 sending a message
to P_2, followed by P_2 carrying out a computation. If during the execution,
any party fails to receive a message or receives one that is ill-formed, it
outputs abort$_i$ (where P_i is the party that failed to send the message). This
holds unless the party is explicitly instructed above to output corrupted$_i$
instead (as in Step 9).

For reference throughout the proof, we provide a high-level diagram of the
protocol in Figure 5.1.

We have motivated the protocol construction above and thus proceed directly to prove its security. Note that we assume that the oblivious transfer
protocol is secure with the same ϵ as above (of course, one can also use an
oblivious transfer protocol that is secure in the presence of malicious adversaries, because this is secure in the presence of covert adversaries for any ϵ).

Theorem 5.2.2 Let ℓ and m be parameters in the protocol that are both
upper-bound by poly(n), and set $\epsilon = (1-1/\ell)(1-2^{-m+1})$. Let f be any probabilistic polynomial-time function. Assume that the encryption scheme used to
generate the garbled circuits has indistinguishable encryptions under chosen-
plaintext attacks (and has an elusive and efficiently verifiable range), and that
the oblivious transfer protocol used is secure in the presence of covert adversaries with ϵ-deterrent according to Definition 2.4.1. Then, Protocol 5.2.1
securely computes f in the presence of covert adversaries with ϵ-deterrent
according to Definition 2.4.1.

Proof. Our analysis of the security of the protocol is in the (OT, ϵ)-hybrid
model, where the parties are assumed to have access to a trusted party computing the oblivious transfer functionality following the ideal model of 2.4.1;
see Section 2.7. Thus the simulator that we describe will play the trusted
party in the oblivious transfer when simulating for the adversary. We separately consider the case where P_2 is corrupted and the case where P_1 is
corrupted.

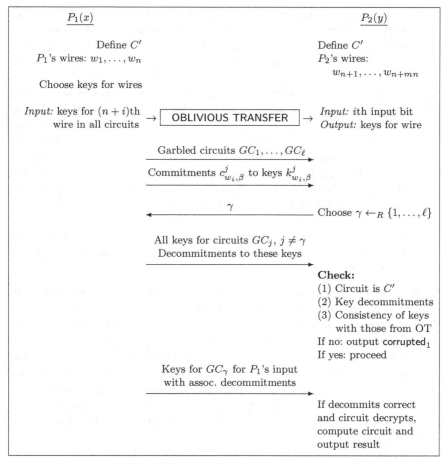

Fig. 5.1 A high-level diagram of the protocol

Party P_2 is corrupted. Intuitively, the security in this case relies on the fact that P_2 can only learn a single set of keys in the oblivious transfers and thus can decrypt the garbled circuit to only a single value as required. Formally, let $\mathcal{A}$ be a probabilistic polynomial-time adversary controlling P_2. The simulator $\mathcal{S}$ fixes $\mathcal{A}$'s random tape to a uniformly distributed tape and works as follows:

1. $\mathcal{S}$ chooses ℓ sets of mn random keys as P_1 would.
2. $\mathcal{S}$ plays the trusted party for the batch oblivious transfer with $\mathcal{A}$ as P_2. $\mathcal{S}$ receives the input that $\mathcal{A}$ sends to the trusted party (as its input as P_2 to the oblivious transfers):

   a. If the input is abort$_2$ or corrupted$_2$, $\mathcal{S}$ sends abort$_2$ or corrupted$_2$ (respectively) to the trusted party computing f, simulates P_1 aborting and halts (outputting whatever $\mathcal{A}$ outputs).

b. If the input is cheat_2, $\mathcal{S}$ sends cheat_2 to the trusted party. If it receives back $\mathsf{corrupted}_2$, it hands $\mathcal{A}$ the message $\mathsf{corrupted}_2$ as if it received it from the trusted party, simulates P_1 aborting and halts (outputting whatever $\mathcal{A}$ outputs). If it receives back $\mathsf{undetected}$ (and thus P_1's input x as well), then $\mathcal{S}$ works as follows. First, it hands $\mathcal{A}$ the string $\mathsf{undetected}$ together with the nm pairs of vectors of random keys that it chose (note that $\mathcal{A}$ expects to receive the inputs of P_1 to the oblivious transfers in the case of $\mathsf{undetected}$). Next, $\mathcal{S}$ uses the input x of P_1 that it received in order to perfectly emulate P_1 in the rest of the execution. That is, it runs P_1's honest strategy with input x while interacting with $\mathcal{A}$ playing P_2 for the rest of the execution and outputs whatever $\mathcal{A}$ outputs. (Note that since P_1 receives no output, $\mathcal{S}$ does not send any output value for P_1 to the trusted party.) The simulation ends here in this case.

c. If the input is a series of bits $\tilde{y}^1, \ldots, \tilde{y}^{mn}$, then $\mathcal{S}$ hands $\mathcal{A}$ the keys from above that are "chosen" by the $\tilde{y}^i$ bits, and proceeds with the simulation below.

3. $\mathcal{S}$ defines $y = \oplus_{i=0}^{m-1}(\tilde{y}^{i \cdot n + 1}, \ldots, \tilde{y}^{i \cdot n + n})$ and sends y to the trusted party computing f. $\mathcal{S}$ receives back some output τ.

4. $\mathcal{S}$ chooses a random value ζ and computes the garbled circuits GC_j for $j \neq \zeta$ correctly (using the appropriate input keys from above as P_1 would). However, for the garbled circuit GC_ζ, the simulator $\mathcal{S}$ does not use the true circuit for computing f but rather a circuit $\widetilde{GC}$ that always evaluates to τ (the value it received from the trusted party), using Lemma 4.3.6. $\mathcal{S}$ uses the appropriate input keys from above also in generating GC_ζ. $\mathcal{S}$ also computes commitments to the keys associated with P_1's input in an honest way.

5. $\mathcal{S}$ sends $GC_1, \ldots, GC_\ell$ and the commitments to $\mathcal{A}$ and receives back an index γ.

6. If $\gamma \neq \zeta$ then $\mathcal{S}$ rewinds $\mathcal{A}$ and returns to Step 4 above (using fresh randomness).

   Otherwise, if $\gamma = \zeta$, $\mathcal{S}$ opens all the commitments and garbled circuits GC_j for $j \neq \gamma$, as the honest P_1 would, and proceeds to the next step.

7. $\mathcal{S}$ hands $\mathcal{A}$ arbitrary keys associated with the input wires of P_1. That is, for $i = 1, \ldots, n$, $\mathcal{S}$ hands $\mathcal{A}$ an arbitrary one of the two keys associated with the input wire w_i in GC_γ (one key per wire), together with its correct decommitment.

8. If at any stage, $\mathcal{S}$ does not receive a response from $\mathcal{A}$, it sends abort_2 to the trusted party (resulting in P_1 outputting abort_2). If the protocol proceeds successfully to the end, $\mathcal{S}$ sends $\mathsf{continue}$ to the trusted party and outputs whatever $\mathcal{A}$ outputs.

Denoting Protocol 5.2.1 as π and $I = \{2\}$ (i.e., party P_2 is corrupted), we prove that:

$$\left\{ \text{IDEALSC}^{\epsilon}_{f,\mathcal{S}(z),2}((x,y),n) \right\} \overset{c}{\equiv} \left\{ \text{HYBRID}^{\text{OT},\epsilon}_{\pi,\mathcal{A}(z),2}((x,y),n) \right\}. \tag{5.3}$$

In order to prove (5.3) we separately consider the cases of abort (including a "corrupted" input), cheat or neither. If $\mathcal{A}$ sends abort_2 or corrupted_2 as the oblivious transfer input, then $\mathcal{S}$ sends abort_2 or corrupted_2 (respectively) to the trusted party computing f. In both cases the honest P_1 outputs the same (abort_2 or corrupted_2) and the view of $\mathcal{A}$ is identical. Thus, the IDEAL and HYBRID output distributions are identical. The exact same argument is true if $\mathcal{A}$ sends cheat_2 and the reply to $\mathcal{S}$ from the trusted party is corrupted_2. In contrast, if $\mathcal{A}$ sends cheat_2 and $\mathcal{S}$ receives back the reply undetected, then the execution does not halt immediately. Rather, $\mathcal{S}$ plays the honest P_1 with its input x (it can do this because P_1 does not use its input before this point). Since $\mathcal{S}$ follows the exact same strategy as P_1 and P_1 receives no output, it is clear that once again the output distributions are identical. We remark that the probability of the trusted party answering corrupted_2 or undetected is the same in the hybrid and ideal executions (i.e., ϵ), and therefore the output distributions in the cases of abort, corrupted or cheat are identical. We denote the event that $\mathcal{A}$ sends an abort, corrupted or cheat message in the oblivious transfers by bad_{OT}. Thus, we have shown that

$$\left\{ \text{IDEALSC}^{\epsilon}_{f,\mathcal{S}(z),2}((x,y),n) \mid \text{bad}_{\text{OT}} \right\} \equiv \left\{ \text{HYBRID}^{\text{OT},\epsilon}_{\pi,\mathcal{A}(z),2}((x,y),n) \mid \text{bad}_{\text{OT}} \right\}.$$

We now show that the IDEAL and HYBRID distributions are computationally indistinguishable in the case where $\mathcal{A}$ sends valid input in the oblivious transfer phase (i.e., in the event $\overline{\text{bad}_{\text{OT}}}$). In order to show this, we consider a modified simulator $\mathcal{S}'$ that is also given the honest party P_1's real input x. Simulator $\mathcal{S}'$ works exactly as $\mathcal{S}$ does, except that it constructs GC_ζ honestly, and not as $\widetilde{GC}$ from Lemma 4.3.6. Furthermore, in Step 7 it sends the keys associated with P_1's input x and not arbitrary keys. It is straightforward to verify that the distribution generated by $\mathcal{S}'$ is identical to the output distribution of a real execution of the protocol between $\mathcal{A}$ and an honest P_1. This is due to the fact that all ℓ circuits received by $\mathcal{A}$ are honestly constructed and the keys that it receives from $\mathcal{S}'$ are associated with P_1's real input. The only difference is the rewinding. However, since ζ is chosen uniformly, this has no effect on the output distribution. Thus:

$$\left\{ \text{IDEALSC}^{\epsilon}_{f,\mathcal{S}'(z,x),2}((x,y),n) \mid \overline{\text{bad}_{\text{OT}}} \right\} \equiv \left\{ \text{HYBRID}^{\text{OT}}_{\pi,\mathcal{A}(z),2}((x,y),n) \mid \overline{\text{bad}_{\text{OT}}} \right\}.$$

Next we prove that conditioned on the event that $\mathsf{bad}_{\mathrm{OT}}$ does not occur, the distributions generated by $\mathcal{S}$ and $\mathcal{S}'$ are computationally indistinguishable. That is,

$$\left\{\mathrm{IDEALSC}_{f,\mathcal{S}(z),2}^{\epsilon}((x,y),n) \mid \overline{\mathsf{bad}}_{\mathrm{OT}}\right\} \overset{\mathrm{c}}{\equiv} \left\{\mathrm{IDEALSC}_{f,\mathcal{S}'(z,x),2}^{\epsilon}((x,y),n) \mid \overline{\mathsf{bad}}_{\mathrm{OT}}\right\}.$$

In order to see this, notice that the only difference between $\mathcal{S}$ and $\mathcal{S}'$ is in the construction of the garbled circuit GC_ζ. By Lemma 4.3.6 it follows immediately that these distributions are computationally indistinguishable. (Note that we do not need to consider the joint distribution of $\mathcal{A}$'s view and P_1's output because P_1 has no output from Protocol 5.2.1.) This yields the above equation. In order to complete the proof of (5.3), note that the probability that the event $\mathsf{bad}_{\mathrm{OT}}$ happens is *identical* in the IDEAL and HYBRID executions. This holds because the oblivious transfer is the first step of the protocol and $\mathcal{A}$'s view in this step with $\mathcal{S}$ is identical to its view in a protocol execution with a trusted party computing the oblivious transfer functionality. Combining this fact with the above equations we derive (5.3).

The simulator $\mathcal{S}$ described above runs in expected polynomial time. In order to see this, note that by Lemma 4.3.6, a fake garbled circuit is indistinguishable from a real one. Therefore, the probability that $\gamma = \zeta$ is at most negligibly far from $1/\ell$ (otherwise, this fact alone can be used to distinguish a fake garbled circuit from a real one). It follows that the expected number of attempts by $\mathcal{S}$ is close to ℓ, and so its expected running time is polynomial (by the assumption on ℓ). By our definition, $\mathcal{S}$ needs to run in strict polynomial time. However, this is easily achieved by having $\mathcal{S}$ halt if it fails after $n\ell$ rewinding attempts. Following the same argument as above, such a failure can occur with at most negligible probability.

We conclude that $\mathcal{S}$ meets the requirements of Definition 2.4.1. (Note that $\mathcal{S}$ only sends cheat_2 due to the oblivious transfer. Thus, if a "fully secure" oblivious transfer protocol were to be used, the protocol would meet the standard definition of security for malicious adversaries for the case where P_2 is corrupted.)

Party P_1 is corrupted. The proof of security in this corruption case is considerably more difficult. Intuitively, security relies on the fact that if P_1 does not construct the circuits correctly or does not provide the same keys in the oblivious transfers and circuit openings, then it will be caught with probability at least ϵ. In contrast, if it does construct the circuits correctly and provide the same keys, then its behavior is effectively the same as that of an honest party and so security is preserved. Formally, let $\mathcal{A}$ be an adversary controlling P_1. The simulator $\mathcal{S}$ works as follows:

1. $\mathcal{S}$ invokes $\mathcal{A}$ and plays the trusted party for the oblivious transfers with $\mathcal{A}$ as P_1. $\mathcal{S}$ receives the input that $\mathcal{A}$ sends to the trusted party (as its input to the oblivious transfers):

a. If the input is abort_1 or $\mathsf{corrupted}_1$, then $\mathcal{S}$ sends abort_1 or $\mathsf{corrupted}_1$ (respectively) to the trusted party computing f, simulates P_2 aborting and halts (outputting whatever $\mathcal{A}$ outputs).

b. If the input is cheat_1, then $\mathcal{S}$ sends cheat_1 to the trusted party. If it receives back $\mathsf{corrupted}_1$, then it hands $\mathcal{A}$ the message $\mathsf{corrupted}_1$ as if it received it from the trusted party, simulates P_2 aborting and halts (outputting whatever $\mathcal{A}$ outputs). If it receives back $\mathsf{undetected}$ (and thus P_2's input y as well), then $\mathcal{S}$ works as follows. First, it hands $\mathcal{A}$ the string $\mathsf{undetected}$ together with the input string $\tilde{y}$ that an honest P_2 upon input y would have used in the oblivious transfers (note that $\mathcal{A}$ expects to receive P_2's input to the oblivious transfers in the case of $\mathsf{undetected}$). We remark that $\mathcal{S}$ can compute $\tilde{y}$ by simply following the instructions of an honest P_2 with input y from the start (nothing yet has depended on P_2's input so there is no problem of consistency). Next, $\mathcal{S}$ uses the derived input $\tilde{y}$ that it computed above in order to perfectly emulate P_2 in the rest of the execution. That is, it continues P_2's honest strategy with input $\tilde{y}$ while interacting with $\mathcal{A}$ playing P_1 for the rest of the execution. Let τ be the output for P_2 that it receives. $\mathcal{S}$ sends τ to the trusted party (for P_2's output) and outputs whatever $\mathcal{A}$ outputs. The simulation ends here in this case.

c. If the input is a series of mn pairs of ℓ-tuples of keys

$$\left([k^1_{w_{n+i},0}, \ldots, k^\ell_{w_{n+i},0}], [k^1_{w_{n+i},1}, \ldots, k^\ell_{w_{n+i},1}] \right)$$

for $i = 1, \ldots, mn$, then $\mathcal{S}$ proceeds below.

2. $\mathcal{S}$ receives from $\mathcal{A}$ a message consisting of ℓ garbled circuits $GC_1, \ldots, GC_\ell$ and a series of commitments.

3. For $j = 1, \ldots, \ell$, simulator $\mathcal{S}$ sends $\mathcal{A}$ the message $\gamma = j$, receives its reply and rewinds $\mathcal{A}$ back to the point before $\mathcal{A}$ receives γ.

4. $\mathcal{S}$ continues the simulation differently, depending on the validity of the circuit openings. In order to describe the cases, we introduce some terminology.

Legitimate circuit: We say that a garbled circuit GC_j is *legitimate* if there exists a value $\gamma \neq j$ such that when $\mathcal{S}$ sends $\mathcal{A}$ the message γ in Step 3 above, $\mathcal{A}$ replies with decommitments that open GC_j to the (correct) auxiliary input circuit C'. Note that if a circuit is legitimate then whenever $\mathcal{A}$ provides valid decommitments in Step 3, the circuit is decrypted to C'. Furthermore, if a circuit is *illegitimate* then for *every* $\gamma \neq j$ in which $\mathcal{A}$ provides valid decommitments, the resulting opened circuit is *not* C'.

Inconsistent key: This notion relates to the question of whether the keys provided by P_1 in the oblivious transfers are the same as those committed to and thus revealed in a circuit opening. We say that a (committed) key $k^j_{w_i,\beta}$ received in an oblivious transfer is inconsistent if it is different from the analogous key committed to by P_1. We stress that the keys obtained in the oblivious transfers (and of course the committed keys) are *fixed* before this point of the simulation and thus this event is well defined.

Inconsistent wire: A wire w_i is inconsistent if there *exists* a circuit GC_j such that either $k^j_{w_i,0}$ or $k^j_{w_i,1}$ is an inconsistent key.

Totally inconsistent input: An original input bit y_i is *totally inconsistent* if all of the wires associated with the shares of y_i are inconsistent (recall that y_i is split over m input wires). Note that the different inconsistent wires need not be inconsistent in the same circuit, nor need they be inconsistent with respect to the same value (0 or 1). Note that the determination about whether a wire is inconsistent is independent of the value γ sent by $\mathcal{S}$ because the oblivious transfers and commitments to keys take place before $\mathcal{S}$ sends γ in Step 3 above.

Before proceeding to describe how $\mathcal{S}$ works, we remark that our strategy below is to have $\mathcal{S}$ use the different possibilities regarding the legitimacy of the circuit and consistency of keys to cause the honest party in an ideal execution to output $\mathsf{corrupted}_1$ with the same probability with which the honest P_2 catches $\mathcal{A}$ cheating in a real execution. Furthermore, $\mathcal{S}$ does this while ensuring that γ is uniformly distributed and the bits chosen as shares of each y_i are also uniformly distributed. In this light, we describe the expected probabilities of catching $\mathcal{A}$ in three cases:

- *There exists an illegitimate circuit GC_{j_0}:* in this case P_2 certainly catches $\mathcal{A}$ cheating unless $\gamma = j_0$. Thus, P_2 catches $\mathcal{A}$ with probability at least $1 - 1/\ell$. We stress that P_2 may catch $\mathcal{A}$ with higher probability depending on whether or not there are other illegitimate circuits of inconsistent inputs.

- *There exists a totally inconsistent wire:* if the inconsistent values of the wire belong to different circuits then P_2 will always catch $\mathcal{A}$. However, if they belong to one circuit GC_{j_0} then $\mathcal{A}$ will be caught if $\gamma \neq j_0$, or if $\gamma = j_0$ and the keys chosen in the oblivious transfer are all consistent (this latter event happens with probability at most 2^{-m+1} because $m-1$ bits of the sharing are chosen randomly). Thus, P_2 catches $\mathcal{A}$ with probability at least $(1 - \ell^{-1})(1 - 2^{m+1})$.

- *None of the above occurs but there are inconsistent keys:* in this case, P_2 catches $\mathcal{A}$ if the inconsistent keys are those chosen and otherwise does not.

We are now ready to proceed. $\mathcal{S}$ works according to the follows cases:

a. *Case 1 – at least one circuit is illegitimate:* Let GC_{j_0} be the first illegitimate circuit. Then, $\mathcal{S}$ sends $w_1 = \mathsf{cheat}_1$ to the trusted party. By the definition of the ideal model, with probability $\epsilon = (1 - 1/\ell)(1 - 2^{-m+1})$ it receives the message $\mathsf{corrupted}_1$, and with probability $1 - \epsilon$ it receives the message $\mathsf{undetected}$ together with P_2's input y:

    i. If S receives the message corrupted$_1$ from the trusted party, then it chooses $\gamma \neq j_0$ at random and sends γ to A. Then, S receives back A's opening for the circuits, including the illegitimate circuit GC_{j_0}, and simulates P_2 aborting due to detected cheating. S then outputs whatever A outputs and halts.

    ii. If S receives the message undetected from the trusted party (together with P_2's input y), then with probability $p = \frac{\ell-1}{1-\epsilon}$ it sets $\gamma = j_0$, and with probability $1 - p$ it chooses $\gamma \neq j_0$ at random. It then sends γ to A, and continues to the end of the execution emulating the honest P_2 with the input y it received from the trusted party. (When computing the circuit, S takes the keys from the oblivious transfer that P_2 would have received when using input y and when acting as the honest P_2 to define the string $\tilde{y}$.) Let τ be the output that S received when playing P_2 in this execution. S sends τ to the trusted party (to be the output of P_2) and outputs whatever A outputs. Note that if the output of P_2 in this emulated execution would have been corrupted$_1$ then S sends $\tau = $ corrupted$_1$ to the trusted party.[4]

    We will show below that the above probabilities result in γ being uniformly distributed in $\{1, \ldots, \ell\}$.

b. *Case 2 – All circuits are legitimate but there is a totally inconsistent input:* Let y_i be the first totally inconsistent input and, for brevity, assume that the inconsistent keys are all for the 0 value on the wires (i.e., there are inconsistent keys $k^{j_1}_{w_{n+(i-1)m+1},0}, \ldots, k^{j_m}_{w_{n+im},0}$ for some $j_1, \ldots, j_m \in \{1, \ldots, \ell\}$). In this case, S sends $w_1 = $ cheat$_1$ to the trusted party. With probability ϵ it receives the message corrupted$_1$, and with probability $1 - \epsilon$ it receives the message undetected together with P_2's input y:

    i. If S receives the message corrupted$_1$ from the trusted party, then it chooses random values for the bits on the wires $w_{n+(i-1)m+1}, \ldots, w_{n+im-1}$, subject to the constraints that not all are 1; i.e., at least one of these wires gets a value with an inconsistent key.[5] Let G_{j_0} be the first circuit for which there exists a wire with value 0 that obtains an inconsistent key. S chooses $\gamma \neq j_0$ at random and sends it to A. Among other things, S receives back A's opening of GC_{j_0}, and simulates P_2's aborting due to detected cheating. (Note that the probability that a real P_2 will make these two choices – choose the

---

[4] We remark that P_2 may output corrupted$_1$ with probability that is higher than ϵ (e.g., if more than one circuit is illegitimate or if inconsistent keys are presented as well). This possibility is dealt with by having S play P_2 and force a corrupted$_1$ output if this would have occurred in the execution.

[5] Recall that the input wires associated with P_2's input bit y_i are $w_{n+(i-1)m+1}, \ldots, w_{n+im}$. Thus, the simulator here fixes the values on all the wires except the last (recall also that the first $m - 1$ values plus P_2's true input bit fully determine the value for the last wire w_{n+im}).

values for the first $m-1$ wires so that not all are 1, and choose $\gamma \neq j_0$
– is exactly ϵ.) S then outputs whatever A outputs and halts.

ii. If S receives the message undetected (and thus the real input y of P_2)
from the trusted party, it first determines the values for the shares
of y_i and for the value γ, as follows:

- With probability $p = \frac{2^{-m+1}}{1-\epsilon}$, for all $t = 1,\ldots,m-1$ it sets
  the value on the wire $w_{n+(i-1)m+t}$ to equal 1 (corresponding to
  not choosing the inconsistent keys), and the value on the wire
  w_{n+im} to equal the XOR of y_i with the values set on the wires
  $w_{n+(i-1)m+1},\ldots,w_{n+(i-1)m+m-1}$. The value γ is chosen at random (out of $1,\ldots,\ell$).
- With probability $1-p$, for all $t = 1,\ldots,m-1$ it sets the
  value on the wire $w_{n+(i-1)m+t}$ to a random value, subject to
  the constraint that not all are 1 (i.e., at least one of the shares
  has an inconsistent key), and it sets the value on the wire
  w_{n+im} to equal the XOR of y_i with the values set on the wires
  $w_{n+(i-1)m+1},\ldots,w_{n+(i-1)m+m-1}$. Let G_{j_0} be the first circuit for
  which there exists a wire with value 0 that obtains an inconsistent
  key. Then S sets $\gamma = j_0$.

The values for shares of all other input bits are chosen at random
(subject to the constraint that their XOR is the input value obtained
from the trusted party, as an honest P_2 would choose). S now sends
γ to A, and completes the execution emulating an honest P_2 using
these shares and γ. It outputs whatever A would output, and sets
P_2's output to whatever P_2 would have received in the executions,
including corrupted$_1$, if this would be the output (this is as described
at the end of Step 4(a)ii above).

c. *Case 3 – All circuits are legitimate and there is no totally inconsistent
input:* For each inconsistent wire (i.e., a wire for which there *exists* an
inconsistent key), if there are any, S chooses a random value, and checks
whether the value it chose corresponds to an inconsistent key. There are
two cases:

i. *Case 3a – S chose bits with inconsistent keys:* In this case, S sends
$w_1 = \text{cheat}_1$ to the trusted party. With probability ϵ it receives the
message corrupted$_1$, and with probability $1-\epsilon$ it receives the message
undetected together with P_2's input y. Let w_{i_0} be the first of the wires
for which the bit chosen has an inconsistent key, and let GC_{j_0} be the
first circuit in which the key is inconsistent:

A. If S receives the message corrupted$_1$ from the trusted party, then
it chooses $\gamma \neq j_0$ at random and sends it to A. S then simulates P_2 aborting due to detected cheating, outputs whatever A
outputs and halts.

B. If S receives undetected together with $y = (y_1,\ldots,y_n)$ from the
trusted party, then first it chooses bits for the (consistent) shares
at random, subject to the constraint that for any input bit y_i,

the XOR of all its shares equals the value of this bit, as provided by the trusted party. In addition:

- With probability $p = \frac{\ell^{-1}}{1-\epsilon}$, simulator $\mathcal{S}$ sets $\gamma = j_0$.
- With probability $1 - p$, simulator $\mathcal{S}$ chooses $\gamma \neq j_0$ at random.

In both cases, $\mathcal{S}$ sends γ to $\mathcal{A}$ and completes the execution emulating an honest P_2 using the above choice of shares, and outputting the values as explained in Step 4(a)ii above (in particular, if the output of the emulated P_2 is corrupted$_1$, then $\mathcal{S}$ causes this to be the output of P_2 in the ideal model).

ii. *Case 3b – $\mathcal{S}$ chose only bits with consistent keys:* $\mathcal{S}$ reaches this point of the simulation if all garbled circuits are legitimate and if either all keys are consistent or it is simulating the case where no inconsistent keys were chosen. Thus, intuitively, the circuit and keys received by $\mathcal{S}$ from $\mathcal{A}$ are the same as from an honest P_1. The simulator $\mathcal{S}$ begins by choosing a random γ and sending it to $\mathcal{A}$. Then, $\mathcal{S}$ receives the opening of the other circuits, as before. In addition, $\mathcal{S}$ receives from $\mathcal{A}$ the set of keys and decommitments for the wires $w_1, \ldots, w_n$ for the unopened circuit GC_γ, in order to evaluate it. If anything in this process is invalid (i.e., any of the circuits is not correctly decrypted, or the decommitments are invalid, or the keys cannot be used in the circuit), then $\mathcal{S}$ sends abort$_1$ or corrupted$_1$ to the trusted party causing P_2 to output abort$_1$ or corrupted$_1$, respectively (the choice of whether to send abort$_1$ or corrupted$_1$ is according to the protocol description and what causes P_2 to output abort$_1$ and what causes it to output corrupted$_1$). Otherwise, $\mathcal{S}$ uses the opening of the circuit GC_γ obtained above, together with the keys obtained in order to derive the input x' used by $\mathcal{A}$. Specifically, in Step 3, the simulator $\mathcal{S}$ receives the opening of all circuits and this reveals the association between the keys on the input wires and the input values. Thus, when $\mathcal{A}$ sends the set of keys associated with its input in circuit GC_γ, simulator $\mathcal{S}$ can determine the exact input x' that is defined by these keys. $\mathcal{S}$ sends the trusted party x' (and continue) and outputs whatever $\mathcal{A}$ outputs.

This concludes the description of $\mathcal{S}$. For reference throughout the analysis below, we present a high-level outline and summary of the simulator in Figures 5.2 and 5.3. We present it in the form of a "protocol" between the simulator $\mathcal{S}$ and the real adversary $\mathcal{A}$.

Denote by bad$_{\mathrm{OT}}$ the event that $\mathcal{A}$ sends abort$_1$, corrupted$_1$ or cheat$_1$ in the oblivious transfers. The analysis of the event bad$_{\mathrm{OT}}$ is identical to the case where P_2 is corrupted and so denoting π as Protocol 5.2.1 and $I = \{1\}$ (i.e., party P_1 is corrupted), we have that

$$\left\{ \mathrm{IDEALSC}^{\epsilon}_{f, \mathcal{S}(z), 1}((x, y), n) \mid \mathsf{bad}_{\mathrm{OT}} \right\} \equiv \left\{ \mathrm{HYBRID}^{\mathrm{OT}}_{\pi, \mathcal{A}(z), 1}((x, y), n) \mid \mathsf{bad}_{\mathrm{OT}} \right\}.$$

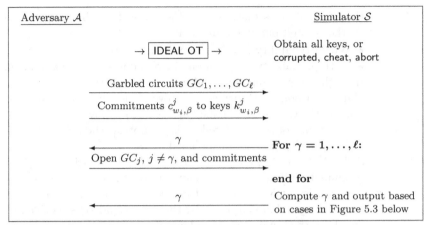

Fig. 5.2 A high-level diagram of the simulator (P_1 corrupted)

Case 1 – at least one illegitimate circuit: Send cheat$_1$ to trusted party. Then:

1. If receive corrupted$_1$: set $\gamma \neq j_0$ at random.
2. If receive undetected: with probability $p = \frac{\ell-1}{1-\epsilon}$ set $\gamma = j_0$; with probability $1 - p$ set $\gamma \neq j_0$ at random; complete execution using real y.

Case 2 – there exists a totally inconsistent input: Send cheat$_1$ to trusted party.

1. If receive corrupted$_1$: choose values for inconsistent input so at least one inconsistent key chosen. Set $\gamma \neq j_0$ at random.
2. If receive undetected: with probability $p = \frac{2^{-m+1}}{1-\epsilon}$ choose values so no inconsistent key chosen and choose $\gamma \leftarrow_R \{1, \ldots, \ell\}$; with probability $1 - p$ choose values so at least one inconsistent key chosen and set $\gamma = j_0$.
   Complete execution using real y.

Case 3 – all other cases: Choose random values for inconsistent wires (if they exist).

1. If a bit with an inconsistent key chosen: send cheat$_1$. If receive corrupted$_1$, set $\gamma \neq j_0$ at random. If receive undetected, choose rest of values under constraint that are consistent with real input of P_2. With probability $p = \frac{\ell-1}{1-\epsilon}$ set $\gamma = j_0$; with probability $1 - p$ choose $\gamma \neq j_0$ at random.
2. If no inconsistent keys chosen: derive input from keys and openings sent by $\mathcal{A}$. Send it to trusted party and conclude simulation (checking for abort or corrupted as in protocol specification).

Fig. 5.3 Cases for the simulator $\mathcal{S}$ (P_1 corrupted)

It remains to analyze the case where $\overline{\text{bad}}_{\text{OT}}$ (i.e., the oblivious transfer is not aborted). We will prove the case following the same case analysis as in the description of the simulator. Before doing so, notice that the only messages that $\mathcal{A}$ receives in a protocol execution are in the oblivious transfers and the challenge value γ. Thus, when analyzing Protocol 5.2.1 in a hybrid model with a trusted party computing the oblivious transfer functionality, its view consists only of the value γ. Thus, in order to show that $\mathcal{A}$'s view in the simulation is indistinguishable from its view in a real execution, it suffices to show that the value γ that $\mathcal{S}$ hands $\mathcal{A}$ is (almost) uniformly distributed in $\{1, \ldots, \ell\}$. We stress that this is not the case when considering the *joint distribution* including P_2's output (because cheating by $\mathcal{A}$ can cause P_2 to output an incorrect value). The focus of the proof below is thus to show that the distribution over the challenge value γ sent by $\mathcal{S}$ during the simulation is uniform, and that the joint distribution over $\mathcal{A}$'s view and the output of P_2 in the simulation is statistically close to a real execution.

1. *Case 1 – at least one circuit is illegitimate:* We first show that the value γ sent by $\mathcal{S}$ in the simulation is uniformly distributed over $\{1, \ldots, \ell\}$, just like the value sent by P_2 in a real execution. In order to see this, we distinguish between the case where $\mathcal{S}$ receives $\mathsf{corrupted}_1$ and the case where it receives $\mathsf{undetected}$. We first prove that $\gamma = j_0$ with probability $1/\ell$ (recall that GC_{j_0} is the first illegitimate circuit):

$$\Pr[\gamma = j_0] = \Pr[\gamma = j_0 \mid \mathsf{corrupted}_1] \cdot \Pr[\mathsf{corrupted}_1]$$
$$+ \Pr[\gamma = j_0 \mid \mathsf{undetected}] \cdot \Pr[\mathsf{undetected}]$$
$$= 0 \cdot \Pr[\mathsf{corrupted}_1] + \frac{\ell - 1}{1 - \epsilon} \cdot \Pr[\mathsf{undetected}]$$
$$= \frac{1}{\ell} \cdot \frac{1}{1 - \epsilon} \cdot (1 - \epsilon) = \frac{1}{\ell}$$

where the second equality is by the simulator's code, and the third follows from the fact that $\Pr[\mathsf{undetected}] = 1 - \epsilon$, by definition. We now proceed to prove that for every $j \neq j_0$ it also holds that $\Pr[\gamma = j] = 1/\ell$. For every $j = 1, \ldots, \ell$ with $j \neq j_0$:

$$\Pr[\gamma = j] = \Pr[\gamma = j \mid \mathsf{corrupted}_1] \cdot \Pr[\mathsf{corrupted}_1]$$
$$+ \Pr[\gamma = j \mid \mathsf{undetected}] \cdot \Pr[\mathsf{undetected}]$$
$$= \Pr[\gamma = j \mid \mathsf{corrupted}_1] \cdot \epsilon + \Pr[\gamma = j \mid \mathsf{undetected}] \cdot (1 - \epsilon)$$
$$= \left(\frac{1}{\ell - 1} \right) \cdot \epsilon + \left(\left(1 - \frac{1}{\ell(1 - \epsilon)} \right) \cdot \frac{1}{\ell - 1} \right) \cdot (1 - \epsilon)$$
$$= \frac{1}{\ell - 1} \cdot \left(\epsilon + \left(1 - \frac{1}{\ell(1 - \epsilon)} \right) \cdot (1 - \epsilon) \right)$$
$$= \frac{1}{\ell - 1} \cdot \left(\epsilon + (1 - \epsilon) - \frac{1 - \epsilon}{\ell(1 - \epsilon)} \right)$$
$$= \frac{1}{\ell - 1} \cdot \left(1 - \frac{1}{\ell} \right) = \frac{1}{\ell}$$

where, once again, the third equality is by the code of the simulator. (Recall that when $\mathsf{undetected}$ is received, then with probability $1 - p$ for $p = \frac{\ell - 1}{(1 - \epsilon)}$

the value γ is uniformly distributed under the constraint that it does not equal j_0. Thus, when undetected occurs, the probability that γ equals a given $j \neq j_0$ is $\frac{1}{\ell-1}$ times $1 - p$.)

We now proceed to show that the joint distribution of $\mathcal{A}$'s view and P_2's output in a real execution (or more exactly, a hybrid execution where the oblivious transfers are computed by a trusted party) is identical to the joint distribution of $\mathcal{S}$ and P_2's output in an ideal execution. We show this separately for the case where $\gamma \neq j_0$ and the case where $\gamma = j_0$. Now, when a real P_2 chooses $\gamma \neq j_0$, then it always outputs corrupted$_1$. Likewise, in an ideal execution where the trusted party sends corrupted$_1$ to P_2, the simulator $\mathcal{S}$ sets $\gamma \neq j_0$. Thus, when $\gamma \neq j_0$, the honest party outputs corrupted$_1$ in both the real and ideal executions. Next consider the case where $\gamma = j_0$. In the simulation by $\mathcal{S}$, this only occurs when $\mathcal{S}$ receives back undetected, in which case $\mathcal{S}$ perfectly emulates a real execution because it is given the honest party's real input y. Thus P_2's output is distributed identically in both the real and ideal executions when $\gamma = j_0$. (Note that P_2 may output corrupted$_1$ in this case as well. However, what is important is that this will happen with exactly the same probability in the real and ideal executions.) Finally recall from above that γ as chosen by $\mathcal{S}$ is uniformly distributed, and thus the two cases (of $\gamma \neq j_0$ and $\gamma = j_0$) occur with the same probability in the real and ideal executions. We therefore conclude that the overall distributions are identical. This completes this case.

2. *Case 2 – All circuits are legitimate but there is a totally inconsistent input:* We analyze this case in a way analogous to that above. Let 'all=1' denote the case where in a *real* execution all of the $m-1$ first wires associated with the totally inconsistent input are given value 1 (and so the inconsistent keys determined for those wires are not revealed because these are all 0). Since the values on these wires are chosen by P_2 uniformly, we have that $\Pr['all=1'] = 2^{-m+1}$. Noting also that γ is chosen by P_2 independently of the values on the wires, we have that in a real execution

$$\Pr[\gamma \neq j_0 \ \& \ \neg'all=1'] = \left(1 - \frac{1}{\ell}\right)\left(1 - \frac{1}{2^{m-1}}\right) = \epsilon$$

where the second equality is by the definition of ϵ (recall that j_0 is the index of the first circuit for which an inconsistent key is chosen by $\mathcal{S}$). Now, the trusted party sends corrupted$_1$ with probability exactly ϵ. Furthermore, in this case, $\mathcal{S}$ generates a transcript for which the event $\gamma \neq j_0$ & $\neg'all=1'$ holds (see item (i) of case (2) of the simulator), and such an event in a real execution results in P_2 certainly outputting corrupted$_1$. We thus have that the corrupted$_1$ event in the ideal model is mapped with probability exactly ϵ to a sub-distribution over the real transcripts in which P_2 outputs corrupted$_1$.

Next we analyze the case where not all values on the wires are 1, but $\gamma = j_0$. In a real execution, we have that this event occurs with the following probability:

$$\Pr[\gamma = j_0 \ \& \ \neg\text{'all=1'}] = \frac{1}{\ell} \cdot \left(1 - 2^{-m+1}\right).$$

By the description of $\mathcal{S}$, this occurs in the simulation with probability $(1 - \epsilon)(1 - p)$ where $p = 2^{-m+1}/(1 - \epsilon)$; see the second bullet of Case (2), subitem (ii), and observe that γ is always set to j_0 in this case. Now,

$$
\begin{aligned}
(1 - \epsilon)(1 - p) &= (1 - \epsilon) \cdot \left(1 - \frac{2^{-m+1}}{1 - \epsilon}\right) \\
&= 1 - \epsilon - 2^{-m+1} \\
&= 1 - \left(1 - 2^{-m+1}\right)\left(1 - \ell^{-1}\right) - 2^{-m+1} \\
&= 1 - \left(1 - \frac{1}{\ell} - 2^{-m+1} + \frac{2^{-m+1}}{\ell}\right) - 2^{-m+1} \\
&= \frac{1}{\ell} - \frac{2^{-m+1}}{\ell} \\
&= \frac{1}{\ell} \cdot \left(1 - 2^{-m+1}\right).
\end{aligned}
$$

Thus, the probability of this event in the simulation by $\mathcal{S}$ is exactly the same as in a real execution. Furthermore, the transcript generated by $\mathcal{S}$ in this case (and the output of P_2) is identical to that in a real execution, because $\mathcal{S}$ runs an emulation using P_2's real input.

Thus far, we have analyzed the output distributions in the events $(\gamma \neq j_0 \ \& \ \neg\text{'all=1'})$ and $(\gamma = j_0 \ \& \ \neg\text{'all=1'})$, and so have covered the case $\neg\text{'all=1'}$. It remains for us to analyze the event 'all=1'. That is, it remains to consider the case where all $m - 1$ wires do equal 1; this case is covered by the simulation in the first bullet of Case (2), subitem (ii). In a real execution, this case occurs with probability 2^{-m+1}. Likewise, in the simulation, $\mathcal{S}$ reaches subitem (ii) with probability $1 - \epsilon$ and then proceeds to the first bullet with probability $p = 2^{-m+1}/(1 - \epsilon)$. Therefore, this case appears with overall probability 2^{-m+1} exactly as in a real execution. Furthermore, as above, the simulation by $\mathcal{S}$ is perfect because it emulates the execution using P_2's real input.

We have shown that for the events $(\gamma \neq j_0 \ \& \ \neg\text{'all=1'})$, $(\gamma = j_0 \ \& \ \neg\text{'all=1'})$, and 'all=1', the joint output distribution generated by $\mathcal{S}$ is identical to that in a real execution. Furthermore, we have shown that these events occur with the same probability in the real and ideal executions. Since these events cover all possibilities, we conclude that the simulation by $\mathcal{S}$ in this case is perfect. (By perfect, we mean that when all circuits are legitimate but there is a totally inconsistent input, the joint output distribution of $\mathcal{S}$ and P_2 in an ideal execution is identical to the joint output distribution of $\mathcal{A}$ and P_2 in a hybrid execution of the protocol where a trusted party is used for the oblivious transfers.)

3. *Case 3 – all circuits are legitimate and there is no totally inconsistent input:* We have the following subcases:

   a. *Case 3a – $\mathcal{S}$ chose values with inconsistent keys:* First observe that $\mathcal{S}$ chooses values with inconsistent keys with exactly the same probability as P_2 in a real execution. This holds because there are no totally inconsistent values and thus the choice of values on the wires with inconsistent keys is uniform. (Note that P_2's strategy for choosing values is equivalent to choosing any subset of $m - 1$ values uniformly and then choosing the last value so that the XOR equals the associated input bit. Since there is at least one wire where both keys are consistent, we can look at this wire as being the one that determines the actual unknown input bit of P_2, and all others are chosen uniformly by $\mathcal{S}$ and P_2. Thus, the probability that $\mathcal{S}$ chooses an inconsistent key is the same as P_2.) We therefore fix the choice of values for the wires and proceed to analyze the transcripts generated by the simulator, *conditioned on this choice of keys.*

   In a real execution in which P_2 chose inconsistent keys, it outputs corrupted$_1$ if the circuit in which the inconsistent keys were chosen is opened (it may also output corrupted$_1$ if the circuit is not opened but this is not relevant here). Now, if the trusted party sends corrupted$_1$, then the simulator ensures that the circuit in which the inconsistent keys were chosen is opened (it does this by choosing γ uniformly under the constraint that $\gamma \neq j_0$ where G_{j_0} is the first circuit with an inconsistent key; see subitem (A) of subitem (i) in Case 3a). In contrast, if the trusted party sends undetected, then $\mathcal{S}$ runs a perfect emulation using P_2's real input; the two subcases (with probability p and $1 - p$) are to ensure that γ is chosen uniformly. Thus, it remains to show that in this case, for every $j = 1, \ldots, \ell$ we have $\Pr[\gamma = j] = 1/\ell$. As above, we separately analyze the probability for $j = j_0$ and $j \neq j_0$. The computation is almost the same as in Case 1 above and we are therefore brief:

   $$\Pr[\gamma = j_0] = \Pr[\gamma = j_0 \mid \text{corrupted}_1] \cdot \epsilon + \Pr[\gamma = j_0 \mid \text{undetected}] \cdot (1 - \epsilon)$$
   $$= 0 \cdot \epsilon + \frac{\ell - 1}{1 - \epsilon} \cdot (1 - \epsilon) = \frac{1}{\ell} .$$

   In addition, for all $j \neq j_0$,

   $$\Pr[\gamma = j] = \Pr[\gamma = j \mid \text{corrupted}_1] \cdot \epsilon + \Pr[\gamma = j \mid \text{undetected}] \cdot (1 - \epsilon)$$
   $$= \left(\frac{1}{\ell - 1} \right) \cdot \epsilon + \left(\left(1 - \frac{1}{\ell(1 - \epsilon)} \right) \cdot \frac{1}{\ell - 1} \right) \cdot (1 - \epsilon) = \frac{1}{\ell} .$$

   Thus, in this case, $\mathcal{S}$ chooses γ uniformly in $\{1, \ldots, \ell\}$. Furthermore, the transcript in each subcase is exactly as in a real execution, as required.

b. *Case 3b – S chose only values with consistent keys:* As above, the probability that S chose only values with consistent keys is identical to the probability that a real P_2 chooses only values with consistent keys. Now, in such a case, all circuits are legitimate, and in addition, all keys that are retrieved by P_2 are consistent (this includes the keys for the opened circuits and for the circuit that is computed). This means that the computation of the circuit using the keys retrieved by P_2 is identical to the computation of an honestly generated circuit. (Note that P_2 may abort or output corrupted$_1$ in this case. However, here we are interested in the result of the computation of the circuit G_γ, *if* it is computed by P_2.) We also note that the keys provided by P_1 that are associated with its own input are provided via decommitments. Thus, P_1 can either not provide valid decommitments, or must provide decommitments that yield keys that result in the circuit being decrypted correctly. This also means that the associations made by S between the input keys of P_1 and the string x' that it sends to the trusted party are correct. We conclude that in this case, the joint output of $\mathcal{A}$ and the real P_2 in a real execution is identical to the joint output of S and P_2 in an ideal execution, as required.

This completes the proof of security in (OT, ϵ)-hybrid model. Applying Theorem 2.7.2 (sequential composition), we have that Protocol 5.2.1 is secure in the real model when using a real oblivious transfer protocol that is secure in the presence of covert adversaries with ϵ-deterrent. ■

5.2.3 Non-halting Detection Accuracy

It is possible to modify Protocol 5.2.1 so that it achieves *non-halting detection accuracy*; see Definition 2.4.2. Before describing how we do this, note that the reason that we need to recognize a halting-abort as cheating in Protocol 5.2.1 is that if P_1 generates one faulty circuit, then it can always just refuse to continue (i.e., abort) in the case where P_2 asks it to open the faulty circuit. This means that if aborting is not considered cheating, then a corrupted P_1 can form a strategy whereby it is never detected cheating, but succeeds in actually cheating with probability $1/\ell$. In order to solve this problem, we construct a method whereby P_1 does not know whether or not it will be caught. We do so by having P_2 receive the circuit openings via a fully secure 1-out-of-ℓ oblivious transfer protocol, rather than having P_1 send it explicitly. This forces P_1 to either abort before learning anything, or to risk being caught with probability $1 - 1/\ell$. In order to describe this in more detail, we restate the circuit opening stage of Protocol 5.2.1 as follows:

1. Party P_1 sends ℓ garbled circuits $GC_1, \ldots, GC_\ell$ to party P_2.
2. P_2 sends a random challenge $\gamma \leftarrow_R \{1, \ldots, \ell\}$.

3. P_1 opens GC_j for all $j \neq \gamma$ by sending decommitments, keys and so on. In addition, it sends the keys associated with its own input in GC_γ.
4. P_2 checks the circuits GC_j for $j \neq \gamma$ and computes GC_γ (using the keys from P_1 in the previous step and the keys it obtained earlier in the oblivious transfers). P_2's output is defined to be the output of GC_γ.

Notice that P_2 only outputs corrupted$_1$ if the checks from the circuit that is opened do not pass. As we have mentioned, there is no logical reason why an adversarial P_1 would ever actually reply with an invalid opening; rather, it would just abort. Consider now the following modification:

1. Party P_1 sends ℓ garbled circuits $GC_1, \ldots, GC_\ell$ to party P_2.
2. P_1 and P_2 participate in a (fully secure) 1-out-of-ℓ oblivious transfer with the following inputs:

   a. P_1 defines its inputs $(x, \ldots, x_\ell)$ as follows. Input x_i consists of the opening of circuits GC_j for $j \neq i$ together with the keys associated with its own input in GC_i.
   b. P_2's input is a random value $\gamma \leftarrow_R \{1, \ldots, \ell\}$.

3. P_2 receives an opening of $\ell - 1$ circuits together with the keys needed to compute the unopened circuit and proceeds as above.

Notice that this modified protocol is essentially equivalent to Protocol 5.2.1 and thus its proof of security is very similar. However, in this case, an adversarial P_1 that constructs one faulty circuit must decide *before* the oblivious transfer if it wishes to abort (in which case there is no successful cheating) or if it wishes to proceed (in which case P_2 will receive an explicitly invalid opening). Note that due to the security of the oblivious transfer, P_1 cannot know what value γ party P_2 inputs, and so cannot avoid being detected.

The price of this modification is that of one fully secure 1-out-of-ℓ oblivious transfer and the replacement of all of the original oblivious transfer protocols with fully secure ones. (Of course, we could use oblivious transfer protocols that are secure in the presence of covert adversaries with non-halting detection accuracy, but we do not know how to construct such a protocol more efficiently than a fully secure one.) A highly efficient oblivious transfer protocol with a constant number of exponentiations per execution is presented in Chapter 7. Using this protocol, we achieve non-halting detection accuracy at a similar cost. As we have mentioned, this is a significant advantage. (We remark that one should not be concerned with the lengths of $x_1, \ldots, x_\ell$ in P_1's input to the oblivious transfer. This is because P_1 can send them encrypted ahead of time with independent symmetric keys $k_1, \ldots, k_\ell$. Then the oblivious transfer takes place only on the keys.)

5.3 Efficiency of the Protocol

We analyze the efficiency of Protocol 5.2.1, as in previous chapters:

- **Number of rounds:** The number of rounds equals six in the oblivious transfer plus an additional three rounds for the rest of the protocol. However, the last message of the oblivious transfer can be sent together with the first message for the rest of the protocol. Thus, overall there are eight rounds of communication.
- **Asymmetric computations:** Party P_2 expands its input of length n to mn, and so mn oblivious transfers are run. The oblivious transfer presented in this chapter is actually not more efficient than those presented in Chapter 7 (we present it for didactic reasons). Thus, we refer to Chapter 7 for exact costs (in short, the cost is $O(1)$ exponentiations per oblivious transfer).
- **Symmetric computations:** The cost here is ℓ times that of the semi-honest protocol in Chapter 3. Thus, we have $12{\cdot}|C|{\cdot}\ell$ symmetric operations.
- **Bandwidth:** The cost here again is approximately ℓ times that of the semi-honest protocol.

We conclude that the cost of achieving security in the presence of covert adversaries is approximately ℓ times the bandwidth and symmetric computation of the semi-honest case, and m times the number of asymmetric computations. For $\epsilon = 1/2$, it suffices to take $\ell = m = 3$ and so we have that the cost is just three times that of the semi-honest case. This is significantly more efficient than the malicious case.

Part III
Specific Constructions

In this final section, we present efficient protocols for problems of specific interest. In Chapters 6 and 7, we study Σ-protocols and oblivious transfer; these are basic building blocks that are used widely when constructing protocols. Then, in Chapters 8 and 9, we present examples of efficient secure protocols that are useful in their own right.

Chapter 6
Sigma Protocols and Efficient Zero-Knowledge[1]

A zero-knowledge proof is an interactive proof with the additional property that the verifier learns nothing beyond the correctness of the statement being proved. The theory of zero-knowledge proofs is beautiful and rich, and is a cornerstone of the foundations of cryptography. In the context of cryptographic protocols, zero-knowledge proofs can be used to enforce "good behavior" by having parties prove that they indeed followed the protocol correctly. These proofs must reveal nothing about the parties' private inputs, and as such must be zero knowledge. Zero-knowledge proofs are often considered an expensive (and somewhat naive) way of enforcing honest behavior, and those who view them in this way consider them to be not very useful when constructing efficient protocols. Although this is true for arbitrary zero-knowledge proofs of $\mathcal{NP}$ statements, there are many languages of interest for which there are extraordinarily efficient zero-knowledge proofs. Indeed, in many cases an efficient zero-knowledge proof is the best way to ensure that malicious parties do not cheat. We will see some good examples of this in Chapter 7 where efficient zero-knowledge proofs are used to achieve oblivious transfer with very low cost.

In this chapter, we assume familiarity with the notions of zero-knowledge and honest verifier zero-knowledge, and refer readers to [30, Chapter 4] for the necessary theoretical background, and for definitions and examples. As everywhere else in this book, here we are interested in achieving *efficiency*.

6.1 An Example

We begin by motivating the notion of a Σ-protocol through an example. Let p be a prime, q a prime divisor of $p - 1$, and g an element of order q in $\mathbb{Z}_p^*$. Suppose that a prover P has chosen a random $w \leftarrow_R \mathbb{Z}_q$ and has published $h = g^w \bmod p$. A verifier V who receives (p, q, g, h) can efficiently

---

[1] Much of this chapter is based on a survey paper by Ivan Damgård [20]. We thank him for graciously allowing us to use his text.

C. Hazay, Y. Lindell, *Efficient Secure Two-Party Protocols,*
Information Security and Cryptography, DOI 10.1007/978-3-642-14303-8_6,
© Springer-Verlag Berlin Heidelberg 2010

verify that p, q are prime, and that g, h both have order q. Since there is only one subgroup of order q in Z_p^*, this automatically means that $h \in \langle g \rangle$ and thus there always exists a value w such that $h = g^w$ (this holds because in a group of prime order all elements apart from the identity are generators and so g is a generator). However, this does not necessarily mean that P *knows* w.

Protocol 6.1.1 by Schnorr provides a very efficient way for P to convince V that it knows the unique value $w \in \mathbb{Z}_q$ such that $h = g^w \bmod p$:

PROTOCOL 6.1.1 (Schnorr's Protocol for Discrete Log)

- **Common input:** The prover P and verifier V both have (p, q, g, h).
- **Private input:** P has a value $w \in \mathbb{Z}_q$ such that $h = g^w \bmod p$.
- **The protocol:**

  1. The prover P chooses a random $r \leftarrow_R \mathbb{Z}_q$ and sends $a = g^r \bmod p$ to V.
  2. V chooses a random *challenge* $e \leftarrow_R \{0, 1\}^t$ and sends it to P, where t is fixed and it holds that $2^t < q$.
  3. P sends $z = r + ew \bmod q$ to V, which checks that $g^z = ah^e \bmod p$, that p, q are prime and that g, h have order q, and accepts if and only if this is the case.

Intuitively, this is a proof of knowledge because if some P^*, having sent a, can answer two *different* challenges e, e' correctly, then this means that it could produce z, z' such that $g^z = ah^e \bmod p$ and also $g^{z'} = ah^{e'} \bmod p$. Dividing one equation by the other, we have that $g^{z-z'} = h^{e-e'} \bmod p$. Now, by the assumption, $e - e' \neq 0 \bmod q$ (otherwise e and e' are not different challenges), and so it has a multiplicative inverse modulo q. Since g, h have order q, by raising both sides to this power, we get $h = g^{(z-z')(e-e')^{-1}} \bmod p$, and so $w = (z - z')(e - e')^{-1} \bmod q$. Observing that z, z', e, e' are all known to the prover, we have that the prover itself can compute w and thus *knows* the required value (except with probability 2^{-t}, which is the probability of answering correctly with a random guess). Thus, Protocol 6.1.1 is a *proof of knowledge*; we prove this formally below in Section 6.3.

In contrast, Protocol 6.1.1 is *not* known to be zero-knowledge. In order to see this, observe that in order for the problem of finding w to be non-trivial in the first place, q must be (exponentially) large. Furthermore, to achieve negligible error in a single run of the protocol, 2^t must be exponentially large too. In this case, standard rewinding techniques for zero-knowledge simulation will fail because it becomes too hard for a simulator to guess the value of e in advance. It is therefore not known if there exists some efficient malicious strategy that the verifier may follow which, perhaps after many executions of the protocol, enables it to obtain w (or learn more than is possible from (p, q, g, h) alone).

On the positive side, the protocol is *honest verifier zero-knowledge*. To simulate the view of an honest verifier V, simply choose a random $z \leftarrow_R \mathbb{Z}_q$

and $e \leftarrow_R \{0,1\}^t$, compute $a = g^z h^{-e} \bmod p$, and output (a, e, z). It is immediate that (a, e, z) has exactly the same probability distribution as a real conversation between the honest prover and the honest verifier. It is even possible to take any given value e and then produce a conversation where e occurs as the challenge: just choose z at random and compute the a that matches. In other words, the simulator does not have to insist on choosing e itself; it can take an e value as input. As we will see, this property turns out to be very useful.

One might argue that honest verifier zero-knowledge is not a very useful property, since it does not make sense in practice to assume that the verifier is honest. However, as we will see, protocols with this weaker property can often be used as building blocks in constructions that are indeed secure against active cheating, and are almost as efficient as the original protocol.

A final practical note: in a real-life scenario, it will often be the case where p and q are fixed for a long period, and so V can check primality once and for all, and will not have to do it in every execution of the protocol.

Working in general groups. From here on, we will consider general groups for proofs such as those above. The translation of Protocol 6.1.1 to this general language is as follows. The common input to the parties is a tuple $(\mathbb{G}, q, g, h)$ where $\mathbb{G}$ denotes a concise representation of a finite group of prime order q, and g and h are generators of $\mathbb{G}$. The proof then remains identical, with all operations now being in the group $\mathbb{G}$, and not necessarily operations $\bmod p$. In addition, the checks by V in the last step regarding the group are different. Specifically, V must be able to efficiently verify that $\mathbb{G}$ is indeed a group of prime order q, and that $g, h \in \mathbb{G}$ (note that in the concrete example above, V checks membership in $\mathbb{G}$ by checking that the element in question is of order q).

6.2 Definitions and Properties

Based on the Schnorr example, we will now define some abstract properties for a protocol that capture the essential properties of the example. It turns out that this abstraction holds for a wide class of protocols that are extremely useful when constructing efficient secure protocols.

Let R be a binary relation. That is, $R \subset \{0,1\}^* \times \{0,1\}^*$, where the only restriction is that if $(x, w) \in R$, then the length of w is at most $p(|x|)$, for some polynomial $p()$. For some $(x, w) \in R$, we may think of x as an instance of some computational problem, and w as the solution to that instance. We call w a *witness* for x.

For example, we could define a relation $\mathcal{R}_{\mathrm{DL}}$ for the discrete log problem by

$$\mathcal{R}_{\mathrm{DL}} = \{((\mathbb{G}, q, g, h), w) \mid h = g^w\},$$

where it is to be understood that $\mathbb{G}$ is of order q with generator g, that q is prime and that $w \in \mathbb{Z}_q$. In this case, R contains the entire set of discrete log problems of the type we considered above, together with their solutions.

We will be concerned with protocols of the form of Protocol 6.2.1.

PROTOCOL 6.2.1 (Protocol Template for Relation R)

- **Common input:** The prover P and verifier V both have x.
- **Private input:** P has a value w such that $(x, w) \in R$.
- **The protocol template:**

  1. P sends V a message a.
  2. V sends P a *random* t-bit string e.
  3. P sends a reply z, and V decides to accept or reject based solely on the data it has seen; i.e., based only on the values (x, a, e, z).

Furthermore, we will assume throughout that both P and V are probabilistic polynomial time machines, and so P's only advantage over V is that it knows the witness w. We say that a transcript (a, e, z) is an accepting transcript for x if the protocol instructs V to accept based on the values (x, a, e, z). We now formally define the notion of a Σ-protocol (the name Σ-protocol comes from the form of the letter Σ that depicts a three-move interaction).

Definition 6.2.2 *A protocol π is a Σ-protocol for relation R if it is a three-round public-coin protocol of the form in Protocol 6.2.1 and the following requirements hold:*

- **Completeness:** *If P and V follow the protocol on input x and private input w to P where $(x, w) \in R$, then V always accepts.*
- **Special soundness:** *There exists a polynomial-time algorithm A that given any x and any pair of accepting transcripts $(a, e, z), (a, e', z')$ for x, where $e \neq e'$, outputs w such that $(x, w) \in R$.*
- **Special honest verifier zero knowledge:** *There exists a probabilistic polynomial-time simulator M, which on input x and e outputs a transcript of the form (a, e, z) with the same probability distribution as transcripts between the honest P and V on common input x. Formally, for every x and w such that $(x, w) \in R$ and every $e \in \{0, 1\}^t$ it holds that*

$$\left\{ M(x, e) \right\} \equiv \left\{ \langle P(x, w), V(x, e) \rangle \right\}$$

where $M(x, e)$ denotes the output of simulator M upon input x and e, and $\langle P(x, w), V(x, e) \rangle$ denotes the output transcript of an execution between P and V, where P has input (x, w), V has input x, and V's random tape (determining its query) equals e.

The value t is called the challenge length.

It should be clear that Protocol 6.1.1 is a Σ-protocol. However, this protocol is not the exception to the rule. Rather, there are many such examples. This makes the general study of Σ-protocols important.

Until now, we have considered proofs of knowledge, and not proofs of membership. Observe that in the case of the discrete log there always exists a w for which $h = g^w$; the question is just whether or not P knows w. However, Σ-protocols often also arise in the setting of proofs of membership. Define L_R to be the set of inputs x for which there exists a w such that $(x, w) \in L_R$. Then the special soundness property implies that a Σ-protocol for R is always an interactive proof system for L_R with error probability 2^{-t}. This is due to the fact that if the special soundness algorithm A must be able to output w given any two accepting transcripts, then there must only be a *single* accepting transcript whenever $x \notin L_R$. Formally,

Proposition 6.2.3 *Let π be a Σ-protocol for a relation R with challenge length t. Then, π is an interactive proof of membership for L_R with soundness error 2^{-t}.*

Proof. Let $x \notin L_R$. We show that no P^* can convince V to accept with probability greater than 2^{-t}, even if P^* is computationally unbounded. Assume by contradiction that P^* can convince V with probability greater than 2^{-t}. This implies that there exists a first message a from P^* and at least two queries e, e' from V that result in accepting transcripts. This is the case because if for every a there exists at most one query e that results in an accepting transcript, then P^* would convince V with probability at most 2^{-t}; i.e., P^* would convince V only if V chose the single query e that P^* can answer, and since $e \leftarrow_R \{0,1\}^t$ this happens with probability only 2^{-t}. Observe now that the special soundness property requires that A output a witness w such that $(x, w) \in R$ (implying that $x \in L_R$) when given *any* pair of accepting transcripts $(a, e, z), (a, e', z')$ with $e \neq e'$. Thus, we conclude that whenever P^* can convince V with probability greater than 2^{-t} it holds that $x \in L_R$, in contradiction to the assumption. ∎

Before proceeding, we give another example of a useful Σ-protocol that is also a proof of membership (i.e., unlike the case of the discrete log, here the fact that the input is in the language is non-trivial to verify). Specifically, we consider the case of proving that a tuple $(\mathbb{G}, q, g, h, u, v)$ is of the Diffie-Hellman form, that is, that there exists a w such that $u = g^w$ and $v = h^w$. By the decisional Diffie-Hellman assumption, the case where $u = g^w$ and $v = h^w$ for some w is computationally indistinguishable from the case where $u = g^w$ and $h = g^{w'}$ for $w \neq w'$. Thus, a proof of this fact is non-trivial. We remark that proofs of this type come up in many settings. For just one example, this exact proof is utilized by Protocol 7.4.1 for securely computing oblivious transfer in the malicious setting; see Section 7.4. Formally, Protocol 6.2.4 below is a proof system for the relation

$$\mathcal{R}_{\text{DH}} = \left\{ ((\mathbb{G}, q, g, h, u, v), w) \mid g, h \in \mathbb{G} \ \& \ u = g^w \ \& \ v = h^w \right\}.$$

PROTOCOL 6.2.4 (Σ Protocol for Diffie-Hellman Tuples)

- **Common input:** The prover P and verifier V both have $(\mathbb{G}, q, g, h, u, v)$.
- **Private input:** P has a value w such that $u = g^w$ and $v = h^w$.
- **The protocol:**

  1. The prover P chooses a random $r \leftarrow_R \mathbb{Z}_q$ and computes $a = g^r$ and $b = h^r$. It then sends (a, b) to V.
  2. V chooses a random challenge $e \leftarrow_R \{0,1\}^t$ where $2^t < q$ and sends it to P.
  3. P sends $z = r + ew \bmod q$ to V, which checks that $g^z = au^e$ and $h^z = bv^e$, that $\mathbb{G}$ is a group of order q with generators g and h, and that $u, v \in \mathbb{G}$, and accepts if and only if this is the case.

Claim 6.2.5 *Protocol 6.2.4 is a Σ-protocol for the relation $\mathcal{R}_{\mathrm{DH}}$.*

Proof. In order to see that completeness holds, observe that when P runs the protocol honestly we have

$$g^z = g^{r+ew} = g^r \cdot g^{we} = g^r \cdot (g^w)^e = a \cdot u^e$$

where the last equality is due to the fact that $a = g^r$ and $u = g^w$. A similar calculation shows that $h^z = bv^e$. Regarding special soundness, let $((a,b), e, z)$ and $((a,b), e', z')$ be two accepting transcripts. We have that $g^z = au^e$ and $g^{z'} = au^{e'}$, as well as $h^z = bv^e$ and $h^{z'} = bv^{e'}$. Dividing both pairs of equations we have that $g^{z-z'} = u^{e-e'}$ and $h^{z-z'} = v^{e-e'}$. Continuing as in Schnorr's protocol, we conclude that $u = g^{(z-z')/(e-e')}$ and $v = h^{(z-z')/(e-e')}$. The machine A can therefore set $w = \frac{z-z'}{e-e'}$ and it holds that $((\mathbb{G}, q, g, h, u, v), w) \in \mathcal{R}_{\mathrm{DH}}$, as required.

It remains to demonstrate special honest verifier zero knowledge. However, this is also almost identical to the simulator in Protocol 6.1.1, and is therefore omitted. ■

Properties of Σ-protocols. We conclude this section with two important, yet easily verified properties of Σ-protocols. We first consider parallel repetition, where the parties run the same protocol multiple times with the same input, and V accepts if and only if it accepts in all repetitions. We stress that the parties use independent randomness for generating their messages in every execution.

Lemma 6.2.6 *The properties of Σ-protocols are invariant under parallel repetition. That is, the parallel repetition ℓ times of a Σ-protocol for R with challenge length t yields a new Σ-protocol for R with challenge length $\ell \cdot t$.*

Combining this with Proposition 6.2.3 we have that parallel repetition reduces the error exponentially fast. It is worthwhile noting that if there are two accepting transcripts of the parallel protocol then this actually means that there are ℓ pairs of accepting transcripts. For this reason, the soundness error is reduced to $2^{-\ell \cdot t}$ as expected.

Next we show that the challenge can be set to any arbitrary length.

Lemma 6.2.7 *If there exists a Σ-protocol π for R with challenge length t, then there exists a Σ-protocol π' for R with challenge length t', for any t'.*

Proof. Let t be the challenge length for the given protocol π. Then a Σ-protocol with challenge length t' shorter than t can be constructed as follows. P sends the first message a as in π. V then sends a random t'-bit string e. Finally, P appends $t - t'$ 0s to e, calls the result e', and computes the answer z to e' as in π. The verifier V checks z as in π as if the challenge were e'. Special soundness trivially still holds. In addition, special zero-knowledge holds because the simulator M is required to work for *every* challenge e, including those the conclude with $t - t'$ 0s.

Regarding longer challenge lengths $t' > t$; this can be achieved by first running π in parallel j times, for j such that $jt \geq t'$ (as in Lemma 6.2.6), and then adjusting jt down to t' as above in the case where $jt > t'$. ∎

6.3 Proofs of Knowledge

In this section we prove that any Σ-protocol with challenge length t is a proof of knowledge with knowledge error 2^{-t}. We begin by reviewing the notion of a proof of knowledge, as defined in [7]. It should be noted that it is not at all straightforward to define this notion. Specifically, what does it mean for a machine to "know" or "not know" something? The aim of a proof of knowledge is to enable a prover to convince a verifier that it knows a witness for some statement. Thus, it must be the case that only a machine that "knows" a witness should be able to convince the verifier. The definition below formalizes this by essentially saying that a machine knows a witness if it can be used to efficiently compute it. Stated differently, if a witness can be efficiently computed given access to P^*, then this means that P^* knows the witness. Indeed, P^* can run the efficient computation on itself and thereby explicitly obtain the witness. We refer to [30, Chapter 4] for more discussion, and proceed directly to the definition. (Note that the definition here is as in [7] and differs from that presented in [30]. Specifically, we consider probabilistic provers as in [7] and not only deterministic ones as considered in [30]. See [8] for a discussion regarding the difference between these definitions.)

Definition 6.3.1 *Let* $\kappa : \{0,1\}^* \to [0,1]$ *be a function. A protocol* (P,V) *is* a proof of knowledge *for the relation* R *with* knowledge error κ *if the following properties are satisfied:*

Completeness: *If* P *and* V *follow the protocol on input* x *and private input* w *to* P *where* $(x,w) \in R$, *then* V *always accepts.*

Knowledge soundness (or validity):   *There exists a constant* $c > 0$ *and a probabilistic oracle machine* K, *called the* knowledge extractor, *such that for every interactive prover function* P^* *and every* $x \in L_R$, *the machine* K *satisfies the following condition. Let* $\epsilon(x)$ *be the probability that* V *accepts on input* x *after interacting with* P^*. *If* $\epsilon(x) > \kappa(x)$, *then upon input* x *and oracle access to* P^*, *the machine* K *outputs a string* w *such that* $(x,w) \in R$ *within an expected number of steps bounded by*

$$\frac{|x|^c}{\epsilon(x) - \kappa(x)}.$$

One can think of the error κ as the probability that one can convince the verifier without knowing a valid w, and the ability to convince the verifier with higher probability means that the prover knows w. Furthermore, the higher the probability that P^* convinces V, the more efficient it is to compute w.

As we have mentioned, we prove here that any Σ-protocol with challenge length t is a proof of knowledge with knowledge error 2^{-t}. Intuitively Σ-protocols are proofs of knowledge because this is essentially what the special soundness property states. Specifically, as we have seen, if P can convince V with probability greater than 2^{-t} then there must be two accepting transcripts, in which case it is possible to apply the extraction machine guaranteed to exist by special soundness. The proof that this indeed holds is however more involved because the witness extractor K has to first *find* two accepting transcripts for such a prover, whereas the special soundness machine only needs to work when somehow magically given such a pair of transcripts. We now prove this theorem.

Theorem 6.3.2 *Let* π *be a* Σ-protocol *for a relation* R *with challenge length* t. *Then* π *is a proof of knowledge with knowledge error* 2^{-t}.

Proof.   Completeness (or non-triviality) is clear by definition. We know prove the *validity* property; i.e., the existence of an extractor K as described above. Let H be the 0/1-matrix with a row for each possible set of random choices ρ by P^*, and one column for each possible challenge value e. An entry $H_{\rho,e}$ equals 1 if V accepts with this random choice and challenge, and 0 otherwise. Using P^* as a black box while setting its internal random coins to be random and choosing a random challenge, we can probe a random entry in H. Furthermore, by rewinding P^* and reusing the same internal random coins as before, we can probe a random entry in the same row. In this light, our goal is to find two 1s in the same row. Using special soundness the resulting two transcripts provide us with sufficient information to efficiently compute a witness w for x.

Let P^* be a prover that convinces V upon input x with probability $\epsilon(x)$, and assume that $\epsilon(x) > 2^{-t}$. This implies that $\epsilon := \epsilon(x)$ equals the fraction of 1-entries in H. However, it is important to note that this gives *no guarantees* about the distribution of 1s in a given row. For instance, there may be some rows with many 1s and others with few 1s (a row with a single 1 cannot be used to find w because it does not give two accepting transcripts). Despite the above, we can make the following observation about the distribution of 1s in H. Define a row to be heavy if it contains a fraction of at least $\epsilon/2$ ones (there are 2^t entries in a row, and so this means that there are $\epsilon \cdot 2^{t-1}$ 1s in a heavy row). By a simple counting argument, it holds that more than half of the 1s are located in heavy rows. In order to see this, let H' be the sub-matrix of H consisting of all rows that are not heavy, and denote by h' the total number of entries in H', and by h the total number of entries in H. By the assumption, the number of 1s in H is $h \cdot \epsilon$, and the number of 1s in H' is smaller than $h' \cdot \epsilon/2$. This implies that the number of 1s in heavy rows is greater than

$$h \cdot \epsilon - h' \cdot \frac{\epsilon}{2} \geq h \cdot \epsilon - h \cdot \frac{\epsilon}{2} = \frac{h \cdot \epsilon}{2},$$

demonstrating that half of the 1s are indeed in heavy rows.

We first show how the extractor works under the assumption that

$$\epsilon \geq 2^{-t+2},$$

so that a heavy row contains at least two 1s (recall that a heavy row has at least $\epsilon/2$ 1s, and so if ϵ is at least 2^{-t+2}, there are at least $2^{-t+2}/2 \cdot 2^t = 2$ 1s in each such row). In this case, we will show that we can find two 1s in the same row in expected time $O(1/\epsilon)$. This will be more than sufficient since $1/\epsilon$ is less than required by the definition, namely $1/(\epsilon - 2^{-t})$.

Our approach will be to first repeatedly probe H at random, until we find a 1 entry, a "first hit". This happens after an expected number of $1/\epsilon$ tries, because the fraction of 1s in H is exactly ϵ. By the observation above, with probability greater than $1/2$, the first hit lies in a heavy row. Now, *if* it does (but note that we cannot check if it does), and if we continue probing at random along this row, the probability of finding another 1 in one attempt equals

$$\frac{\epsilon/2 \cdot 2^t - 1}{2^t}$$

because there are $\frac{\epsilon}{2} \cdot 2^t$ 1s in this row (by the assumption that it is heavy) and we need to find a different 1. This implies that the expected number T of tries to find the second hit satisfies

$$T = \frac{2^t}{\epsilon/2 \cdot 2^t - 1} = \frac{2^t}{\epsilon/2 \cdot (2^t - 2/\epsilon)} = \frac{2}{\epsilon} \cdot \frac{2^t}{2^t - 2/\epsilon} \leq \frac{2}{\epsilon} \cdot \frac{2^t}{2^t - 2^{t-1}} = \frac{4}{\epsilon}$$

tries, where the inequality follows from the assumption that $\epsilon \geq 2^{-t+2}$ above.

We therefore conclude that if the extractor's first hit was in a heavy row, then it finds two different accepting transcripts, as required for obtaining the witness, within an expected $O(1/\epsilon)$ tries. However, it may not be the case that the extractor's first hit was in a heavy row, and in such a case, we may spend too much time finding another 1 (if it exists at all). To remedy this, we include an "emergency break", resulting in the following algorithm (which is still not the final extractor, as we will see below):

1. As above, probe random entries in H until the first 1 is found (the first hit). The row in which this is found is called the current row.
2. Next, start the following two processes in *parallel*, and stop when either one stops:

   - **Process Pr_1:** Probe random entries in the current row, until another 1 entry is found (the second hit).
   - **Process Pr_2:** Repeatedly flip a coin that comes out heads with probability ϵ/d, for some *constant* d (we show how to choose d below) until you get heads. This can be achieved by probing a random entry in H and choosing a random number out of $1, 2, ..., d$. Then, output heads if both the entry in H is a 1 and the number chosen is 1.

We now analyze the expected running time of the above algorithm and its probability of success. Regarding the running time, observe that whenever one of the processes stops the entire algorithm halts. Thus, it suffices to analyze the expected running time of Pr_2 and this provides an upper bound on the expected running time of the entire algorithm. Now, since Pr_2 halts when the first heads appears, we have that its expected running time is d/e, which for a constant d equals $O(1/\epsilon)$ as required. We now show that for an appropriate choice of the constant d, the algorithm succeeds in extracting a witness with probability at least $1/8$. Observe that the algorithm only succeeds in extracting a witness if Pr_1 finishes first. Therefore, we have to choose d so that Pr_1 has enough time to finish before Pr_2, if indeed the first hit is in a heavy row.

The probability that Pr_2 finishes after exactly k attempts is $\epsilon/d \cdot (1 - \epsilon/d)^{k-1}$. Using the crude estimate $(1 - \epsilon/d)^{k-1} \leq 1$, we have that the probability that Pr_2 finishes *within* k attempts is less than or equal to $k\epsilon/d$. Thus, by setting $k = 8/\epsilon$ and $d = 16$ we have that the probability that Pr_2 finishes within $8/\epsilon$ attempts is less than or equal to

$$\frac{k\epsilon}{d} = \frac{\frac{8}{\epsilon} \cdot \epsilon}{16} = \frac{1}{2}.$$

We conclude that setting $d = 16$, it holds that Pr_2 finishes after more than $8/\epsilon$ tries with probability *at least* $1/2$.

We are now ready to show that the algorithm succeeds in extracting a witness with probability at least $1/8$. In order to see this, first recall that *if* the first hit is in a heavy row, then the expected number of steps of Pr_1 is at most $T = \epsilon/4$ and so by Markov's inequality, the probability that Pr_1 takes $2T \le 8/\epsilon$ or more steps is upper-bounded by $1/2$. Stated differently, if the first hit is in a heavy row then with probability at least $1/2$ process Pr_1 concludes in less than $8/\epsilon$ steps. Since the random variables determining the stopping condition of Pr_1 and Pr_2 are independent, we have that if the first hit is in a heavy row then Pr_1 finishes before Pr_2 with probability at least $1/2 \cdot 1/2 = 1/4$. Next, recall that the first hit is in a heavy row with probability at least $1/2$. We have that the probability that the algorithm succeeds equals at least the probability that the first hit is a heavy row times the probability that Pr_1 finishes before Pr_2, *conditioned* on the first hit being a heavy row. Thus, the algorithm succeeds with probability $1/2 \cdot 1/4 = 1/8$, as required.

We now describe the actual knowledge extractor (however, still only for the case where $\epsilon \ge 2^{-t+2}$). The knowledge extractor K works by repeating the above algorithm until it succeeds. Since the expected number of repetitions is constant (at most eight), we have that K succeeds in finding a witness in an expected number of steps that is bound by $O(1/\epsilon)$, as desired.

It still remains to consider the case where $2^{-t} < \epsilon < 2^{-t+2}$; recall that above we assumed that $\epsilon \ge 2^{-t+2}$ but the extractor K must work whenever $\epsilon \ge 2^{-t}$. We treat this case by a separate algorithm, using the fact that when ϵ is so small, we actually have enough time to probe an entire row. The knowledge extractor K runs this algorithm in parallel with the above one.

Define δ to be such that $\epsilon = (1 + \delta)2^{-t}$. Note that for the values of ϵ that we consider here it holds that $0 < \delta < 3$. Let R be the number of rows in H. Then we have that the number of 1s in H is at least $(1 + \delta) \cdot R$, where the total number of entries in H equals $R \cdot 2^t$. Since there are only R rows, it follows that at most $R-1$ of the $(1 + \delta)R$ ones can be alone in a row, and thus at least δR of them must be in rows with at least two 1s. We call such a row semi-heavy. The algorithm here works as follows:

1. Probe random entries until a 1 is found; call this row the current row.
2. Search the entire current row for another 1 entry. If no such entry is found, then go back to Step 1.

It is clear that the algorithm successfully finds a witness. However, the question is whether it does so in time $O(1/\epsilon - 2^{-t})$. In order to analyze this, note that the fraction of 1s in semi-heavy rows is $\delta/(1 + \delta)$ among all 1s (because the fraction of 1s overall is $(1 + \delta)$ and a δ fraction of these are in semi-heavy rows). In addition, the fraction of 1s in semi-heavy rows is at least $\delta/2^t$ among all entries.

Now, the expected number of probes to find a 1 is

$$\frac{1}{\epsilon} = \frac{2^t}{(1 + \delta)}. \tag{6.1}$$

Furthermore, by the fact that the fraction of 1s in semi-heavy rows among all entries is $\delta/2^t$, the expected number of probes to find a 1 in a semi-heavy row is $2^t/\delta$. We therefore expect to find a 1 in a semi-heavy row after finding $(1+\delta)/\delta$ 1s. This implies that the expected number of times that the algorithm carries out Steps 1 and 2 equals $(1+\delta)/\delta$. In each such repetition, the expected number of probes in Step 1 equals $2^t/(1+\delta)$; see (6.1). Then, once a 1 is found, the number of probes in Step 2 is exactly 2^t. We therefore have that the expected cost is

$$\frac{1+\delta}{\delta} \cdot \left(\frac{2^t}{1+\delta} + 2^t \right) = 2^t \cdot \left(\frac{1}{\delta} + \frac{1+\delta}{\delta} \right) = 2^t \cdot \frac{2+\delta}{\delta} < 5 \cdot \frac{2^t}{\delta}.$$

However,

$$\frac{1}{\epsilon - 2^{-t}} = \frac{1}{(1+\delta)2^{-t} - 2^{-t}} = \frac{2^t}{\delta},$$

proving that the algorithm runs in time $O(1/(\epsilon - 2^{-t}))$, as required. This completes the proof that any Σ-protocol with challenge length t is a proof of knowledge with knowledge error 2^{-t}. ∎

6.4 Proving Compound Statements

In this section, we show how basic Σ-protocols can be used to prove compound statements. This is a very useful tool in cryptographic protocols and is a good demonstration of the usefulness of Σ-protocols. It is easy to prove the AND of two statements: simply have the prover prove both in parallel with the verify sending a single challenge e for both proofs. It is easy to see that the result is a Σ-protocol for a compound relation comprised of the AND of two statements. We leave the proof of this as an exercise.

In contrast, proving the OR of two statements is more involved. Specifically, the problem we wish to solve is as follows. A prover and verifier P and V hold a pair (x_0, x_1) for their common input. The prover P wishes to prove to V that it knows a witness w such that either $(x_0, w) \in R$ or $(x_1, w) \in R$, *without revealing which is the case*. We use this methodology in Section 7.5 in an oblivious transfer protocol where one party has to prove to another that one of two tuples is a Diffie-Hellman tuple, without revealing which. We remark that similar ideas can be used to extend the construction below to prove that k out of n statements are true, again without revealing which; see [19]. In addition, the construction below remains unchanged when the statements are for different relations R_0 and R_1 and the goal is for P to prove that it knows a witness w such that either $(x_0, w) \in R_0$ or $(x_1, w) \in R_1$. We present the case of a single relation for the sake of clarity.

Assume that we are given a Σ-protocol π for a relation R. Assume also that the pair (x_0, x_1) is common input to P and V, and that w is the private input of P, where either $(x_0, w) \in R$ or $(x_1, w) \in R$. Let b be such that $(x_b, w) \in R$. Roughly speaking, the idea is to have the prover complete two instances of π, one with input x_0 and the other with input x_1. Of course, the problem is that the prover can only do this for x_b, because it only knows a witness for this statement (indeed, the other statement x_{1-b} may not even be in the language L_R). Thus, the prover "fakes" the other statement by running the simulator M. This does not seem to solve the problem because in order to simulate, M needs to be able to determine e_{1-b} by itself, or at least to know e_{1-b} before generating a_{1-b}. Thus, the verifier V will know in which execution its challenge is used, and in which execution the prover set the challenge. This problem is overcome in the following fascinating way. The prover runs M to obtain $(a_{1-b}, e_{1-b}, z_{1-b})$. The prover P then sends V the first message a_{1-b} from the simulated transcript along with the first message a_b of the real execution. After receiving a challenge s from V, the prover P sets the challenge for the real execution to be $e_b = s \oplus e_{1-b}$ and completes both executions using challenges e_0 and e_1, respectively. (Observe that this means that P sends z_{1-b} from the simulated transcript in the simulated execution, as required.) Now, soundness still holds because a_{1-b} *binds* P to the challenge e_{1-b} (by the special soundness of π, if P does not know w such that $(x_{1-b}, w) \in R$ then it cannot complete the proof with any $e'_{1-b} \neq e_{1-b}$). Thus, the first message a_{1-b} together with the challenge s from the verifier actually defines a *unique* and *random* challenge $e_b = s \oplus e_{1-b}$ for the prover in the real execution. Thus, the prover must know w such that $(x_b, w) \in R$ in order to complete this execution. Interestingly, the verifier cannot distinguish which of the executions is real and which is simulated, because the transcript generated by M looks the same as a real one, and because $s \oplus e_{1-b}$ is distributed identically to $s \oplus e_b$. We therefore have that P can only complete the proof if it has a witness to one of the statements, and yet V cannot know which witness P has. Protocol 6.4.1 contains the full details of the OR protocol.

PROTOCOL 6.4.1 (OR Protocol for Relation R Based on π)

- **Common input:** The prover P and verifier V both have a pair (x_0, x_1).
- **Private input:** P has a value w and a bit b such that $(x_b, w) \in R$.
- **The protocol:**
  1. P computes the first message a_b in π, using (x_b, w) as input.
     P chooses e_{1-b} at random and runs the simulator M on input (x_{1-b}, e_{1-b}); let $(a_{1-b}, e_{1-b}, z_{1-b})$ be the output of M.
     P sends (a_0, a_1) to V.
  2. V chooses a random t-bit string s and sends it to P.
  3. P sets $e_b = s \oplus e_{1-b}$ and computes the answer z_b in π to challenge e_b using (x_b, a_b, e_b, w) as input. P sends (e_0, z_0, e_1, z_1) to V.
  4. V checks that $e_0 \oplus e_1 = s$ and that both transcripts (a_0, e_0, z_0) and (a_1, e_1, z_1) are accepting in π, on inputs x_0 and x_1, respectively.

Let $R_{\mathrm{OR}} = \{((x_0, x_1), w) \mid (x_0, w) \in R \text{ or } (x_1, w) \in R\}$. Then we have:

Theorem 6.4.2 *Protocol 6.4.1 is a Σ-protocol for the relation R_{OR}. Moreover, for any verifier V^*, the probability distribution over transcripts between P and V^*, where w is such that $(x_b, w) \in R$, is independent of b.*

Proof. It is clear that the protocol is of the right form. To verify special soundness, let there be two accepting transcripts $(a_0, a_1, s, e_0, e_1, z_0, z_1)$ and $(a_0, a_1, s', e'_0, e'_1, z'_0, z'_1)$ where $s \neq s'$. Since the transcripts are accepting it holds that $e_0 \oplus e_1 = s$ and $e'_0 \oplus e'_1 = s'$. Since $s \neq s'$ this implies that for some $c \in \{0, 1\}$ it must hold that $e_c \neq e'_c$. By the special soundness of π this implies that from (a_c, e_c, z_c) and (a_c, e'_c, z'_c) it is possible to efficiently compute w such that $(x_c, w) \in R$, implying in turn that $((x_0, x_1), w) \in R_{\mathrm{OR}}$.

Special honest verifier zero-knowledge is immediate: given s, choose e_0 and e_1 at random subject to $s = e_0 \oplus e_1$ and run M twice, once with input (x_0, e_0) and once with input (x_1, e_1).

It remains to prove that the probability distribution over transcripts is independent of b. Let V^* be an arbitrary verifier. Then, observe that the distribution over transcripts between P and V^* can be specified as follows. A transcript is of the form $(a_0, a_1, s, e_0, e_1, z_0, z_1)$, where a_0, a_1 are distributed as an honest prover in π would choose them (this is immediate for a_b and holds for a_{1-b} by the perfect honest verifier zero-knowledge of π). Then s has whatever distribution V^* outputs, given (x_0, x_1, a_0, a_1), and e_0, e_1 are random subject to $s = e_0 \oplus e_1$. Finally, z_0 and z_1 both have whatever distribution the honest prover in π outputs upon the respective transcript prefixes (x_0, a_0, e_0) and (x_1, a_1, e_1). As above, this is immediate for z_b and holds for z_{1-b} by the perfect honest verifier zero-knowledge of π. We therefore conclude that the distribution is independent of b.   ∎

By the last claim in this theorem, Protocol 6.4.1 is what is known as *witness indistinguishable* [27]. There are several different values of w that a prover may know that would enable it to complete the protocol successfully. However, there is no way that a verifier (even a malicious one) can know which of the possible witnesses the prover knows. This is a first sign that it is possible to obtain security properties that hold for arbitrary verifiers, even when starting from a protocol that is only honest verifier zero knowledge.

6.5 Zero-Knowledge from Σ-Protocols

In this section, we show how to construct efficient zero-knowledge protocols from any Σ-protocol. We remark that this can be done in a number of ways. We will first present a construction using any perfectly-hiding commitment scheme. This construction achieves zero knowledge, but is *not* a proof of knowledge. We will then show what needs to be modified so that the result is a zero-knowledge *proof of knowledge*.

6.5.1 The Basic Zero-Knowledge Construction

In order to motivate the construction, observe that the reason why a Σ-protocol is not zero-knowledge for arbitrary verifiers is that a simulator cannot predict the challenge e with probability greater than 2^{-t}. Thus, a verifier that outputs a different essentially random challenge for every first message a (say, by applying a pseudorandom function to a) cannot be simulated. This problem can be solved by simply having the verifier commit to its challenge e before the execution begins. This is the methodology used by [34] for the case of computational zero-knowledge. However, the simulation strategy and its analysis is far more simple here because the underlying Σ-protocol has a *perfect* zero-knowledge property (albeit for honest verifiers). Let com be a perfectly-hiding commitment protocol for committing to binary strings of length t. See Protocol 6.5.1 for the details of the construction.

PROTOCOL 6.5.1 (Zero-Knowledge Proof for L_R Based on π)

- **Common input:** The prover P and verifier V both have x.
- **Private input:** P has a value w such that $(x, w) \in R$.
- **The protocol:**
    1. V chooses a random t-bit string e and interacts with P via the commitment protocol com in order to commit to e.
    2. P computes the first message a in π, using (x, w) as input, and sends it to V.
    3. V decommits to e to P.
    4. P verifies the decommitment and aborts if it is not valid. Otherwise, it computes the answer z to challenge e according to the instructions in π, and sends z to V.
    5. V accepts if and only if transcript (a, e, z) is accepting in π on input x.

Theorem 6.5.2 *If* com *is a perfectly-hiding commitment protocol and π is a Σ-protocol for relation R, then Protocol 6.5.1 is a zero-knowledge proof for L_R with soundness error 2^{-t}.*

Proof. Completeness is immediate. Soundness follows from Proposition 6.2.3 and the hiding property of the commitment scheme. Specifically, by the perfect hiding property of the commitment protocol, a cheating prover knows nothing of e before it sends a. Thus, there is only a single challenge that it can answer, and the probability that it equals e is at most 2^{-t}.

In order to prove zero knowledge, we construct a black-box simulator $\mathcal{S}$, as follows:

1. $\mathcal{S}$ invokes the polynomial-time verifier V^* upon input x and interacts with it in the commitment protocol.
2. $\mathcal{S}$ runs the Σ-protocol honest verifier simulator M on a random $\tilde{e}$ in order to obtain a first message a', and hands it to V^*.

3. S receives back V^*'s decommitment. If it is invalid, S outputs what V^* upon receiving $(x, a, \bot)$ and halts. Otherwise, let e be the decommitted value. Then, S continues as follows:

   a. S invokes the simulator M upon input e and obtains back (a, z).
   b. S hands a to V^* and receives back its decommitment. If the decommitment is to e, then S outputs whatever V^* outputs upon receiving (a, e, z) and halts. If the decommitment is to some $e' \neq e$ then S outputs fail. If the decommitment is not valid, S returns to the previous step (and repeats with independent randomness).

We first observe that the computational binding of the commitment scheme ensures that the probability that S outputs fail is at most negligible. The reduction for this is identical to that in [34] and is omitted (one just needs to use S^{V^*} to break the computational binding, while truncating the simulation to run in strict polynomial time). Next, we claim that the distribution over the output of S given that it does not output fail is identical to the distribution over real transcripts. This follows from the perfect zero-knowledge property of M. Formally, let $S^{V^*}(x)$ denote the output of the simulator upon input x, and let $\langle P, V^* \rangle(x)$ denote the output of V^* after a real execution. Then, for any distinguisher D, we have

$$\left| \Pr[D(S^{V^*}(x)) = 1] - \Pr[D(\langle P, V^* \rangle(x)) = 1] \right|$$

$$= \left| \Pr[D(S^{V^*}(x)) = 1 \wedge S^{V^*}(x) \neq \mathsf{fail}] - \Pr[D(\langle P, V^* \rangle(x)) = 1] \right.$$

$$\left. + \Pr[D(S^{V^*}(x)) = 1 \wedge S^{V^*}(x) = \mathsf{fail}] \right|$$

$$\leq \left| \Pr[D(S^{V^*}(x)) = 1 \wedge S^{V^*}(x) \neq \mathsf{fail}] - \Pr[D(\langle P, V^* \rangle(x)) = 1] \right|$$

$$+ \Pr[S^{V^*}(x) = \mathsf{fail}].$$

Now,

$$\left| \Pr[D(S^{V^*}(x)) = 1 \wedge S^{V^*}(x) \neq \mathsf{fail}] - \Pr[D(\langle P, V^* \rangle(x)) = 1] \right|$$

$$= \left| \Pr[D(S^{V^*}(x)) = 1 \mid S^{V^*}(x) \neq \mathsf{fail}] \cdot \Pr[S^{V^*}(x) \neq \mathsf{fail}] \right.$$

$$\left. - \Pr[D(\langle P, V^* \rangle(x)) = 1] \right|$$

$$= \left| \Pr[D(S^{V^*}(x)) = 1 \mid S^{V^*}(x) \neq \mathsf{fail}] \cdot \left(1 - \Pr[S^{V^*}(x) = \mathsf{fail}] \right) \right.$$

$$\left. - \Pr[D(\langle P, V^* \rangle(x)) = 1] \right|$$

$$\leq \left| \Pr[D(S^{V^*}(x)) = 1 \mid S^{V^*}(x) \neq \mathsf{fail}] - \Pr[D(\langle P, V^* \rangle(x)) = 1] \right|$$

$$+ \Pr[D(S^{V^*}(x)) = 1 \mid S^{V^*}(x) \neq \mathsf{fail}] \cdot \Pr[S^{V^*}(x) = \mathsf{fail}].$$

Combining the above with the fact that $\Pr[\mathcal{S}^{V^*}(x) = \mathsf{fail}]$ is negligible, we have that there exists a negligible function μ such that

$$\left| \Pr[D(\mathcal{S}^{V^*}(x)) = 1] - \Pr[D(\langle P, V^* \rangle(x)) = 1] \right|$$

$$\leq \left| \Pr[D(\mathcal{S}^{V^*}(x)) = 1 \mid \mathcal{S}^{V^*}(x) \neq \mathsf{fail}] - \Pr[D(\langle P, V^* \rangle(x)) = 1] \right| + \mu(|x|).$$

We are now ready to apply the above intuition that the output of $\mathcal{S}$ conditioned on it not outputting fail is identical to a real execution transcript. That is, we claim that

$$\Pr[D(\mathcal{S}^{V^*}(x)) = 1 \mid \mathcal{S}^{V^*}(x) \neq \mathsf{fail}] = \Pr[D(\langle P, V^* \rangle(x)) = 1].$$

In order to see this, note that if the execution terminates with the transcript $(x, a, \bot)$, then the distributions are identical to the distribution of a as generated by M. Otherwise, in the first pass $\mathcal{S}$ receives a decommitment to e and repeats, generating an independent sample of (a, e, z) from M until this same event repeats. Furthermore, the probability that V^* aborts in a real execution is identical to the probability that it aborts in the simulation. Combining the above together, we have that the distributions are identical. We conclude that

$$\left| \Pr[D(\mathcal{S}^{V^*}(x)) = 1] - \Pr[D(\langle P, V^* \rangle(x)) = 1] \right| \leq \mu(|x|).$$

It remains to show that $\mathcal{S}$ runs in expected polynomial time. Let p be the probability that V^* sends a valid decommitment (to any value) upon receiving a from $\mathcal{S}$. The key observation is that since M is a perfect zero-knowledge simulator, it follows that the distribution over a when M is invoked upon x and a random $\tilde{e}$ is *identical* to the distribution over a when M is invoked upon x and the value e that was revealed in the first pass. This implies that V^* sends a valid decommitment in the rewinding phase with probability exactly p. We stress that $\mathcal{S}$ halts as soon as V^* sends a valid decommitment in this stage, irrespective of whether it is to the same e or some $e' \neq e$. Thus, the expected running time of $\mathcal{S}$ is

$$\mathrm{poly}(|x|) \cdot \left((1 - p) + p \cdot \frac{1}{p} \right) = \mathrm{poly}(|x|),$$

completing the proof. ∎

Efficient perfectly-hiding commitments. The complexity of Protocol 6.5.1 depends on the cost of running a perfectly-hiding commitment protocol. In order to appreciate the additional cost involved, we describe the highly efficient Pedersen's commitment scheme that is secure under the discrete log assumption; see Protocol 6.5.3.

In order to see that the above commitment scheme is perfectly hiding, observe that $c = g^r \cdot \alpha^x = g^{r+ax}$. Now, for every $x' \in \mathbb{Z}_q$ there exists a value

---

PROTOCOL 6.5.3 (The Pedersen Commitment Protocol)

- **Input:** The committer C and receiver R both hold 1^n, and the committer C has a value $x \in \{0,1\}^n$ interpreted as an integer between 0 and 2^n.

- **The commit phase:**
  1. The receiver R chooses $(\mathbb{G}, q, g)$ where $\mathbb{G}$ is a group of order q with generator g and $q > 2^n$. R then chooses a random $a \leftarrow \mathbb{Z}_q$, computes $\alpha = g^a$ and sends $(\mathbb{G}, q, g, \alpha)$ to C.
  2. The committer C verifies that $\mathbb{G}$ is a group of order q, that g is a generator and that $\alpha \in \mathbb{G}$. Then, it chooses a random $r \leftarrow \mathbb{Z}_q$, computes $c = g^r \cdot \alpha^x$ and sends c to R.

- **The decommit phase:**
  The committer C sends (r, x) to R, which verifies that $c = g^r \cdot \alpha^x$.

---

r' such that $r' + ax' = r + ax \bmod q$; specifically, take $r' = r + ax - ax'$. Thus, since r is chosen randomly, the distribution over commitments to x is identical to the distribution over commitments to x'. In order to prove computational binding, observe that given two decommitments $(x, r), (x', r)$ to $x, x' \in \mathbb{Z}_q$ where $x \neq x'$, it is possible to compute the discrete log of α, exactly as in Schnorr's protocol. Specifically, by the assumption that both decommitments are valid, it holds that $g^x \alpha^r = g^{x'} \alpha^{r'}$ and so $\alpha = g^{(x'-x)/(r-r') \bmod q}$. This similarity to Schnorr's protocol is not coincidental, and indeed it is possible to construct efficient commitments from *any* Σ-protocol; see Section 6.6 below.

Overall efficiency. The overall cost of Protocol 6.5.1 is exactly that of a *single* invocation of the underlying Σ-protocol, together with a *single* invocation of the commit and decommit phase of the perfectly-hiding commitment protocol. Using Pedersen's commitment, for example, we have that the overall cost of achieving zero-knowledge is just five exponentiations in $\mathbb{G}$, in addition to the cost of the Σ-protocol. (We remark that the parameters $(\mathbb{G}, q, g)$ can be reused many times, or can even be fixed and verified just once. We therefore ignore the cost of choosing and verifying the parameters.)

In addition to the above computational cost, there are an additional two rounds of communication (making an overall of five), and the additional communication of two group elements and two values in $\mathbb{Z}_q$. (As above, we ignore the cost of the parameters because they are typically fixed.)

6.5.2 Zero-Knowledge Proofs of Knowledge

As we have seen, any Σ-protocol is a proof of knowledge. Unfortunately, however, the zero-knowledge protocol of Section 6.5.1 appears to *not* be a proof of knowledge. The reason for this is that an extractor cannot send the prover two different queries $e \neq e'$ for the same a, because it is committed to e before the prover P^* sends a. Thus, the entire approach of Section 6.3 fails. This can be solved by using a perfectly-hiding *trapdoor* commitment scheme instead. Such a commitment scheme has the property that there exists a trapdoor that, when known, enables the committer to generate special commitment values that are distributed exactly like regular commitments, but can be opened to *any* value in the decommitment phase. Then, in the last step of the protocol, after the verifier has already decommitted to e and so is no longer relevant, the prover can send the trapdoor. Although meaningless in a real proof, this solves the problem described above because the knowledge extractor can obtain the trapdoor and then rewind P^* in order to provide it with different values of e for the same a, as needed to extract.

Let com be a perfectly-hiding trapdoor commitment protocol; and denote the trapdoor by trap. We assume that the receiver in the commitment protocol obtains the trapdoor, and that the committer can efficiently verify that the trapdoor is valid if it receives it later. We do not provide a formal definition of this, and later show that Pedersen's commitment described in Protocol 6.5.3 has this additional property. See Protocol 6.5.4 for the details of the construction.

PROTOCOL 6.5.4 (ZK Proof of Knowledge for R Based on π)

- **Common input:** The prover P and verifier V both have x.
- **Private input:** P has a value w such that $(x, w) \in R$.
- **The protocol:**
  1. V chooses a random t-bit challenge e and interacts with P via the commitment protocol com in order to commit to e.
  2. P computes the first message a in π, using (x, w) as input, and sends it to V.
  3. V reveals e to P by decommitting.
  4. P verifies the decommitment and aborts if it is not valid. Otherwise, it computes the answer z to challenge e according to the instructions in π, and sends z and the trapdoor trap to V.
  5. V accepts if and only if the trapdoor trap is valid and the transcript (a, e, z) is accepting in π on input x.

Theorem 6.5.5 *If* com *is a perfectly-hiding trapdoor commitment protocol and π is a Σ-protocol for relation R, then Protocol 6.5.4 is a zero-knowledge proof of knowledge for R with knowledge error 2^{-t}.*

Proof. The fact that the protocol is zero knowledge follows from the proof that Protocol 6.5.1 is zero knowledge. In particular, we define the same simulator S as in the proof of Theorem 6.5.2. The only subtlety that arises is in the part of the analysis in which we show that S does not output fail. Recall that if S outputs fail with non-negligible probability then it is possible to construct a machine that uses $S^{V^*}(x)$ in order to break the computational binding of the commitment scheme. However, when the trapdoor is known there is no guarantee whatsoever of binding and thus the machine used to break the computational binding is not given the trapdoor. This means that it cannot hand V^* the last message of the protocol that includes the trapdoor trap. However, observe that the instructions of $\mathcal{S}$ are such that it only hands the last message z to V^* *after* receiving a decommitment to e for the second time. That is, if $\mathcal{S}^{V^*}(x)$ outputs fail with non-negligible probability, then this event can be reproduced without $\mathcal{S}$ ever handing V^* the last message z and trap. Thus, the proof that $\mathcal{S}$ outputs fail with negligible probability goes through.

In order to demonstrate that the protocol is a proof of knowledge, we need to show that the knowledge extractor of the Σ-protocol π, as described in the proof of Theorem 6.3.2, can be applied here. The main observation is that a "hit" (or 1 in the matrix H), as described in the proof of Theorem 6.3.2, occurs when V is convinced. By the specification of Protocol 6.5.4, whenever this occurs, the knowledge extractor K obtains the valid trapdoor trap. Thus, after the first hit, K can start the entire extraction process again. However, this time it sends a "special" commitment that can be opened to any value at a later time. This enables K to send any challenge e that it wishes, relative to the same prover message a. In the terminology of the proof of Theorem 6.3.2, this means that K can sample multiple entries in the same row. Finally, observe that until the first hit occurs, K just randomly samples from the matrix. This can be achieved by K beginning the execution from scratch with a commitment to a new random e, and so the trapdoor is not needed at this point. We therefore conclude that the extractor K described in the proof of Theorem 6.3.2 can be applied here, as required. ∎

Pedersen's commitment is a trapdoor commitment. Observe that if the committer in Protocol 6.5.3 knows the exponent a chosen by the receiver to compute α, then it can decommit c to any value it wishes. In particular, C can commit to x by sending $c = g^r \cdot \alpha^x$ and can later decommit to any x' by computing $r' = r + ax - ax'$. This then implies that $r' + ax' = r + ax$ and so $g^r \cdot \alpha^x = c = g^{r'} \cdot \alpha^{x'}$. Stated differently, (r', x') is a valid decommitment to c, and so a is the desired trapdoor.[2]

Efficiency. When considering the concrete instantiation of Protocol 6.5.4 with Pedersen's commitments, we have that the additional cost of making the protocol a proof of knowledge is just a single additional exponentiation (this exponentiation is the verification by V of the trapdoor). It therefore follows that any Σ-protocol can be transformed into a zero-knowledge proof of knowledge with an additional six group exponentiations only. As above, the number of rounds of communication is five, and the additional communication costs are three group elements and one value in $\mathbb{Z}_q$.

---

[2] Observe that the trapdoor property of Pedersen's commitment is actually stronger than defined above. Specifically, it is possible for the committer to construct a regular commitment to a value x, and then open it to whatever value it wishes even when it receives the trapdoor *after* the commitment phase concludes.

Concrete efficiency – discrete log and Diffie-Hellman. Later on this book, we will use zero-knowledge proofs of knowledge for the discrete log and Diffie-Hellman tuple relations. In order to enable a full and concrete analysis of the efficiency of these protocols, we now give an exact analysis of the cost of applying Protocol 6.5.4 to the Σ-protocols for the discrete log (Protocol 6.1.1) and Diffie-Hellman tuples (Protocol 6.2.4).

- **ZKPOK for discrete log:** The complexity of Protocol 6.1.1 is just three exponentiations, and the communication of one group element and two elements of $\mathbb{Z}_q$ (for simplicity we count e as an element of $\mathbb{Z}_q$). Thus, the overall cost of the zero-knowledge proof of knowledge for this relation is five rounds of communication, nine exponentiations, and the exchange of four group elements and four elements of $\mathbb{Z}_q$.
- **ZKPOK for DH tuples:** The complexity of Protocol 6.2.4 is six exponentiations, and the communication of two group elements and two elements of $\mathbb{Z}_q$. Thus, the overall cost of the zero-knowledge proof of knowledge for this relation is five rounds of communication, 12 exponentiations, and the exchange of five group elements and four elements of $\mathbb{Z}_q$.

As can easily be seen here, zero-knowledge proofs of knowledge do *not* have to be expensive, and Σ-protocols are a very useful tool for obtaining efficient proofs of this type.

6.5.3 The ZKPOK Ideal Functionality

When constructing secure protocols, a very useful tool for proving security is to use the modular composition theorem outlined in Section 2.7. However, in order to do this for zero-knowledge, we must show that such a protocol is secure by the ideal/real definitions of secure computation; see Section 2.3. Let R be a relation and define the zero-knowledge functionality $\mathcal{F}_{\text{ZK}}^R$ by

$$\mathcal{F}_{\text{ZK}}^R((x, w), w) = (\lambda, R(x, w)).$$

Intuitively, any zero-knowledge proof of knowledge for R securely realizes $\mathcal{F}_{\text{ZK}}^R$ because simulation takes care of the case where V is corrupted and witness extraction takes care of the case where P is corrupted. However, for technical reasons that will become apparent in the proof below, this is not so simple. Although the proof of this fact is not related to Σ-protocols specifically, it makes the task of proving the security of protocols that use zero-knowledge proofs of knowledge much easier. Indeed, we will use this numerous times in the coming chapters. For this reason, we prove the theorem here (we remark that although it is folklore that this holds, we are not aware of any full proof having appeared before). In addition, the technical problem that arises when trying to prove this theorem occurs often in the setting of secure computation. Thus, the technique used to solve it, as introduced by [34], is important to study.

Before proving the theorem we remark on one technical change that must be made to the zero-knowledge proof of knowledge protocol. Specifically, in the setting of secure computation, the prover may be invoked with input (x, w) such that $(x, w) \notin R$. In this case, the prover should just send 0 to the verifier and do nothing else. Thus, we add an instruction to the protocol to have the prover first verify that $(x, w) \in R$. If so, it proceeds with the protocol, and if not it sends 0 to the verifier and halts. Observe that in order for this to be possible, the relation R must be in $\mathcal{NP}$. (We also need the following property: for every x it is possible to efficiently find a value w such that $(x, w) \notin R$. This property holds for all "interesting" relations that we know of. Note that if it does not hold, then a random witness is almost always a valid one, and so running such a proof is meaningless.)

Theorem 6.5.6 *Let π be a zero-knowledge proof of knowledge with negligible knowledge error for an $\mathcal{NP}$-relation R. Then, π securely computes the zero-knowledge functionality $\mathcal{F}_{\mathrm{ZK}}^R$ in the presence of malicious adversaries.*

Proof. Let $\mathcal{A}$ be an adversary. We separately consider the case where $\mathcal{A}$ corrupts the prover P and the case where $\mathcal{A}$ corrupts the verifier V. In the case where the verifier is corrupted, the simulator $\mathcal{S}$ for $\mathcal{F}_{\mathrm{ZK}}^R$ receives a bit $b \in \{0, 1\}$ from the trusted party. If $b = 0$, then $\mathcal{S}$ hands 0 to $\mathcal{A}$ as if coming from P and halts. Otherwise, if $b = 1$, then $\mathcal{S}$ runs the zero-knowledge simulator that is guaranteed to exist for π with $\mathcal{A}$ as the verifier. In the case where $b = 0$, the adversary $\mathcal{A}$ sees exactly what an honest prover would send. In the case where $b = 1$, by the security properties of the zero-knowledge simulator, the output generated by $\mathcal{S}$ is computationally indistinguishable from the output of $\mathcal{A}$ in a real execution with P. This therefore completes this corruption case.

We now consider the case where P is corrupted by $\mathcal{A}$. Intuitively, the simulator $\mathcal{S}$ works as follows:

1. $\mathcal{S}$ plays the honest verifier with $\mathcal{A}$ as the prover.

   a. If $\mathcal{A}$ sends 0 to the verifier in this execution, then $\mathcal{S}$ sends the trusted party computing $\mathcal{F}_{\mathrm{ZK}}^R$ an invalid witness w such that $(x, w) \notin R$. Then, $\mathcal{S}$ outputs whatever $\mathcal{A}$ outputs and halts.

   b. If $\mathcal{S}$, playing the honest verifier, is not convinced by the proof with $\mathcal{A}$, then it sends abort$_P$ to the trusted party computing $\mathcal{F}_{\mathrm{ZK}}^R$, outputs whatever $\mathcal{A}$ outputs and halts.

   c. If $\mathcal{S}$ is convinced by the proof with $\mathcal{A}$, then it records the output that $\mathcal{A}$ outputs and proceeds to the next step.

2. $\mathcal{S}$ runs the knowledge extractor K that is guaranteed to exist for π on the prover $\mathcal{A}$, and receives back a witness w such that $(x, w) \in R$. $\mathcal{S}$ then sends w to the trusted party computing $\mathcal{F}_{\mathrm{ZK}}^R$ and outputs the output of $\mathcal{A}$ recorded above.

The intuition behind this simulation is clear. In the case where $\mathcal{A}$ would not convince the verifier in a real execution, the same behavior (output of the value 0 or abort$_P$) is achieved in the simulation. However, in the case where $\mathcal{A}$ would convince the verifier, the simulator $\mathcal{S}$ has to send a valid witness w to the trusted party computing $\mathcal{F}_{\mathrm{ZK}}^R$. It therefore runs the knowledge extractor K to do this. However, K runs in expected time

$$\frac{|x|^c}{\epsilon(x) - \kappa(x)}$$

where $\epsilon(x)$ is the probability that $\mathcal{A}$ would convince the verifier and $\kappa(x)$ is the (negligible) knowledge error. Now, since $\mathcal{S}$ only runs the extractor in the case where $\mathcal{A}$ convinces it in the proof while $\mathcal{S}$ plays the honest verifier, we have that $\mathcal{S}$ only runs the extractor with probability $\epsilon(x)$. Thus, the expected running time of $\mathcal{S}$ is

$$\epsilon(x) \cdot \frac{|x|^c}{\epsilon(x) - \kappa(x)} = |x|^c \cdot \frac{\epsilon(x)}{\epsilon(x) - \kappa(x)}$$

It may be tempting at this point to conclude that the above is polynomial because $\kappa(x)$ is negligible, and so $\epsilon(x) - \kappa(x)$ is almost the same as $\epsilon(x)$. This is true for "large" values of $\epsilon(x)$. For example, if $\epsilon(x) > 2\kappa(x)$ then $\epsilon(x) - \kappa(x) > \epsilon(x)/2$. This then implies that $\epsilon(x)/(\epsilon(x) - \kappa(x)) < 2$. Unfortunately, however, this is *not* true in general. For example, consider the case where $\kappa(x) = 2^{-|x|}$ and $\epsilon(x) = \kappa(x) + 2^{-|x|/2} = 2^{-|x|} + 2^{-|x|/2}$. Then,

$$\frac{\epsilon(x)}{\epsilon(x) - \kappa(x)} = \frac{2^{-|x|} + 2^{-|x|/2}}{2^{-|x|/2}} = 2^{|x|/2} + 1,$$

which is exponential in $|x|$. In addition to the above problem, the guarantee regarding the running time of K and its success only holds if $\epsilon(x) > \kappa(x)$. Thus, if K runs for time $\epsilon(x)^{-2}$ whenever $\epsilon(x) \leq \kappa(x)$, we once again have a similar problem. For example, consider the case where $\kappa(x) = \epsilon(x) = 2^{-|x|}$. Then, the expected running time of $\mathcal{S}$ for such an $\mathcal{A}$ is

$$(1 - \epsilon(x)) \cdot \mathrm{poly}(|x|) + \epsilon(x) \cdot \frac{1}{\epsilon(x)^2} > \frac{1}{\epsilon(x)} = 2^{|x|}.$$

This technical problem was observed and solved by [34] in the context of zero-knowledge. We now show how to use their technique here.

Both of the problems described above are solved by ensuring that the extractor never runs "too long". Specifically, if $\mathcal{S}$ is convinced of the proof by $\mathcal{A}$, and so proceeds to the second step of the simulation, then it first estimates the value of $\epsilon(x)$, where $\epsilon(x)$ denotes the probability that $\mathcal{A}$ successfully proves that it knows a witness w such that $(x, w) \in R$. This is done by repeating the verification until $m(x)$ successful verifications occur for a large enough polynomial $m(\cdot)$. Then, an estimate $\tilde{\epsilon}$ of ϵ is taken to be m/T, where

T is the overall number of attempts until m successful verifications occurred. As we will see, this suffices to ensure that the probability that $\tilde{\epsilon}$ is not within a constant factor of $\epsilon(x)$ is at most $2^{-|x|}$. We show this using the following bound:

Lemma 6.5.7 (Tail inequality for geometric variables [43]): *Let $X_1, \ldots, X_m$ be m independent random variables with geometric distribution with probability ϵ (i.e., for every i, $Pr[X_i = j] = (1 - \epsilon)^{j-1} \cdot \epsilon$). Let $X = \sum_{i=1}^{m} X_i$ and let $\mu = E[X] = m/\epsilon$. Then, for every δ,*

$$\Pr[X \geq (1 + \delta)\mu] \leq e^{-\frac{m\delta^2}{2(1+\delta)}}.$$

Proof. In order to prove this lemma, we define a new random variable Y_α for any $\alpha \in \mathbb{N}$ as follows. Consider an infinite series of independent Bernoulli trials with probability ϵ (i.e., the probability of any given trial being 1 is ϵ). Then, write the results of these trials as a binary string and let Y_α be the number of 1s appearing in the prefix of length α. It is clear that

$$\mu_\alpha = E[Y_\alpha] = \alpha \cdot \epsilon.$$

Furthermore,
$$\Pr[X \geq (1 + \delta)\mu] = \Pr[Y_\alpha < m]$$

for $\alpha = (1+\delta)\mu$. In order to see why this holds, observe that one can describe the random variable $X = \sum_{i=1}^{m} X_i$ by writing an infinite series of Bernoulli trials with probability ϵ (as above), and then taking X to be the index of the mth 1 to appear in the string. Looking at it in this way, we have that X is greater than or equal to $(1 + \delta)\mu$ if and only if $Y_{(1+\delta)\mu} < m$ (because if $Y_{(1+\delta)\mu} < m$ then this means that m successful trials have not yet been obtained). Observe now that $\mu_\alpha = \alpha \cdot \epsilon$, $\alpha = (1 + \delta)\mu$, and $\mu = m/\epsilon$. Thus, $\mu_\alpha = (1 + \delta) \cdot m$. This implies that

$$\left(1 - \tfrac{\delta}{1+\delta}\right) \cdot \mu_\alpha = \left(1 - \tfrac{\delta}{1+\delta}\right) \cdot (1+\delta) \cdot m = (1+\delta) \cdot m - \delta \cdot m = m \, ,$$

and so
$$\Pr[Y_\alpha < m] = \Pr\left[Y_\alpha < \left(1 - \tfrac{\delta}{1+\delta}\right) \cdot \mu_\alpha\right].$$

Applying the Chernoff bound[3], we have that

$$\Pr\left[Y_\alpha < m\right] = \Pr\left[Y_\alpha < \left(1 - \tfrac{\delta}{1+\delta}\right)\mu_\alpha\right] < e^{-\frac{\mu_\alpha}{2} \cdot \left(\frac{\delta}{1+\delta}\right)^2}.$$

Once again using the fact that $\mu_\alpha = (1 + \delta) \cdot m$ we conclude that

---

[3] We use the following version of the Chernoff bound. Let $X_1, \ldots, X_m$ be independent Bernoulli trials where $\Pr[X_i = 1] = \epsilon$ for every i, and let $X = \sum_{i=1}^{m} X_i$ and $\mu = E[X] = m\epsilon$. Then, for every δ it holds that $\Pr[X < (1 - \beta)\mu] < e^{-\frac{\mu}{2} \cdot \beta^2}$.

$$\Pr[X \geq (1+\delta)\mu] = \Pr\left[Y_\alpha < m\right] < e^{-\frac{(1+\delta)m}{2}\cdot\left(\frac{\delta}{1+\delta}\right)^2} = e^{-\frac{m\delta^2}{2(1+\delta)}}$$

as required.   ∎

Define X_i to be the random variable that equals the number of attempts needed to obtain the ith successful verification (not including the attempts up until the $(i-1)$th verification), and let $\delta = \pm 1/2$. Clearly, each X_i has a geometric distribution with probability ϵ. It therefore follows that

$$\Pr\left[X \leq \frac{m}{2\epsilon} \vee X \geq \frac{3m}{2\epsilon}\right] \leq 2 \cdot \Pr\left[X \geq \frac{3}{2}\cdot\frac{m}{\epsilon}\right] \leq 2 \cdot e^{-\frac{m}{12}}.$$

Stated in words, the probability that the *estimate* $\tilde{\epsilon} = m/X$ is not between $2\epsilon/3$ and 2ϵ is at most $2e^{-m/12}$. Thus, if $m(x) = 12|x|$ it follows that the probability that $\tilde{\epsilon}$ is not within the above bounds is at most $2^{-|x|}$, as required.

Next, $\mathcal{S}$ repeats the following up to $|x|$ times: $\mathcal{S}$ runs the extractor K and answers all of K's oracle queries with the $\mathcal{A}$ as the prover. However, $\mathcal{S}$ limits the number of steps taken by K to $|x|^{c+1}/\tilde{\epsilon}$ steps, where c is the constant from the *knowledge soundness* (or *validity*) condition in the definition of proofs of knowledge (every extractor K has a single constant c associated with it and so $\mathcal{S}$ can use the appropriate c). Note that a "call" to $\mathcal{A}$ as the prover is counted by $\mathcal{S}$ as a single step. Now, if within this time K outputs a witness w, then $\mathcal{S}$ sends w to the trusted party computing $\mathcal{F}_{\mathrm{ZK}}^R$ and outputs the output of $\mathcal{A}$ that it first recorded. (We note that $\mathcal{S}$ does not need to check if w is a valid witness because by the definition of K, it only outputs valid witnesses.) If K does not output a witness within this time, then $\mathcal{S}$ aborts this attempt and tries again. As mentioned above, this is repeated up to $|x|$ times; we stress that in each attempt, K is given independent coins by $\mathcal{S}$. If the extractor K did not output a witness in any of the $|x|$ attempts, then $\mathcal{S}$ halts and outputs fail. We will show that this strategy ensures that the probability that $\mathcal{S}$ outputs fail is negligible. Therefore, the probability that the initial verification of the proof succeeded, yet $\mathcal{S}$ does not output a valid witness, is negligible.

In addition to the above, $\mathcal{S}$ keeps a count of the overall running time of K and if it reaches $2^{|x|}$ steps, it halts, outputting fail. (This additional time-out is needed to ensure that $\mathcal{S}$ does not run too long in the case where the estimate $\tilde{\epsilon}$ is not within a constant factor of $\epsilon(x)$. Recall that this "bad event" can only happen with probability $2^{-|x|}$.)

We first claim that $\mathcal{S}$ runs in expected polynomial time.

Claim 6.5.8 *Simulator $\mathcal{S}$ runs in expected time that is polynomial in $|x|$.*

Proof. Recall that $\mathcal{S}$ initially verifies the proof provided by $\mathcal{A}$. Since $\mathcal{S}$ merely plays an honest verifier, this takes a strict polynomial number of steps. Next, $\mathcal{S}$ obtains an estimate $\tilde{\epsilon}$ of $\epsilon(x)$. This involves repeating the verification until $m(|x|)$ successes are obtained. Therefore, the expected number of repetitions in order to obtain $\tilde{\epsilon}$ equals exactly $m(|x|)/\epsilon(x)$ (since the expected number of trials for a single success is $1/\epsilon(x)$). After the estimation $\tilde{\epsilon}$ has been obtained, $\mathcal{S}$ runs the extractor K for a maximum of $|x|$ times, each time for at most $|x|^{c+1}/\tilde{\epsilon}$ steps.

Given the above, we are ready to compute the expected running time of $\mathcal{S}$. In order to do this, we differentiate between two cases. In the first case, we consider what happens if $\tilde{\epsilon}$ is *not* within a constant factor of $\epsilon(x)$. The only thing we can say about $\mathcal{S}$'s running time in this case is that it is bound by $2^{|x|}$ (since this is an overall bound on its running time). However, since this event happens with probability at most $2^{-|x|}$, this case adds only a polynomial number of steps to the overall expected running time. We now consider the second case, where $\tilde{\epsilon}$ *is* within a constant factor of $\epsilon(x)$. In this case, we can bound the expected running time of $\mathcal{S}$ by

$$\text{poly}(|x|) \cdot \epsilon(x) \cdot \left(\frac{m(|x|)}{\epsilon(x)} + \frac{|x| \cdot |x|^{c+1}}{\tilde{\epsilon}} \right) = \text{poly}(|x|) \cdot \frac{\epsilon(x)}{\tilde{\epsilon}} = \text{poly}(|x|)$$

and this concludes the analysis. ∎

It is clear that the output of $\mathcal{S}$ is distributed exactly like the output of $\mathcal{A}$ in a real execution. This is because $\mathcal{S}$ just plays the honest verifier with $\mathcal{A}$ as the prover, and so the view of $\mathcal{A}$ in this simulation is identical to a real execution. Thus, the only problem that arises is if $\mathcal{S}$ accepts $\mathcal{A}$'s proof, but fails to obtain a valid witness. Notice that whenever $\mathcal{A}$'s proof is accepting, $\mathcal{S}$ runs the extractor K and either obtains a proper witness w or outputs fail. That is, in the case of accepting proofs, if $\mathcal{S}$ does not output fail, then it outputs a proper witness. Therefore, it suffices to show that the probability that $\mathcal{S}$ outputs fail is negligible.

Claim 6.5.9 *The probability that $\mathcal{S}$ outputs* fail *is negligible in $|x|$.*

Proof. Notice that the probability that $\mathcal{S}$ outputs fail is less than or equal to the probability that the extractor K does not succeed in outputting a witness w in any of the $|x|$ extraction attempts *plus* the probability that K runs for $2^{|x|}$ steps.

We first claim that the probability that K runs for $2^{|x|}$ steps is negligible. We have already shown in Claim 6.5.8 that $\mathcal{S}$ (and thus K) runs in expected polynomial time. Therefore, the probability that an execution will deviate so far from its expectation and run for $2^{|x|}$ steps is negligible. (It is enough to use Markov's inequality to establish this fact.)

We now continue by considering the probability that in all $|x|$ extraction attempts, the extractor K does not output a witness within $|x|^{c+1}/\tilde{\epsilon}$ steps. Consider the following two possible cases (recall that $\epsilon(x)$ equals the probability that $\mathcal{A}$ succeeds in proving the proof, and that κ is the negligible knowledge error function of the proof system):

1. *Case 1:* $\epsilon(x) \leq 2\kappa(x)$: In this case, $\mathcal{A}$ succeeds in proving the proof with only negligible probability. This means that the probability that $\mathcal{S}$ even

reaches the stage that it runs K is negligible (and thus $\mathcal{S}$ outputs fail with negligible probability only).

2. *Case 2:* $\epsilon(x) > 2\kappa(x)$: Recall that by the definition of proofs of knowledge, the constant c is such that the expected number of steps taken by K to output a witness is at most $|x|^c/(\epsilon(x) - \kappa(x))$. Now, since in this case $\epsilon(x) > 2\kappa(x)$, it holds that the expected number of steps required by K is less than $2|x|^c/\epsilon(x)$. Assuming that $\tilde{\epsilon}$ is within a constant factor of $\epsilon(x)$, we have that the expected number of steps is bound by $O(|x|^c/\tilde{\epsilon})$. Therefore, by Markov's inequality, the probability that K runs longer than $|x|^{c+1}/\tilde{\epsilon}$ steps in any given extraction attempt is at most $O(1/|x|)$. It follows that the probability that K runs longer than this time in $|x|$ independent attempts is negligible in $|x|$ (specifically, it is bound by $O(1/|x|)^{|x|}$). This covers the case where $\tilde{\epsilon}$ is within a constant factor of $\epsilon(x)$. However, the probability that $\tilde{\epsilon}$ is *not* within a constant factor of $\epsilon(x)$ is also negligible. Putting this together, we have that $\mathcal{S}$ outputs fail with negligible probability only.

∎

Combining the above two claims, together with the fact that the simulation by $\mathcal{S}$ is perfect when it does not output fail, we conclude that $\mathcal{S}$ is a valid simulator for the case where P is corrupted. Thus, π securely computes the $\mathcal{F}_{\mathrm{ZK}}^R$ functionality. ∎

6.6 Efficient Commitment Schemes from Σ-Protocols

In our constructions of zero-knowledge proofs and zero-knowledge proofs of knowledge in Sections 6.5.1 and 6.5.2, we relied on the existence of perfectly-hiding commitments and perfectly-hiding trapdoor commitments. In this section, we present a *general* construction of (trapdoor) commitment schemes from Σ-protocols. The advantage of the construction here is that it does not rely on any specific hardness assumption.

Hard relations. We typically use zero-knowledge for relations R with the property that given x it is "hard" to find a witness w such that $(x, w) \in R$. Otherwise, the verifier V could just find a witness by itself and would not need the prover. This is not entirely accurate. However, in the context of protocols for secure computation, this seems to almost always be the case. We formally define the notion of a hard relation in this light.

Definition 6.6.1 *An $\mathcal{NP}$-relation R is said to be hard if*

- **Efficient generation of instances:** *There exists a probabilistic polynomial-time algorithm G, called the* generator, *that on input 1^n outputs a pair $(x, w) \in R$ where $|x| = n$. Denote the first element of G's output (i.e., the x portion) by G_1.*

- **Hardness of generated instances:** *For every probabilistic polynomial-time algorithm $\mathcal{A}$ there exists a negligible function μ such that*

$$\Pr[(x, \mathcal{A}(G_1(1^n)) \in R] \le \mu(n).$$

In words, given x as output by G, no efficient $\mathcal{A}$ can find a witness w such that $(x, w) \in R$ with more than negligible probability.

Consider the following example of a hard relation. Let $\mathbb{G}$ be a group of order q with generator g, and let $R_{\mathrm{DL}} = \{(h, a) \mid h = g^a\}$. Then, consider an algorithm G that chooses a random $a \leftarrow \mathbb{Z}_q$ and outputs (g^a, a). Then, the relation R_{DL} is hard if and only if the standard discrete logarithm problem is hard in the group $\mathbb{G}$.

Commitments from hard relations. Assume that R is a hard relation with a generator G, and let π be a Σ-protocol for R. Assume also that it is easy to check membership in L_R. That is, given x it is easy to decide if there exists a w such that $(x, w) \in R$ (of course, since R is hard, it is hard to find such a w).

With this setup, we can build a perfectly hiding commitment scheme, which is as efficient as π and enables commitment to many bits at once; see Protocol 6.6.2.

PROTOCOL 6.6.2 (Commitment from Σ-Protocol)

- **Input:** The committer C and receiver R both hold 1^n, and the committer C has a value $e \in \{0, 1\}^t$.
- **The commit phase:**

  1. The receiver R runs (in private) the generator G on input 1^n to obtain $(x, w) \in R$, and sends x to C.
  2. C verifies that $x \in L_R$; if not, it aborts. If so, in order to commit to $e \in \{0, 1\}^t$, the committer C runs the Σ-protocol simulator M on input (x, e) and obtains a transcript (a, e, z). C then sends a to R.

- **The decommit phase:** In order to decommit, the committer C sends the remainder of the transcript (e, z) to R, which accepts e as the decommitted value if and only if (a, e, z) is an accepting transcript in π with respect to input x.

Observe that the generation of (x, w) and verification that $x \in L_R$ need to be run only once, and can be reused for many commitments.

Theorem 6.6.3 *Protocol 6.6.2 is a perfectly-hiding commitment scheme with computational binding.*

Proof. We first prove that the protocol is perfectly hiding. In order to see this, observe that in a real execution of the Σ-protocol π, the first message a by the prover is independent of the challenge e. Now, since $x \in L_R$, the simulation by M is perfect, by the definition of a Σ-protocol. Thus, the value a generated by M (upon input (x, e)) is independent of e. We now prove computational binding. Assume that some probabilistic polynomial-time cheating committer C^* can output a, and then later decommit to two different values $e \neq e'$, with non-negligible probability. This implies that C^* outputs a and two continuations $(e, z), (e', z')$ such that (a, e, z) and (a, e', z') are both accepting transcripts. By the special soundness property of Σ-protocols, we know that w can be computed efficiently from these transcripts. Thus, it is possible to construct an efficient algorithm that is given x and computes w such that $(x, w) \in R$ with non-negligible probability. This contradicts the assumption that R is a hard relation. ∎

Trapdoor commitments. In Section 6.5.2 we used an additional property of commitments in order to obtain a zero-knowledge *proof of knowledge*. Specifically, we needed there to be a trapdoor trap that was generated during the commit phase and known only to the receiver. The desired property is that given trap, the committer could generate a special commitment that looks exactly like a regular commitment, but that can be opened to any value it wishes. Observe that Protocol 6.6.2 is actually a *trapdoor commitment scheme* because if C has a witness w then it can run the *honest prover strategy* (and not the simulator M) in order to generate the commitment value a. Since C knows w, by perfect completeness, for any e it can generate the proper prover response z. In addition, by the special honest verifier zero-knowledge property, the commitment value a generated in this way is distributed exactly like a regular commitment value. We have the following:

Theorem 6.6.4 *Protocol 6.6.2 is a perfectly-hiding trapdoor commitment scheme with computational binding.*

6.7 Summary

Σ-protocols are an important and useful tool for building efficient secure protocols. The underlying structure behind many (if not most) zero-knowledge proofs for concrete number-theoretic languages is actually that of a Σ-protocol. The advantage in making this explicit is that it is possible to worry only about the relatively simpler problem of constructing the basic Σ-protocol, and then the transformations necessary for achieving zero-knowledge proofs of knowledge and for proving compound statements follow easily without sacrificing efficiency. We remark also that proving that a proof is a Σ-protocol is *much* easier than proving that it is a zero-knowledge proof of knowledge. In this chapter we have reviewed only a few of the results regarding Σ-protocols. Many other efficient transformations exist. For just one example, Σ-protocols have been used as a tool to significantly improve the efficiency of proving a large number of statements simultaneously [18].

Chapter 7
Oblivious Transfer and Applications

In this chapter we carry out an in-depth study of the problem of constructing efficient protocols for oblivious transfer (as introduced in Section 3.2.2). Oblivious transfer is one of the most important building blocks in cryptography, and is very useful for constructing secure protocols. We demonstrate this by showing how to achieve secure pseudorandom function evaluation using oblivious transfer.

The 1-out-of-2 oblivious transfer (OT) functionality, denoted by $\mathcal{F}_{\mathrm{OT}}$, is defined by $((x_0, x_1), \sigma) \mapsto (\lambda, x_\sigma)$ where λ denotes the empty string. That is, a sender has a pair of inputs (x_0, x_1) and a receiver has a bit σ. The aim of the protocol is for the receiver to receive x_σ (and x_σ only), without revealing anything about σ to the sender. Oblivious transfer is one of the basic building blocks of cryptographic protocols and its efficiency is a bottleneck in many protocols using it; see Chapter 4 for just one example. In this chapter we present three protocols for three different notions of security: privacy only, one-sided simulation and full simulation, as defined in Sections 2.6.1, 2.6.2 and 2.3, respectively.

Our starting point is the private and highly efficient protocol of [62] which relies on the hardness of the decisional Diffie-Hellman (DDH) problem. Thereafter, in Section 7.3 we show how to slightly modify this protocol so that it achieves one-sided simulatability. Then, in Section 7.4 we present another protocol for this task that is fully secure in the presence of malicious adversaries. The security of this protocol follows from the same hardness assumption (e.g., DDH) and the (exact) computation and communication costs are approximately doubled. We conclude with a construction for *batch* oblivious transfer which is of great importance for protocols such as that of Chapter 4, where many OT executions are run together. This protocol is considerably more efficient than just taking the previous protocol and running it many times in parallel.

In addition to the protocols for OT, we present the protocol of [28] for securely computing the pseudorandom function evaluation functionality, denoted by $\mathcal{F}_{\mathrm{PRF}}$, and defined by $(k, x) \mapsto (\lambda, F_{\mathrm{PRF}}(k, x))$. The aim of the pro-

C. Hazay, Y. Lindell, *Efficient Secure Two-Party Protocols*,
Information Security and Cryptography, DOI 10.1007/978-3-642-14303-8_7,
© Springer-Verlag Berlin Heidelberg 2010

tocol is for party P_2 to learn $F_{\mathrm{PRF}}(k, x)$ (and $F_{\mathrm{PRF}}(k, x)$) only), without revealing anything about x to P_1. We consider the concrete pseudorandom function of [64] and present the secure (slightly modified) construction of [28] for this specific function. We will use $\mathcal{F}_{\mathrm{PRF}}$ in later constructions. We remark that with one exception, the protocols do not utilize any specific properties of this function and can employ any other (possibly more efficient) secure protocol that implements the pseudorandom function evaluation functionality. All the protocols in this chapter achieve security in the presence of malicious adversaries.

7.1 Notational Conventions for Protocols

We now describe some notational conventions that are used in this and the subsequent chapters. The aim of our notation is to enable the reader to quickly identify if a protocol is secure in the semi-honest or covert model, or whether it provides privacy only, one-sided simulatability or full security in the presence of malicious adversaries. In addition, since our more complex protocols use a number of subprotocols, we include notation to identify the function being computed. The general template is

$$\pi_f^{\mathrm{S}}$$

where f is the function being computed and S is the level of security obtained. For example, protocols for computing the oblivious transfer functionality in the presence of semi-honest and covert models are denoted by $\pi_{\mathrm{OT}}^{\mathrm{SH}}$ and $\pi_{\mathrm{OT}}^{\mathrm{CO}}$, respectively. Regarding malicious adversaries, protocols achieving privacy only and one-sided simulatability are denoted by $\pi_{\mathrm{OT}}^{\mathrm{P}}$ and $\pi_{\mathrm{OT}}^{\mathrm{OS}}$, respectively. Protocols achieving *full simulation* are not given any superscript, and so π_{OT} denotes an OT protocol achieving full security in the presence of malicious adversaries.

7.2 Oblivious Transfer – Privacy Only

7.2.1 A Protocol Based on the DDH Assumption

We now present the protocol of [62] that computes $\mathcal{F}_{\mathrm{OT}}$ with privacy in the presence of malicious adversaries; see Definition 2.6.1 in Section 2.6.1 for the definition of private oblivious transfer. This protocol involves a sender S and a receiver R and is implemented in two rounds. Its high efficiency is due to the fact that the parties are not required to prove that they followed the protocol specification, and this is also why only privacy is achieved.

We assume that the parties have a probabilistic polynomial-time algorithm V that checks membership in $\mathbb{G}$ (i.e., for every h, $V(h) = 1$ if and only if $h \in \mathbb{G}$). This is easily achieved in typical discrete log groups.

PROTOCOL 7.2.1 (Private Oblivious Transfer π^P_{OT})

- **Inputs:** The sender has a pair of strings $x_0, x_1 \in \mathbb{G}$ and the receiver has a bit $\sigma \in \{0, 1\}$.
- **Auxiliary inputs:** Both parties have the security parameter 1^n and the description of a group $\mathbb{G}$ of *prime order*, including a generator g for the group and its order q. The group can be chosen by P_2 if not given as auxiliary input.
- **The protocol:**

  1. The receiver R chooses $\alpha, \beta, \gamma \leftarrow_R \{1, \ldots, q\}$ and computes $\bar{a}$ as follows:
     a. If $\sigma = 0$ then $\bar{a} = (g^\alpha, g^\beta, g^{\alpha\beta}, g^\gamma)$.
     b. If $\sigma = 1$ then $\bar{a} = (g^\alpha, g^\beta, g^\gamma, g^{\alpha\beta})$.
     R sends $\bar{a}$ to S.
  2. Denote the tuple $\bar{a}$ received by S by (x, y, z_0, z_1). Then, S checks that $x, y, z_0, z_1 \in \mathbb{G}$ and that $z_0 \neq z_1$. If not, it aborts outputting $\perp$. Otherwise, S chooses random $u_0, u_1, v_0, v_1 \leftarrow_R \{1, \ldots, q\}$ and computes the following four values:

     $$w_0 = x^{u_0} \cdot g^{v_0}, \qquad k_0 = (z_0)^{u_0} \cdot y^{v_0},$$
     $$w_1 = x^{u_1} \cdot g^{v_1}, \qquad k_1 = (z_1)^{u_1} \cdot y^{v_1}.$$

     S then encrypts x_0 under k_0 and x_1 under k_1. For the sake of simplicity, assume that one-time pad type encryption is used. That is, assume that x_0 and x_1 are mapped to elements of $\mathbb{G}$. Then, S computes $c_0 = x_0 \cdot k_0$ and $c_1 = x_1 \cdot k_1$ where multiplication is in the group $\mathbb{G}$.
     S sends R the pairs (w_0, c_0) and (w_1, c_1).
  3. R computes $k_\sigma = (w_\sigma)^\beta$ and outputs $x_\sigma = c_\sigma \cdot (k_\sigma)^{-1}$.

The security of Protocol 7.2.1 rests on the decisional Diffie-Hellman (DDH) assumption that states that tuples of the form $(g^\alpha, g^\beta, g^\gamma)$ where $\alpha, \beta, \gamma \leftarrow_R \{1, \ldots, q\}$ are indistinguishable from tuples of the form $(g^\alpha, g^\beta, g^{\alpha\beta})$ where $\alpha, \beta \leftarrow_R \{1, \ldots, q\}$ (recall that q is the order of the group $\mathbb{G}$ that we are working in).[1] This implies that an adversarial sender S^* cannot discern whether the message sent by R is $(g^\alpha, g^\beta, g^{\alpha\beta}, g^\gamma)$ or $(g^\alpha, g^\beta, g^\gamma, g^{\alpha\beta})$ and so R's input is hidden from S^*. The motivation for privacy in the case where R^* is malicious is more difficult and it follows from the fact that – informally speaking – the exponentiations computed by S in Step 2 completely randomize the

---

[1] Formally, we consider the game where a PPT distinguisher D_{DDH} is given a description of a prime order group $\mathbb{G}$, a generator g and a tuple $\langle x, y, z \rangle$ and outputs a bit b. We say that the DDH problem is hard relative to $\mathbb{G}$ if the probability D_{DDH} outputs b when given a Diffie-Hellman tuple is essentially the same as when given a tuple of random elements. See [49] for a formal definition of the DDH problem.

triple $(g^\alpha, g^\beta, g^\gamma)$ when $\gamma \neq \alpha \cdot \beta$, and so $w_{1-\sigma}$ is independent of $x_{1-\sigma}$. Interestingly, it is still possible for R to derive the key k_σ that results from the randomization of a Diffie-Hellman tuple of the form $(g^\alpha, g^\beta, g^{\alpha\beta})$. None of these facts are evident from the protocol itself but are demonstrated below in the proof. We therefore proceed directly to prove the following theorem:

Theorem 7.2.2 *Assume that the decisional Diffie-Hellman problem is hard in $\mathbb{G}$ with generator g. Then, Protocol 7.2.1 is a private oblivious transfer, as in Definition 2.6.1.*

Proof. The first requirement of Definition 2.6.1 is that of non-triviality, and we prove this first. Let x_0, x_1 be S's input and let σ be R's input. The pair (w_σ, c_σ) sent by S to R is defined as $w_\sigma = x^{u_\sigma} \cdot g^{v_\sigma}$ and $c_\sigma = x_\sigma \cdot k_\sigma$ where $k_\sigma = (z_\sigma)^{u_\sigma} \cdot y^{v_\sigma}$. Non-triviality follows from the fact that

$$(w_\sigma)^\beta = x^{u_\sigma \cdot \beta} \cdot g^{v_\sigma \cdot \beta} = g^{(\alpha \cdot \beta) \cdot u_\sigma} \cdot g^{\beta \cdot v_\sigma} = (z_\sigma)^{u_\sigma} \cdot y^{v_\sigma} = k_\sigma$$

where the second equality is because $x = g^\alpha$ and the third equality is due to the fact that $z_\sigma = g^{\alpha\beta}$. Thus, R recovers the correct key k_σ and can compute $(k_\sigma)^{-1}$, implying that it obtains the correct value when computing $x_\sigma = c_\sigma \cdot (k_\sigma)^{-1}$.

Next, we prove the privacy requirement for the case of a malicious S^*. Recall that this requirement is that S^*'s view when R has input 0 is indistinguishable from its view when R has input 1. Now, the view of an adversarial sender S^* in Protocol 7.2.1 consists merely of R's first message $\bar{a}$. By the DDH assumption, we have that

$$\left\{(g^\alpha, g^\beta, g^{\alpha\beta})\right\}_{\alpha,\beta \leftarrow_R \{1,\dots,q\}} \overset{c}{\equiv} \left\{(g^\alpha, g^\beta, g^\gamma)\right\}_{\alpha,\beta,\gamma \leftarrow_R \{1,\dots,q\}}.$$

Now, assume by contradiction that there exists a probabilistic polynomial-time distinguisher D and a non-negligible function ϵ such that for every n

$$\left|\Pr[D(g^\alpha, g^\beta, g^{\alpha\beta}, g^\gamma) = 1] - \Pr[D(g^\alpha, g^\beta, g^\gamma, g^{\alpha\beta}) = 1]\right| \geq \epsilon(n)$$

where $\alpha, \beta, \gamma \leftarrow_R \{1, \dots, q\}$. Then, by subtracting and adding

$$\Pr[D(g^\alpha, g^\beta, g^\gamma, g^\delta) = 1]$$

we have,

$$\left|\Pr[D(g^\alpha, g^\beta, g^{\alpha\beta}, g^\gamma) = 1] - \Pr[D(g^\alpha, g^\beta, g^\gamma, g^{\alpha\beta}) = 1]\right|$$
$$\leq \left|\Pr[D(g^\alpha, g^\beta, g^{\alpha\beta}, g^\gamma) = 1] - \Pr[D(g^\alpha, g^\beta, g^\gamma, g^\delta) = 1]\right|$$
$$+ \left|\Pr[D(g^\alpha, g^\beta, g^\gamma, g^\delta) = 1] - \Pr[D(g^\alpha, g^\beta, g^\gamma, g^{\alpha\beta}) = 1]\right|$$

where $\alpha, \beta, \gamma, \delta \leftarrow_R \{1, \dots, q\}$. Therefore, by the contradicting assumption,

$$\left|\Pr[D(g^{\alpha},g^{\beta},g^{\alpha\beta},g^{\gamma})=1]-\Pr[D(g^{\alpha},g^{\beta},g^{\gamma},g^{\delta})=1]\right|\geq\frac{\epsilon(n)}{2} \quad (7.1)$$

or

$$\left|\Pr[D(g^{\alpha},g^{\beta},g^{\gamma},g^{\delta})=1]-\Pr[D(g^{\alpha},g^{\beta},g^{\gamma},g^{\alpha\beta})=1]\right|\geq\frac{\epsilon(n)}{2}. \quad (7.2)$$

Assume that (7.1) holds. We construct a distinguisher D' for the DDH problem that works as follows. Upon input $\bar{a}=(x,y,z)$, the distinguisher D' chooses a random $\delta\leftarrow_R\{1,\ldots,q\}$ and hands D the tuple $\bar{a}'=(x,y,z,g^{\delta})$. The key observation is that on the one hand, if $\bar{a}=(g^{\alpha},g^{\beta},g^{\gamma})$ then $\bar{a}'=(g^{\alpha},g^{\beta},g^{\gamma},g^{\delta})$. On the other hand, if $\bar{a}=(g^{\alpha},g^{\beta},g^{\alpha\beta})$ then $\bar{a}'=(g^{\alpha},g^{\beta},g^{\alpha\beta},g^{\delta})$. Noting that in this last tuple γ does not appear, and γ and δ are distributed identically, we have that $\bar{a}'=(g^{\alpha},g^{\beta},g^{\alpha\beta},g^{\gamma})$. Thus,

$$\begin{aligned}
&\left|\Pr[D'(g^{\alpha},g^{\beta},g^{\alpha\beta})=1]-\Pr[D'(g^{\alpha},g^{\beta},g^{\gamma})=1]\right|\\
&=\left|\Pr[D(g^{\alpha},g^{\beta},g^{\alpha\beta},g^{\gamma})=1]-\Pr[D(g^{\alpha},g^{\beta},g^{\gamma},g^{\delta})=1]\right|\\
&\geq\frac{\epsilon(n)}{2}
\end{aligned}$$

in contradiction to the DDH assumption. A similar analysis follows in the case where (7.2) holds. It therefore follows that ϵ must be a negligible function. The proof of R's privacy is concluded by noting that $(g^{\alpha},g^{\beta},g^{\alpha\beta},g^{\gamma})$ is exactly the distribution over R's message when $\sigma=0$ and $(g^{\alpha},g^{\beta},g^{\gamma},g^{\alpha\beta})$ is exactly the distribution over R's message when $\sigma=1$. Thus, the privacy of R follows from the DDH assumption over the group in question.

It remains to prove privacy in the case of a malicious R^*. Let R^* be a non-uniform deterministic receiver with auxiliary input z (our proof will hold even for R^* that is computationally unbounded). Let $\bar{a}=(x,y,z_0,z_1)$ denote $R^*(z)$'s first message, and let α and β be such that $x=g^{\alpha}$ and $y=g^{\beta}$ (we cannot efficiently compute α and β from R^*'s message, but they are well defined because S checks that $x,y,z_0,z_1\in\mathbb{G}$). If $z_0=z_1$ then S sends nothing and so clearly the requirement in Definition 2.6.1 holds. Otherwise, let $\tau\in\{0,1\}$ be such that $z_{\tau}\neq g^{\alpha\beta}$ (note that since $z_0\neq z_1$ it cannot be that both $z_0=g^{\alpha\beta}$ and $z_1=g^{\alpha\beta}$). The sender S's security is based on the following claim:

Claim 7.2.3 *Let $x=g^{\alpha}$, $y=g^{\beta}$ and $z_{\tau}=g^{\gamma}\neq g^{\alpha\beta}$. Then, given α, β and γ, the pair of values (w_{τ},k_{τ}), where $w_{\tau}=x^{u_{\tau}}\cdot g^{v_{\tau}}$ and $k_{\tau}=(z_{\tau})^{u_{\tau}}\cdot y^{v_{\tau}}$, is uniformly distributed when u_{τ},v_{τ} are chosen uniformly in $\{1,\ldots,q\}$.*

Proof. We prove that for every $(\mu_0,\mu_1)\in\mathbb{G}\times\mathbb{G}$,

$$\Pr[w_{\tau}=\mu_0\wedge k_{\tau}=\mu_1]=\frac{1}{|\mathbb{G}|^2}, \quad (7.3)$$

where the probability is taken over random choices of u_τ, v_τ, and w_τ, k_τ are computed according to the honest sender's instruction based on the message $\bar{a}$ from R^* (this is equivalent to the stated claim). Recall first that

$$w_\tau = x^{u_\tau} \cdot g^{v_\tau} = g^{\alpha u_\tau} \cdot g^{v_\tau} = g^{\alpha u_\tau + v_\tau},$$

and that

$$k_\tau = (z_\tau)^{u_\tau} \cdot y^{v_\tau} = g^{\gamma u_\tau} \cdot g^{\beta v_\tau} = g^{\gamma u_\tau + \beta v_\tau}$$

where the above holds because $x = g^\alpha$, $y = g^\beta$ and $z_\tau = g^\gamma$. Let ϵ_0 and ϵ_1 be such that $\mu_0 = g^{\epsilon_0}$ and $\mu_1 = g^{\epsilon_1}$. Then, since u_τ and v_τ are uniformly distributed in $\{1, \ldots, q\}$, we have that (7.3) holds if and only if there is a single solution to the equations

$$\alpha \cdot u_\tau + v_\tau = \epsilon_0 \quad \text{and} \quad \gamma \cdot u_\tau + \beta \cdot v_\tau = \epsilon_1.$$

Now, there exists a single solution to these equations if and only if the matrix

$$\begin{pmatrix} \alpha & 1 \\ \gamma & \beta \end{pmatrix}$$

is invertible, which is the case here because its determinant is $\alpha \cdot \beta - \gamma$ and by the assumption $\alpha \cdot \beta \neq \gamma$ and so $\alpha \cdot \beta - \gamma \neq 0$. This completes the proof. ∎

Proving this claim, it follows that k_τ is uniformly distributed, even given w_τ (and $k_{1-\tau}, w_{1-\tau}$). Thus, for every $x, x_\tau \in \mathbb{G}$, $k_\tau \cdot x_\tau$ is distributed identically to $k_\tau \cdot x$. This completes the proof of the sender's privacy. ∎

Exact efficiency. Recall that we measure the protocols in this book according to three parameters; round complexity, bandwidth and asymmetric computations (i.e., the number of modular exponentiations computed relative to some finite group) According to these, Protocol 7.2.1 is extremely efficient. Specifically, the number of rounds of communication is only two. Furthermore, the total number of modular exponentiations is thirteen (eight by the sender and five by the receiver), and each party exchanges four group elements.

7.2.2 A Protocol from Homomorphic Encryption

In this section we show how to construct a private oblivious transfer protocol from any homomorphic encryption scheme, as in Definition 5.1.1. Recall that this is an encryption scheme such that for any public key pk the plaintext space $\mathcal{M}$ and ciphertext space $\mathcal{C}$ are *additive groups*, and it is possible to efficiently compute $E_{pk}(m_1 + m_2)$ given pk, $c_1 = E_{pk}(m_1)$ and $c_2 = E_{pk}(m_2)$.

Such encryption schemes can be constructed under the quadratic residuosity, decisional Diffie-Hellman and other assumptions; see [2, 42] for some references. The idea behind the protocol is very simple. The receiver encrypts its input bit, and the server uses the homomorphic properties of the encryption scheme in order to construct a pair of ciphertexts so that if $\sigma = 0$ then the receiver can decrypt the first ciphertext and the second is random, and vice versa if $\sigma = 1$. As we will see in the proof, this idea works as long as the sender can verify that the ciphertext it receives from the receiver encrypts a value in the plaintext group $\mathcal{M}$.

Formally, we call a homomorphic encryption scheme efficiently verifiable if the following conditions all hold:

1. the public-key pk fully defines $\mathcal{M}$ and there is an efficient 1–1 mapping from $\{0,1\}^n$ to $\mathcal{M}$ for pk that is generated with security parameter 1^n,
2. given a ciphertext c and public key pk it is possible to efficiently verify that c encrypts a plaintext $\alpha \in \mathcal{M}$ such that $\alpha = 0$ or α has a *multiplicative inverse* (recall that $\mathcal{M}$ is an additive group and so only elements α with order that is relatively prime to the group order q have this property), and
3. it is possible to efficiently verify that a public-key pk is "valid", meaning that the homomorphic properties always work.

We remark that the assumptions of efficient verifiability can be removed in a number of ways. First, it is possible to prove the required properties in zero-knowledge. Efficient proofs for such statements do exist, and so this can sometimes be carried out without too much additional cost. In addition, if the plaintext group $\mathcal{M}$ is of *prime order*, then every ciphertext that encrypts a value in the group has the required property. In this case, it suffices to be able to efficiently verify (or prove in zero-knowledge) that a ciphertext c encrypts any value in the group. See Protocol 7.2.4 for the full description.

PROTOCOL 7.2.4 (Private Oblivious Transfer $\pi'^{\mathrm{P}}_{\mathrm{OT}}$)

- **Inputs:** The sender has a pair of strings $x_0, x_1 \in \{0,1\}^n$ and the receiver has a bit $\sigma \in \{0,1\}$.
- **Auxiliary inputs:** Both parties have the security parameter 1^n.
- **The protocol:**

  1. The receiver R chooses a pair of keys $(pk, sk) \leftarrow G(1^n)$, computes $c = E_{pk}(\sigma)$ and sends c and p_k to S.
  2. The sender S verifies that pk is a valid public key and that c encrypts either 0 or a value with a multiplicative inverse in the plaintext group $\mathcal{M}$. If both checks pass, then S maps x_0 and x_1 into $\mathcal{M}$ and then uses the homomorphic property of the encryption scheme, and its knowledge of x_0 and x_1, to compute two random encryptions $c_0 = E_{pk}((1-\sigma) \cdot x_0 + r_0 \cdot \sigma)$ and $c_1 = E_{pk}(\sigma \cdot x_1 + r_1 \cdot (1-\sigma))$ where $r_0, r_1 \leftarrow_R \mathcal{M}$ are random elements in the plaintext group.
  3. R computes and outputs $s_r = D_{sk}(c')$.

We have the following theorem:

Theorem 7.2.5 *Assume that the encryption scheme (G, E, D) is indistinguishable under chosen-plaintext attacks and is efficiently verifiable. Then, Protocol 7.2.4 is a private oblivious transfer, as in Definition 2.6.1.*

Proof. We begin by proving non-triviality. In order to see this, observe that if $\sigma = 0$ then $c_0 = E_{pk}(1 \cdot x_0 + 0 \cdot s_1) = E_{pk}(x_0)$ and if $\sigma = 1$ then $c_1 = E_{pk}(1 \cdot x_1 + (1 - 1) \cdot r_1) = E_{pk}(x_1)$.

We now prove privacy in the case of a malicious sender S^*. Privacy in this case is derived from the fact that S^*'s entire view consists of an encryption $c = E_{pk}(\sigma)$. Furthermore, the encryption key and ciphertext are generated honestly by R. Thus, the ability to distinguish between an execution where R has input 0 and where R has input 1 is equivalent to the ability to distinguish between the distributions $\{pk, E_{pk}(0)\}$ and $\{pk, E_{pk}(1)\}$, in contradiction to the semantic security of the encryption scheme.

It remains to prove privacy in the case of a malicious receiver. Let R^* be an arbitrary sender, and let z be an auxiliary input. Then $R^*(z)$'s first message consists of fixed public key pk and ciphertext c. There are three cases:

1. *The checks regarding the validity of the public key and the fact that c encrypts a value in $\mathcal{M}$ do not pass:* In this case, the sender does nothing and so clearly the distributions over inputs (x_0, x_1) and (x_0, x) are identical for all $x_0, x_1, x \in \{0, 1\}^n$.
2. *The checks pass and $c = E_{pk}(0)$:* In this case, $c_1 = E_{pk}(r_1)$ and so R^*'s view is independent of x_1. That is, R^*'s view when S has input (x_0, x_1) is *identical* to its view when S has input (x_0, x).
3. *The checks pass and $c = E_{pk}(1)$:* In this case, $c_0 = E_{pk}(r_0)$ and so R^*'s view is independent of x_0. Thus, as above, R^*'s view when S has input (x_0, x_1) is identical to its view when S has input (x, x_1).
4. *The checks pass and $c \notin \{E_{pk}(0), E_{pk}(1)\}$:* Let $\alpha \in \mathcal{M}$ be such that $c = E_{pk}(\alpha)$. Then, $c_0 = E_{pk}((1-\alpha)x_0 + r_0 \cdot \alpha)$. Now, for every $x \in \mathcal{M}$, we claim that there exists a value r_0' such that $(1 - \alpha)x + r_0' \cdot \alpha = (1 - \alpha)x_0 + r_0 \cdot \alpha$. This holds by taking
$$r_0' = \frac{(1 - \alpha) \cdot x_0 + r_0 \cdot \alpha - (1 - \alpha) \cdot x}{\alpha}.$$

The key point is that such an r_0' exists as long as α has a multiplicative inverse in $\mathcal{M}$ which is guaranteed by the efficient verifiability and checks carried out by S. Since r_0 is chosen randomly, the above demonstrates that the distribution of the view of R^* when S has input (x_0, x_1) is *identical* to its view when S has input (x, x_1).

This completes the proof.   ∎

On the necessity of efficient verifiability. The fact that efficient verifiability is needed for the proof to go through is clear. However, it is natural to ask whether this is a real concern for known homomorphic encryption schemes. We show that this is indeed the case, and in particular, it must be possible to check that a given ciphertext encrypts a value with a multiplicative inverse in the plaintext group $\mathcal{M}$. That is, if the homomorphic encryption scheme of Paillier [68] is used and if the receiver sends an encryption of a value α that is not relatively prime to the modulus N, then it is actually possible for the receiver to cheat. Recall that the plaintext group $\mathcal{M}$ in the case of Paillier equals Z_N, where $N = pq$. Consider a malicious R^* that sends $c = E_{pk}(\alpha)$ where $\alpha = 0 \bmod p$ and $\alpha = 1 \bmod q$. (Such a value can be found efficiently using the extended Euclidean algorithm. Specifically, it is possible to find X and Y such that $Xp + Yq = 1$, and now define $\alpha = Xp = 1 - Yq$. It then follows that $\alpha = 0 \bmod p$ and $\alpha = 1 \bmod q$, as required.) Now, S sends c_0 that is an encryption of $\gamma_0 = (1 - \alpha) \cdot x_0 + r_0 \cdot \alpha$. Using the fact that $\alpha = Xp$ and $1 - \alpha = Yq$ we have that $\gamma_0 = Yq \cdot x_0 + r_0 \cdot Xp$. Thus $\gamma_0 = Yq \cdot x_0 \bmod p$. However, R^* knows Yq and so can obtain $x_0 \bmod p$. Likewise, c_1 is an encryption of $\gamma_1 = \alpha \cdot x_1 + r_1 \cdot (1 - \alpha) = Xp \cdot x_1 + r_1 \cdot Yq$. Thus, $\gamma_1 = Xp \cdot x_1 \bmod q$, and R^* can learn $x_1 \bmod q$. We conclude that R^* learns $x_0 \bmod p$ and $x_1 \bmod q$, in contradiction to the privacy requirement.

This problem can be solved by having R^* prove in zero-knowledge that it encrypted a value that is relatively prime to N. This ensures that the encrypted value has a multiplicative inverse, as required. Alternatively, it is possible for R^* to simply prove that it encrypted either 0 or 1; fortunately, efficient zero-knowledge proofs for this latter task are known.

Efficiency. Protocol 7.2.4 has two rounds of communication. The computational cost involves R choosing a key pair (pk, sk) and computing one encryption and one decryption. In addition, S carries out ten homomorphic operations (six scalar multiplications and four additions).

7.3 Oblivious Transfer – One-Sided Simulation

In this section we present an oblivious transfer protocol that is secure with one-sided simulation, by introducing a minor change to Protocol 7.2.1. Recall that by Definition 2.6.2, simulation must be possible for the party that receives output. We modify Protocol 7.2.1 as follows. In Step 1 where the receiver computes and sends the tuple $\bar{a} = (g^\alpha, g^\beta, \cdot, \cdot)$, we add a zero-knowledge proof of knowledge for α. That is, the receiver proves that it knows α by running a proof for the relation

$$\mathcal{R}_{\mathrm{DL}} = \left\{ ((\mathbb{G}, q, g, h), \alpha) \mid h = g^\alpha \right\}$$

where it is to be understood that $\mathbb{G}$ is of order q, that q is prime and that $\alpha \in \mathbb{Z}_q$. This, in turn, enables the simulator to extract the input bit of the malicious receiver by simply verifying whether it is the third or the fourth element of $\bar{a}$ that equals $(g^\beta)^\alpha = g^{\alpha\beta}$. A protocol for this task is presented in Chapter 6, by applying the transformation of Protocol 6.5.4 (zero-knowledge proof of knowledge from Σ-protocols) to Protocol 6.1.1 (Σ-protocol for $\mathcal{R}_{\mathrm{DL}}$). This protocol requires nine additional exponentiations overall, five rounds of communication, and the exchange of six additional group elements. As in Protocol 7.2.1, we assume that the parties have a probabilistic polynomial-time algorithm V that checks membership in $\mathbb{G}$ (i.e., for every h, $V(h) = 1$ if and only if $h \in \mathbb{G}$), and thus S verifies the membership of all elements within $\bar{a}$. Let $\pi_{\mathrm{OT}}^{\mathrm{OS}}$ denote Protocol 7.2.1 with the above additions. Then we have,

Theorem 7.3.1 *Assume that the DDH problem is hard in $\mathbb{G}$ with generator g. Then, Protocol $\pi_{\mathrm{OT}}^{\mathrm{OS}}$ securely computes $\mathcal{F}_{\mathrm{OT}}$ with one-sided simulation.*

Proof (sketch). Note first that the proof for the case of a malicious sender S^* is identical to the proof of Theorem 7.2.2, with the exception that the distinguisher D for DDH invokes the simulator for the zero-knowledge proof of $\mathcal{R}_{\mathrm{DL}}$ since it does not know α.

The proof for the case of a malicious receiver R^* is in a hybrid model where a trusted party is used to compute an ideal functionality for the zero-knowledge proof of knowledge for $\mathcal{R}_{\mathrm{DH}}$; see Chapter 6. We construct a simulator $\mathcal{S}_{\mathrm{REC}}$ for a malicious receiver R^* as follows.

1. $\mathcal{S}_{\mathrm{REC}}$ invokes R^* upon its input and receives from $\mathcal{A}$ its tuple $\bar{a}$ and its input $(\mathbb{G}, q, g, h, \alpha)$ to the trusted party computing the zero-knowledge functionality $\mathcal{F}_{\mathrm{ZK}}^{\mathrm{DL}}$. If the conditions for outputting $(1, \lambda)$ are not met then $\mathcal{S}_{\mathrm{REC}}$ sends $\perp$ to the trusted party for $\mathcal{F}_{\mathrm{OT}}$ and aborts.
2. Otherwise, $\mathcal{S}_{\mathrm{REC}}$ sets $\sigma = 0$ if $\bar{a} = (g^\alpha, \tilde{h}, \tilde{h}^\alpha, \cdot)$ and $\sigma = 1$ otherwise. Then, $\mathcal{S}_{\mathrm{REC}}$ sends σ to the trusted party computing the oblivious transfer functionality and receives back the string x_σ.
3. $\mathcal{S}_{\mathrm{REC}}$ computes (w_σ, c_σ) as an honest S would, using the value x_σ. In contrast, it computes $(w_{1-\sigma}, c_{1-\sigma})$ using $x_{1-\sigma} = 1$.
4. $\mathcal{S}_{\mathrm{REC}}$ outputs whatever $\mathcal{A}$ outputs and halts.

We continue by proving that

$$\left\{ \mathrm{IDEAL}_{\mathcal{F}_{\mathrm{OT}}, \mathcal{S}_{\mathrm{REC}}(z), R}((x_0, x_1), \sigma, n) \right\} \stackrel{\mathrm{c}}{\equiv} \left\{ \mathrm{HYBRID}_{\pi_{\mathrm{OT}}^{\mathrm{OS}}, \mathcal{A}(z), R}^{\mathrm{ZK}}((x_0, x_1), \sigma, n) \right\}$$

Since the honest S does not have any output, it suffices to show that the distribution over R^*'s output is computationally indistinguishable in a hybrid execution with S and in the simulation with $\mathcal{S}_{\mathrm{REC}}$ (we actually show that the distributions are identical). It is clear that until the last message, $\mathcal{A}$'s view is identical (all the messages until the last are independent of S's input). Furthermore, the only difference is in the way $(w_{1-\sigma}, c_{1-\sigma})$ are constructed. Thus, all we need to show is that for *every* $x_{1-\sigma}$, the distribution over $(w_{1-\sigma}, c_{1-\sigma})$ is the same when constructed by an honest S and $\mathcal{S}_{\text{REC}}$. We do this separately for $\sigma \in \{0, 1\}$ and the case where σ is neither 1 nor 0. We distinguish between the values generated by the simulator $\mathcal{S}_{\text{REC}}$ and those generated by a real S by adding a tilde to the values generated by $\mathcal{S}_{\text{REC}}$.

1. *Case $\sigma \in \{0, 1\}$:* For simplicity we take the concrete case where $\sigma = 0$ (the case where $\sigma = 1$ follows symmetrically). In this case we know that $\bar{a} = (x, y, z_0, z_1)$ where $z_0 = y^\alpha$. We have that the distributions generated over (w_1, c_1) are

   - *generated by the simulator $\mathcal{S}_{\text{REC}}$:* $\tilde{w}_1 = x^{\tilde{u}_1} \cdot g^{\tilde{v}_1}$ and $\tilde{c}_1 = z_1^{\tilde{u}_1} \cdot y^{\tilde{v}_1}$ (recall that $\mathcal{S}_{\text{REC}}$ uses $x_1 = 1$),
   - *generated by the honest sender S:* $w_1 = x^{u_1} \cdot g^{v_1}$ and $c_1 = x_1 \cdot z_1^{u_1} \cdot y^{v_1}$.

   To show that the distributions over (w_1, c_1) and $(\tilde{w}_1, \tilde{c}_1)$ are *identical*, we first let $x = g^r$, $y = g^s$, $z_1 = g^t$ and $x_1 = g^\ell$. (Such values are guaranteed because $x, y, z_1 \in \mathbb{G}$ are checked by $\mathcal{S}_{\text{REC}}$ and the honest S.) Next, let $\tilde{v}_1 = v_1 - r\ell/(t - sr)$ and $\tilde{u}_1 = u_1 + \ell/(t - sr)$. For *fixed* r, s, t and ℓ and uniformly distributed u_1, v_1, the values $\tilde{u}_1, \tilde{v}_1$ chosen in this way are also uniformly distributed. Plugging these values in, we have that

   $$\tilde{w}_1 = x^{\tilde{u}_1} \cdot g^{\tilde{v}_1} = g^{ru_1 + r\ell/(t-sr) + v_1 - r\ell/(t-sr)} = x^{u_1} \cdot g^{v_1}$$

   where the second equality is because $x = g^r$. Furthermore,

   $$\tilde{c}_1 = z_1^{\tilde{u}_1} \cdot y^{\tilde{v}_1} = g^{tu_1 + t\ell/(t-sr) + sv_1 - sr\ell/(t-sr)} = x_1 \cdot z_1^{u_1} \cdot y^{v_1}$$

   because $z = g^t$ and $y = g^s$. Observe now that these are the *exact values* w_1 and c_1 generated by an honest S. Thus, the distribution viewed by $\mathcal{A}$ in this case is the same in the hybrid and simulated executions.

2. *Case $\sigma \notin \{0, 1\}$:* In this case the simulator fixes $\sigma = 1$; thus the computation of (w_1, c_1) is as in the hybrid execution. The proof in which the view of the malicious receiver is identical in both executions, given that $\mathcal{S}_{\text{REC}}$ enters $x_0 = 1$, is as in the previous case.

This concludes the proof.   ∎

Exact efficiency. The complexity of this protocol is the same as Protocol 7.2.1 with the addition of the zero-knowledge proof of knowledge of the discrete log, and the verification that all elements of $\bar{a}$ are in $\mathbb{G}$. Therefore, the number of rounds of communication is six (Protocol 7.2.1 has two rounds and the zero-knowledge proof of knowledge has five, but the first message of the zero-knowledge protocol can be sent together with the receiver's first message in the OT protocol). Furthermore, the total number of modular exponentiations is 22, and overall 14 group elements are exchanged.

7.4 Oblivious Transfer – Full Simulation

In this section we present a protocol for oblivious transfer that is fully secure under the DDH assumption and has only a *constant* number of exponentiations and a *constant* number of rounds. The following protocol uses a zero-knowledge proof of knowledge that a given tuple is a Diffie-Hellman tuple. A protocol for this task is presented in Chapter 6, by applying the transformation of Protocol 6.5.4 (zero-knowledge proof of knowledge from Σ-protocols) to Protocol 6.2.4 (Σ-protocol for the relation of Diffie-Hellman tuples). This protocol has five rounds of communication and 12 exponentiations.

7.4.1 1-out-of-2 Oblivious Transfer

In order to motivate the protocol, we begin by discussing the reason that the protocol π_{OT}^{OS} for one-sided simulation does not achieve full simulation. Clearly, the problem is with respect to simulating the case of a corrupted sender (because a corrupted receiver can be simulated). Recall that if R sends $\bar{a} = (x, y, z_0, z_1)$ where $x = g^\alpha$ and $y = g^\beta$, then for $z_\tau = g^\gamma \neq g^{\alpha\beta}$ it holds that w_τ, k_τ are uniformly distributed and independent of the input x_τ; see Claim 7.2.3. Furthermore, since the sender checks that $z_0 \neq z_1$ it is only possible that at most one of z_0, z_1 equals $g^{\alpha\beta}$. Now, consider a simulator that constructs and sends $\bar{a}$ to an adversary $\mathcal{A}$ that has corrupted the sender S. The simulator must hand $\mathcal{A}$ a message $\bar{a} = (x, y, z_0, z_1)$ and it too can only make one of z_0, z_1 equal $g^{\alpha\beta}$. Therefore, like an honest receiver, the simulator can also only obtain one of the sender's inputs x_0, x_1. The natural way to solve this problem is for the simulator to *rewind* $\mathcal{A}$. Specifically, it first hands $\mathcal{A}$ the message $\bar{a} = (g^\alpha, g^\beta, g^{\alpha\beta}, g^\gamma)$ and learns x_0, and then rewinds $\mathcal{A}$ and hands it the message $\bar{a}' = (g^\alpha, g^\beta, g^\gamma, g^{\alpha\beta})$ so that it learns x_1. The problem with this strategy, however, is that the adversary $\mathcal{A}$ may make its inputs x_0, x_1 depend on the first message that it sees (e.g., it may apply a pseudorandom function to the first message and take the result as the input). In this case, it uses different inputs for every different first message that it receives from the simulator. This is a problem because the simulator does not know the honest party's input bit σ and so cannot know whether to write $\bar{a}$ in the output view or $\bar{a}'$. (Observe that if the honest receiver has input $\sigma = 0$ then its output will be consistent with the view of the adversarial sender $\mathcal{A}$ only if $\bar{a}$ appears in the output transcript. In contrast, if the honest receiver has input $\sigma = 1$ then its output will be consistent with the view of the adversarial sender $\mathcal{A}$ only if $\bar{a}'$ appears in the output transcript. This discrepancy can be detected by the distinguisher in the case where $\mathcal{A}$'s output includes the inputs that it used.)

One way to overcome this problem is to have the receiver send two separate tuples $\bar{a} = (x = g^\alpha, y = g^\beta, z_0, z_1)$ and $\bar{a}' = (x' = g^{\alpha'}, y' = g^{\beta'}, z_0', z_1')$, while proving in zero-knowledge that either $z_0 = g^{\alpha\beta}$ or $z_1' = g^{\alpha'\beta'}$ but *not both*. This ensures that a corrupted receiver will only be able to learn one of x_0, x_1. However, a simulator that can rewind the adversary in the zero-knowledge proof and "cheat" can set both $z_0 = g^{\alpha\beta}$ and $z_1' = g^{\alpha'\beta'}$, thereby learning both values x_0 and x_1. Importantly, both values are obtained at the same time, on the *same execution transcript*. Therefore, the above problem is solved. For the sake of optimization, we actually do something slightly different and combine the two tuples $\bar{a}, \bar{a}'$ into one. More exactly, we have the receiver construct a tuple (h_0, h_1, a, b_0, b_1) as follows,

1. if $\sigma = 0$ it sets

$$h_0 = g^{\alpha_0}, h_1 = g^{\alpha_1}, a = g^r, b_0 = g^{\alpha_0 r}, b_1 = g^{\alpha_1 r},$$

2. if $\sigma = 1$ it sets

$$h_0 = g^{\alpha_0}, h_1 = g^{\alpha_1}, a = g^r, b_0 = g^{\alpha_0 r + 1}, b_1 = g^{\alpha_1 r + 1},$$

and prove that the tuple is correctly constructed. Note that this holds if $(h_0/h_1, a, b_0/b_1)$ is a Diffie-Hellman tuple. (Actually, it is a Diffie-Hellman tuple as long as $b_0 = g^{\alpha_0 r + i}$ and $b_1 = g^{\alpha_1 r + i}$ for any i. However, if $i \notin \{0, 1\}$ then an adversarial receiver will learn nothing about x_0 or x_1.) The prover then uses a similar transformation as in Protocols $\pi_{\mathrm{OT}}^{\mathrm{P}}$ and $\pi_{\mathrm{OT}}^{\mathrm{OS}}$ to ensure that only one of x_0, x_1 is learned. Note that the transformation used is such that when applied to a DH tuple the receiver can learn the input, and when applied to a non-DH tuple the result is uniformly distributed (and perfectly hides the input). Observe now that when $\sigma = 0$ it holds that (h_0, a, b_0) and (h_1, a, b_1) are DH tuples, and when $\sigma = 1$ it holds that $(h_0, a, b_0/g)$ and $(h_1, a, b_1/g)$ are DH tuples. We stress that when $\sigma = 0$ we have that $(h_1, a, b_1/g)$ is *not* a DH tuple, and when $\sigma = 1$ we have that (h_0, a, b_0) is *not* a DH tuple. Therefore, the prover applies the transformation to (h_0, a, b_0) in order to "encrypt" x_0, and to $(h_1, a, b_1/g)$ in order to "encrypt" x_1. By what we have shown, this means that when $\sigma = 0$ the receiver can learn x_0 but x_1 is completely hidden, and vice versa. Of course, all of the above holds unless the receiver sets $b_0 = g^{\alpha_0 r}$ and $b_1 = g^{\alpha_1 r + 1}$, in which case it could learn both x_0 and x_1. However, the zero-knowledge proof described above prevents an adversarial receiver from doing this. In contrast, as described above, the simulator *can* cheat in the zero-knowledge proof and set the b_0, b_1 values so that it can learn both x_0 and x_1. See Protocol 7.4.1.

PROTOCOL 7.4.1 (Fully Simulatable Oblivious Transfer π_{OT})

- **Inputs:** The sender has a pair of strings $x_0, x_1 \in \mathbb{G}$ and the receiver has a bit $\sigma \in \{0, 1\}$.
- **Auxiliary inputs:** Both parties have the security parameter 1^n and the description of a group $\mathbb{G}$ of *prime order*, including a generator g for the group and its order q.[2]
- **The protocol:**

  1. R chooses $\alpha_0, \alpha_1, r \leftarrow_R \{1, \ldots, q\}$ and computes $h_0 = g^{\alpha_0}$, $h_1 = g^{\alpha_1}$ and $a = g^r$. It also computes $b_0 = h_0^r \cdot g^\sigma$ and $b_1 = h_1^r \cdot g^\sigma$.
     R sends (h_0, h_1, a, b_0, b_1) to S.
  2. S checks that all of $h_0, h_1, a, b_0, b_1 \in \mathbb{G}$, and if not, it aborts.
  3. Let $h = h_0/h_1$ and $b = b_0/b_1$. Then, R proves to S that $(\mathbb{G}, q, g, h, a, b)$ is a Diffie-Hellman tuple, using a zero-knowledge proof of knowledge. Formally, R proves the relation:

  $$\mathcal{R}_{\mathrm{DH}} = \left\{ ((\mathbb{G}, q, g, h, a, b), r) \mid a = g^r \ \& \ b = h^r \right\}.$$

  4. If S accepted the proof in the previous step, it chooses $u_0, v_0, u_1, v_1 \leftarrow_R \{1, \ldots, q\}$ and sends (e_0, e_1) computed as follows:
     a. $e_0 = (w_0, z_0)$ where $w_0 = a^{u_0} \cdot g^{v_0}$ and $z_0 = b_0^{u_0} \cdot h_0^{v_0} \cdot x_0$.
     b. $e_1 = (w_1, z_1)$ where $w_1 = a^{u_1} \cdot g^{v_1}$ and $z_1 = \left(\frac{b_1}{g}\right)^{u_1} \cdot h_1^{v_1} \cdot x_1$.
  5. R outputs $\frac{z_\sigma}{w_\sigma^{\alpha_\sigma}}$ and S outputs nothing.

In order to understand the protocol better, we first show that if S and R are honest, then R indeed outputs x_σ. We prove this separately for $\sigma = 0$ and $\sigma = 1$. Recall that $h_\sigma = g^{\alpha_\sigma}$, $a = g^r$ and $b_\sigma = h_\sigma^r \cdot g^\sigma = g^{r\alpha_\sigma + \sigma}$.

1. *Case $\sigma = 0$:* We have that

$$\frac{z_0}{w_0^{\alpha_0}} = \frac{b_0^{u_0} \cdot h_0^{v_0} \cdot x_0}{a^{u_0\alpha_0} \cdot g^{v_0\alpha_0}} = \frac{g^{u_0 r \alpha_0} \cdot g^{v_0 \alpha_0} \cdot x_0}{g^{u_0 \alpha_0 r} \cdot g^{v_0 \alpha_0}} = x_0.$$

2. *Case $\sigma = 1$:* We have that

$$\frac{z_1}{w_1^\alpha} = \frac{\left(\frac{b_1}{g}\right)^{u_1} \cdot h_1^{v_1} \cdot x_1}{a^{u_1\alpha_1} \cdot g^{v_1\alpha_1}} = \frac{g^{u_1 r \alpha_1} \cdot g^{v_1 \alpha_1} \cdot x_1}{g^{u_1 \alpha_1 r} \cdot g^{v_1 \alpha_1}} = x_1$$

where the second equality follows from the fact that $b_1 = h_1^r \cdot g$ and so $\left(\frac{b_1}{g}\right)^{u_1} = h_1^{u_1 r} = g^{u_1 r \alpha_1}$.

We now proceed to prove the security of the protocol:

---

[2] We stress that the group must be of prime order in order to ensure that every element in $\mathbb{G}$ is a generator; we use this fact in the proof of security. In addition, we note that the group can be chosen by R if not given as auxiliary input; in this case, it needs to be possible for S to check that all values are chosen correctly.

Theorem 7.4.2 *Assume that the DDH problem is hard in* $\mathbb{G}$ *with generator* g. *Then, Protocol 7.4.1 securely computes* $\mathcal{F}_{\mathrm{OT}}$ *in the presence of malicious adversaries.*

Proof. We separately prove security in the case where S is corrupted and the case where R is corrupted. We note that the proof is in a hybrid model where a trusted party is used to compute an ideal functionality for the zero-knowledge proof of knowledge for $\mathcal{R}_{\mathrm{DH}}$.

The sender S is corrupted. Let $\mathcal{A}$ be an adversary controlling S. We construct a simulator $\mathcal{S}_{\mathrm{SEND}}$ as follows:

1. $\mathcal{S}_{\mathrm{SEND}}$ invokes $\mathcal{A}$ upon its input and computes h_0 and h_1 as the honest R would. Then it computes $a = g^r$, $b_0 = h_0^r$ and $b_1 = h_1^r \cdot g$. (Note that this message is computed differently than an honest R.)
   $\mathcal{S}_{\mathrm{SEND}}$ hands (h_0, h_1, a, b_0, b_1) to $\mathcal{A}$.
2. $\mathcal{S}_{\mathrm{SEND}}$ receives from $\mathcal{A}$ its input to the trusted party computing the zero-knowledge functionality $\mathcal{F}_{\mathrm{ZK}}^{\mathrm{DH}}$; this input is just $(\mathbb{G}, q, g, h_0/h_1, a, b_0/b_1)$. If the input is the values handed by $\mathcal{S}_{\mathrm{SEND}}$ to $\mathcal{A}$, it simulates the trusted party returning 1; otherwise it simulates the trusted party returning 0.
3. $\mathcal{S}_{\mathrm{SEND}}$ receives from $\mathcal{A}$ two encryptions e_0 and e_1. It then computes $x_0 = \frac{z_0}{w_0^{\alpha_0}}$ and $x_1 = \frac{z_1}{w_1^{\alpha_1}}$ and sends (x_0, x_1) to the trusted party computing the oblivious transfer functionality.
4. $\mathcal{S}_{\mathrm{SEND}}$ outputs whatever $\mathcal{A}$ outputs and halts.

We now show that the joint output distribution of $\mathcal{A}$ and R in a real protocol execution is computationally indistinguishable from the output of $\mathcal{S}_{\mathrm{SEND}}$ and R in the ideal-world simulation. Note that the only difference is that $\mathcal{S}_{\mathrm{SEND}}$ generates the values b_0 and b_1 incorrectly (because b_0 equals h_0^r multiplied by 1 whereas b_1 equals h_1^r multiplied by g). We prove this separately for the case where R's input is $\sigma = 0$ and the case where it is $\sigma = 1$. Denoting Protocol 7.4.1 by π_{OT}, we begin by proving that

$$\left\{ \mathrm{IDEAL}_{\mathcal{F}_{\mathrm{OT}}, \mathcal{S}_{\mathrm{SEND}}(z), S}((x_0, x_1), \sigma, n) \right\} \overset{\mathrm{c}}{\equiv} \left\{ \mathrm{HYBRID}_{\pi_{\mathrm{OT}}, \mathcal{A}(z), S}^{\mathrm{ZK}}((x_0, x_1), \sigma, n) \right\}.$$

We construct a distinguisher D_{DDH} for the DDH problem that distinguishes a Diffie-Hellman tuple from a non-Diffie-Hellman tuple with the same probability that it is possible to distinguish the IDEAL and HYBRID executions. Before doing this we note that if the DDH problem is hard, then it is also hard to distinguish between Diffie-Hellman tuples of the form $(\mathbb{G}, q, g, h, g^r, h^r)$ and non-Diffie-Hellman tuples of the specific form $(g, h, g^r, h^r \cdot g)$.[3] D_{DDH} receives

---

[3] This can be shown as follows. We will use the following distributions: $X_0 = \{(\mathbb{G}, q, g, h, g^r, h^r)\}$, $X_1 = \{(\mathbb{G}, q, g, h, g^r, gh^r)\}$ and $X_2 = \{(\mathbb{G}, q, g, h, g^r, h^s)\}$, where r and s are chosen randomly. First note that for every distinguisher D_{DDH} we have that $|\Pr[D(X_0) = 1] - \Pr[D(X_1) = 1]| \leq |\Pr[D(X_0) = 1] - \Pr[D(X_2) = 1]| + |\Pr[D(X_2) = 1] - \Pr[D(X_1) = 1]|$. The fact that $|\Pr[D(X_0) = 1] - \Pr[D(X_2) = 1]|$ is negligible is exactly the DDH assumption. Regarding $|\Pr[D(X_2) = 1] - \Pr[D(X_1) = 1]|$ this follows

a tuple $(\mathbb{G}, q, g, h, s, t)$ and wishes to determine if there exists an r such that $s = g^r$ and $t = h^r$. First, D_{DDH} chooses α_0 and computes $h_0 = g^{\alpha_0}$ and $b_0 = s^{\alpha_0}$. Then it sets $h_1 = h$ and $b_1 = t$. Note that if $(\mathbb{G}, q, g, h, s, t)$ is a Diffie-Hellman tuple then D_{DDH} generates (h_0, h_1, a, b_0, b_1) exactly as an honest R with input $\sigma = 0$ would. (In order to see this, let r be such that $s = g^r$. Then, $b_1 = t = h^r = h_1^r$. Furthermore, $b_0 = s^{\alpha_0} = h_0^r$, as required.) In contrast, if $(\mathbb{G}, q, g, h, s, t)$ is such that for some r, $s = g^r$ and $t = gh^r$, then D_{DDH} generates (h_0, h_1, a, b_0, b_1) exactly as the simulator S_{SEND} would. (This follows because $b_1 = gh_1^r$ whereas $b_0 = s^{\alpha_0} = h_0^r$.) Next, D_{DDH} continues the simulation as S_{SEND} by handing A the values h_0, h_1, a, b_0, b_1 and receiving back the encryptions e_0, e_1. Finally, it computes $x_0 = \frac{z_0}{w_0^{\alpha_0}}$, sets this as R's output and combines it with A's output (generating a *joint output distribution* of A and R). Now, if D_{DDH} receives a non-Diffie-Hellman tuple (with the fourth element being gh^r), then the output of D_{DDH} is *identical* to the output distribution of the ideal-world execution with S_{SEND} and an honest R. Furthermore, if D_{DDH} receives a Diffie-Hellman tuple, then the output distribution is identical to a hybrid execution between A and R (note that R would derive x_0 in exactly the same way as D_{DDH} derived it from e_0). By the assumed hardness of the DDH problem, this implies that the IDEAL and HYBRID distributions are computationally indistinguishable, as required. The proof for the case of $\sigma = 1$ is almost identical.

The receiver R is corrupted. Let A be an adversary controlling R. We construct a simulator S_{REC} as follows:

1. S_{REC} invokes A upon its input and receives the vector (h_0, h_1, a, b_0, b_1).
2. S_{REC} checks that all $h_0, h_1, a, b_0, b_1 \in \mathbb{G}$; if not, it sends $\perp$ to the trusted party for $\mathcal{F}_{\text{OT}}$, simulates S aborting, outputs whatever A outputs and halts.
3. Otherwise, S_{REC} obtains from A its input $((\mathbb{G}, q, g, h, a, b), r)$ for the trusted party computing the zero-knowledge functionality $\mathcal{F}_{\text{ZK}}^{\text{DH}}$. If the conditions for outputting $(1, \lambda)$ are not met then S_{REC} sends $\perp$ to the trusted party for $\mathcal{F}_{\text{OT}}$ and aborts. Otherwise, S_{REC} proceeds.
4. S_{REC} computes $\ell = b_0/h_0^r$. If $\ell = 1$, S_{REC} sets $\sigma = 0$; if $\ell \neq 1$ (including the case where $\ell = g$), S_{REC} sets $\sigma = 1$. Then, S_{REC} sends σ to the trusted party computing the oblivious transfer functionality and receives back the string x_σ.
5. S_{REC} computes $e_\sigma = (w_\sigma, z_\sigma)$ as an honest S would, using the value x_σ. In contrast, it computes $e_{1-\sigma}$ using $x_{1-\sigma} = 1$.
6. S_{REC} outputs whatever A outputs and halts.

---

from the fact that by dividing the fourth element of the input by g, an instance of X_1 becomes an instance of X_0, whereas an instance of X_2 remains an instance of X_2. Thus, if $|\Pr[D(X_2) = 1] - \Pr[D(X_1) = 1]|$ is non-negligible then we can construct a distinguisher D' so that $|\Pr[D'(X_0) = 1] - \Pr[D'(X_2) = 1]|$ is non-negligible, in contradiction to the DDH assumption.

We continue by proving that

$$\left\{ \text{IDEAL}_{\mathcal{F}_{\text{OT}}, \mathcal{S}_{\text{REC}}(z), R}((x_0, x_1), \sigma, n) \right\} \stackrel{c}{\equiv} \left\{ \text{HYBRID}^{\text{ZK}}_{\pi_{\text{OT}}, \mathcal{A}(z), R}((x_0, x_1), \sigma, n) \right\}.$$

Since the honest S does not have any output, it suffices to show that the distribution over $\mathcal{A}$'s output is computationally indistinguishable in a hybrid execution with S and in the simulation with $\mathcal{S}_{\text{REC}}$ (we actually show that the distributions are identical). It is clear that until the last message, $\mathcal{A}$'s view is identical (all the messages until the last are independent of S's input). Furthermore, the only difference is in the way $e_{1-\sigma}$ is constructed. Thus, all we need to show is that for *every* $x_{1-\sigma}$, the distribution over $e_{1-\sigma}$ is the same when constructed by an honest S and $\mathcal{S}_{\text{REC}}$. We do this separately for $\ell = 1$, $\ell = g$ and $\ell \notin \{1, g\}$ (where ℓ is the value obtained by $\mathcal{S}_{\text{REC}}$ by computing b_0 / h_0^r). We distinguish between the values generated by the simulator $\mathcal{S}_{\text{REC}}$ and those generated by a real S by adding a tilde to the values generated by $\mathcal{S}_{\text{REC}}$.

1. *Case $\ell = 1$:* In this case we know that for some r it holds that $a = g^r$ and $b_0 / b_1 = (h_0 / h_1)^r$ (this is given due to the zero-knowledge protocol). Combining this with the fact that $b_0 / h_0^r = 1$ we have that

$$b_1 = b_0 \cdot \left(\frac{h_1}{h_0} \right)^r = (h_1)^r.$$

Letting α be such that $h_1 = g^\alpha$, we have that $b_1 = (h_1)^r = (g^\alpha)^r$. (Such an α is guaranteed to exist because $h_1 \in \mathbb{G}$ as checked by $\mathcal{S}_{\text{REC}}$ and the honest S.) This implies that $\frac{b_1}{g} = g^{r\alpha} \cdot g^{-1}$. Finally, for every $x_1 \in \mathbb{G}$ it holds that there exists a value s such that $x_1 = g^s$; this holds because g is a generator and x_1 is in $\mathbb{G}$). Recalling that $w_1 = b^{u_1} \cdot g^{v_1}$ and $z_1 = \left(\frac{b_1}{g} \right)^{u_1} \cdot h_1^{v_1} \cdot x_1$ and that $\mathcal{S}_{\text{REC}}$ uses $x_1 = 1$, we have that the distributions generated over e_1 are

- *generated by $\mathcal{S}_{\text{REC}}$:*
$$\tilde{w}_1 = a^{\tilde{u}_1} \cdot g^{\tilde{v}_1} = g^{r\tilde{u}_1 + \tilde{v}_1}$$

  and

$$\tilde{z}_1 = \left(\frac{b_1}{g} \right)^{\tilde{u}_1} \cdot h_1^{\tilde{v}_1} = g^{r\alpha\tilde{u}_1 + \tilde{v}_1\alpha} \cdot g^{-\tilde{u}_1};$$

- *generated by S:*
$$w_1 = a^{u_1} \cdot g^{v_1} = g^{ru_1 + v_1}$$

  and

$$z_1 = \left(\frac{b_1}{g} \right)^{u_1} \cdot h_1^{v_1} \cdot x_1 = g^{r\alpha u_1 + v_1\alpha} \cdot g^s \cdot g^{-u_1}.$$

We claim that the above distributions are *identical*. In order to see this, let $\tilde{u}_1 = u_1 - s$ and $\tilde{v}_1 = v_1 + rs$. For a fixed r and s and a uniformly distributed u_1 and v_1, the values $\tilde{u}_1$ and $\tilde{v}_1$ chosen in this way are also uniformly distributed. Plugging these values in, we have that

$$\tilde{w}_1 = g^{r\tilde{u}_1+\tilde{v}_1} = g^{r(u_1-s)+v_1+rs} = g^{ru_1+v_1}$$

and

$$\tilde{z}_1 = g^{r\alpha\tilde{u}_1+\tilde{v}_1\alpha-\tilde{u}_1} = g^{r\alpha(u_1-s)+\alpha(v_1+rs)-(u_1-s)}$$
$$= g^{r\alpha u_1+\alpha v_1-u_1+s} = g^{r\alpha u_1+\alpha v_1} \cdot g^s \cdot g^{-u_1},$$

which are the *exact values* w_1 and z_1 generated by an honest S. Thus, the distribution viewed by $\mathcal{A}$ in this case is the same in the hybrid and simulated executions.

2. *Case $\ell = g$:* In this case we know that for some r it holds that $a = g^r$ and $b_0 = gh_0^r$ (this is given due to the zero-knowledge protocol and the fact that $b_0/h_0^r = g$, as in the previous case). Now, as above there exists a value α such that $h_0 = g^\alpha$ (because $h_0 \in \mathbb{G}$) and we know that $a = g^r$, implying that $b_0 = gh_0^r = ga^\alpha = g \cdot g^{r\alpha}$. Furthermore, for every $x_0 \in \mathbb{G}$ there exists a value s such that $x_0 = g^s$.
   Recalling that $w_0 = a^{u_0} \cdot g^{v_0}$ and $z_0 = b_0^{u_0} \cdot h_0^{v_0} \cdot x_0$ and that $\mathcal{S}_{\mathrm{REC}}$ uses $x_0 = 1$, we have that the distributions generated over e_0 are

   * *generated by $\mathcal{S}_{\mathrm{REC}}$:*
     $$\tilde{w}_0 = a^{\tilde{u}_0} \cdot g^{\tilde{v}_0} = g^{r\tilde{u}_0+\tilde{v}_0}$$

     and
     $$\tilde{z}_0 = b_0^{\tilde{u}_0} \cdot h_0^{\tilde{v}_0} = g^{r\alpha\tilde{u}_0+\tilde{v}_0\alpha} \cdot g^{\tilde{u}_0};$$

   * *generated by S:*
     $$w_0 = a^{u_0} \cdot g^{v_0} = g^{ru_0+v_0}$$

     and
     $$z_0 = b_0^{u_0} \cdot h_0^{v_0} \cdot x_0 = g^{r\alpha u_0+v_0\alpha} \cdot g^s \cdot g^{u_0}.$$

   As above, let $\tilde{u}_0 = u_0 + s$ and $\tilde{v}_0 = v_0 - rs$; again these are uniformly distributed if u_0 and v_0 are uniformly distributed. Then,

   $$\tilde{w}_0 = g^{r\tilde{u}_0+\tilde{v}_0} = g^{r(u_0+s)+v_0-rs} = g^{ru_0+v_0}$$

   and

   $$\tilde{z}_0 = g^{r\alpha\tilde{u}_0+\tilde{v}_0\alpha} \cdot g^{\tilde{u}_0} = g^{r\alpha(u_0+s)+\alpha(v_0-rs)+(u_0+s)} = g^{r\alpha u_0+\alpha v_0} \cdot g^s \cdot g^{u_0},$$

   which are the exact values generated by an honest S. Thus, the distribution viewed by $\mathcal{A}$ in this case is the same in the hybrid and simulated executions.

3. *Case $\ell \notin \{1, g\}$:* In this case, $\mathcal{S}_{\mathrm{REC}}$ sets $\sigma = 1$ and so generates the distribution for e_1 exactly as an honest S would. In contrast to above, here we only know that $a = g^r$ and $b_0 = \ell \cdot h_0^r$ for *some* ℓ. This means that for α such that $h_0 = g^\alpha$ we have that $b_0 = \ell \cdot g^{r\alpha}$. However, since $b_0 \in \mathbb{G}$ (as

checked by $\mathcal{S}_{\text{REC}}$) and $h_0 = g^{r\alpha}$ is also in $\mathbb{G}$, it follows that $\ell \in \mathbb{G}$; let t be such that $\ell = g^t$ and let s be such that $x_0 = g^s$. As above, we write the distributions generated by $\mathcal{S}_{\text{REC}}$ and an honest S:

- *Generated by $\mathcal{S}_{\text{REC}}$:*
$$\tilde{w}_0 = g^{r\tilde{u}_0 + \tilde{v}_0}$$

and
$$\tilde{z}_0 = g^{r\alpha\tilde{u}_0 + \tilde{v}_0\alpha} \cdot \ell^{\tilde{u}_0} = g^{r\alpha\tilde{u}_0 + \tilde{v}_0\alpha} \cdot g^{t\tilde{u}_0};$$

- *Generated by S:*
$$w_0 = g^{ru_0 + v_0}$$

and
$$z_0 = g^{r\alpha u_0 + v_0\alpha} \cdot g^b \cdot \ell^{u_0} = g^{r\alpha u_0 + v_0\alpha} \cdot g^s \cdot g^{tu_0}.$$

Let $\tilde{u}_0 = u_0 + s/t$ and $\tilde{v}_0 = v_0 - rs/t$. The crucial point here is that $\mathbb{G}$ is a group of *prime order*, and thus t has an inverse modulo q, meaning that s/t is well defined (recall that we can work modulo q in the exponent). Thus, we can define $\tilde{u}_0$ and $\tilde{v}_0$ in this way, and as above, if v_0 and u_0 are uniformly distributed then so are $\tilde{v}_0$ and $\tilde{u}_0$. Plugging these values in as above, we have that $\tilde{w}_0, \tilde{z}_0$ are distributed identically to w_0, z_0, as required.

This completes the proof of the theorem.   ∎

Exact efficiency. The number of rounds of communication in the protocol, including the zero-knowledge subprotocol, is exactly six (each party sends three messages). We note that the first message from R to S is combined with the first message of the zero-knowledge protocol (also sent by R). Regarding exponentiations, we have the following:

1. *The sender S:* In the zero-knowledge protocol S, playing the verifier, computes exactly seven exponentiations. In the rest of the protocol, S computes eight exponentiations (for computing e_0 and e_1). Overall, S has 15 exponentiations.
2. *The receiver R:* In the zero-knowledge protocol R, playing the prover, computes exactly five exponentiations. In the rest of the protocol, R computes six exponentiations (five for computing h_0, h_1, a, b_0, b_1 and one for decrypting e_σ). Overall, R has 11 exponentiations.

Finally, we count the bandwidth in terms of the number of group elements sent by each party:

1. *The sender S:* In the zero-knowledge protocol S, playing the verifier, sends three group elements. In the rest of the protocol, S sends four group elements (w_0, z_0, w_1 and z_1). Overall, S sends seven group elements.
2. *The receiver R:* In the zero-knowledge protocol R, playing the prover, sends five group elements. In the rest of the protocol, R sends five group elements (h_0, h_1, a, b_0 and b_1). Overall, R sends ten group elements.

Summing up, the protocol has six rounds of communication, an overall number of exponentiations for both parties of 26 (almost equally divided between them), and an overall bandwidth of 17 group elements sent. We remark that if the random oracle model can be tolerated, then the zero-knowledge proof of knowledge can be collapsed into a single round (using the Fiat-Shamir paradigm [26]) yielding a *two-round* protocol (this also saves five exponentiations).

Recall that Protocol 7.2.1 that achieves privacy only has 13 exponentiations, an overall bandwidth of eight group elements, and two rounds of communication. Therefore Protocol 7.4.1 achieves *full simulation* at the cost of approximately twice the complexity (and three times the number of rounds).

7.4.2 Batch Oblivious Transfer

In many settings, it is necessary to run many oblivious transfers at the same time. We have seen this, for example, in the protocols of Chapters 4 and 5. Two issues arise in such a case. The first issue relates to the question of whether or not a stand-alone OT protocol remains secure when run many times in parallel. In short, the answer is that it depends; some protocols do and some do not. An alternative is therefore to construct a single protocol that directly handles many oblivious transfers, and prove it secure.[4] Beyond the issue of security, this alternative also has the potential to yield greater efficiency. For example, if an oblivious transfer protocol requires a zero-knowledge proof, then m executions would require m copies of that proof to be run. However, it may be possible to run a single proof, even when considering many executions. As we will see, this is indeed the case with the above protocol.

We call a multi-execution protocol of this type batch oblivious transfer, and define the batch OT functionality as follows:

$$\mathcal{F}_{\mathrm{BOT}} : ((x_1^0, x_1^1), \ldots, (x_m^0, x_m^1), (\sigma_1, \ldots, \sigma_m)) \to (\lambda, (x_1^{\sigma_1}, \ldots, x_m^{\sigma_m})).$$

We note that private and one-sided batch oblivious transfer can be achieved by directly invoking $\pi_{\mathrm{OT}}^{\mathrm{P}}$ and $\pi_{\mathrm{OT}}^{\mathrm{OS}}$ in parallel (this holds as long as the zero-knowledge proof of knowledge for $\mathcal{R}_{\mathrm{DL}}$ composes in parallel, which can be achieved by first constructing the parallel Σ protocol; see Section 6.4). Further optimizations can be made by fixing the first element in the tuple $\bar{a} = (g^\alpha, \cdot, \cdot, \cdot)$. We omit the details here and continue with a detailed analysis for fully simulatable batch oblivious transfer in the malicious setting.

---

[4] This is not the same as parallel composition, because in parallel composition the honest parties run each execution as if it is the only one being run. In contrast, we propose running a single protocol which coordinates all the executions together.

We construct our batch OT protocol by modifying Protocol 7.4.1 as follows. Conceptually, we run many copies of Protocol 7.4.1 in parallel, except for the following changes:

1. *We use the same h_0, h_1 values in each execution:* this is of no consequence because the DDH assumption holds even if the same base is used in many tuples. That is, $(g^\alpha, g^\beta, g^{\alpha\beta}), (g^\alpha, g^\gamma, g^{\alpha\gamma})$ are indistinguishable from $(g^\alpha, g^\beta, g^{\delta_1}), (g^\alpha, g^\gamma, g^{\delta_2})$, even though g^α appears in both. This can be proven quite easily by a reduction to the basic DDH problem.

2. *We provide a single zero-knowledge proof of Diffie-Hellman for all executions:* We achieve this by combining all of the values sent by the receiver into a single tuple, so that if all the values are correctly constructed then the resulting tuple is of the Diffie-Hellman type, but if even one set of values is incorrectly constructed then the resulting tuple is *not* of the Diffie-Hellman type.

Before proceeding to the protocol, we describe how all the values are combined into a single tuple for which a Diffie-Hellman proof can be used. We demonstrate this for the special case of two executions. The receiver prepares a single h_0, h_1 as in Protocol 7.4.1. It then prepares (a^1, b_0^1, b_1^1) and (a^2, b_0^2, b_1^2) separately depending on its input σ_1 in the first execution and σ_2 in the second execution; in both cases it prepares the values again as in Protocol 7.4.1. As a first attempt, these values can be combined by computing

$$a = a^1 \cdot a^2 \quad \text{and} \quad b = \frac{b_0^1}{b_1^1} \cdot \frac{b_0^2}{b_1^2}. \tag{7.4}$$

Observe that if the receiver prepares the tuples correctly, then the tuple $(h_0/h_1, a, b)$ is a Diffie-Hellman tuple. This is because the product of the second and third elements of Diffie-Hellman tuples with the same first element yields a Diffie-Hellman tuple. (That is, given $(g^\alpha, g^\beta, g^{\alpha\beta})$ and $(g^\alpha, g^\gamma, g^{\alpha\gamma})$, it holds that $(g^\alpha, g^{\beta+\gamma}, g^{\alpha\beta+\alpha\gamma})$ is also a Diffie-Hellman tuple.) Recall that in the basic protocol, each $(h_0/h_1, a^j, b_0^j/b_1^j)$ is a Diffie-Hellman tuple, and so the computation of a and b above yields a Diffie-Hellman tuple. One is tempted to conclude that if however at least one of $(h_0/h_1, a^1, b_0^1/b_1^1)$ or $(h_0/h_1, a^2, b_0^2/b_1^2)$ is *not* a Diffie-Hellman tuple, then $(h_0/h_1, a, b)$ where a and b are as in (7.4) is *not* a Diffie-Hellman tuple. However, this is not necessarily the case. For example, if $a^1 = g^{r_1}, a^2 = g^{r_2}$ then a malicious receiver can set $b_0^1 = h_0^{r_1}, b_1^1 = h_1^{r_1} \cdot g$ and $b_0^2 = h_0^{r_1} \cdot g, b_1^2 = h_1^{r_1}$. In this case, it holds that

$$a = \frac{a^1}{a^2} = g^{r_1 - r_2}$$

and

$$b = \frac{b_0^1}{b_1^1} \cdot \frac{b_0^2}{b_1^2} = \frac{h_0^{r_1}}{h_1^{r_1} \cdot g} \cdot \frac{h_0^{r_2} \cdot g}{h_1^{r_2}} = \frac{h_0^{r_1}}{h_1^{r_1}} \cdot \frac{h_0^{r_2}}{h_1^{r_2}} = \left(\frac{h_0}{h_1}\right)^{r_1 - r_2}$$

and thus $(h_0/h_1, a, b)$ is a Diffie-Hellman tuple. Furthermore, observe that in the first execution, the receiver learns *both* x_1^0 and x_1^1 (i.e., both inputs of S in the first execution), and so this is not secure. In order to prevent this attack, the sender chooses random value ρ_1, ρ_2 and requires the receiver to prove that $(h_0/h_1, a, b)$ is a Diffie-Hellman tuple when

$$a = (a^1)^{\rho_1} \cdot (a^2)^{\rho_2} \quad \text{and} \quad b = \left(\frac{b_0^1}{b_1^1}\right)^{\rho_1} \cdot \left(\frac{b_0^2}{b_1^2}\right)^{\rho_2} .$$

The reason why this helps is because a malicious receiver could fool the sender above by preparing $b_0^1, b_1^1, b_0^2, b_1^2$ so that the same deviation from the Diffie-Hellman values is found in both b_0^1/b_1^1 and b_0^2/b_1^2 (but inverted). In this way, when multiplied together, the deviation disappears and the result is a Diffie-Hellman tuple. However, now the receiver can only succeed if this deviation disappears after each quotient is raised to the power of an independent random value. However, this occurs with only very small probability because the random exponentiations cause the deviations from the Diffie-Hellman values to be random. The full description is found in Protocol 7.4.3.

PROTOCOL 7.4.3 (Batch Oblivious Transfer π_{BOT})

- **Inputs:** The sender has a list of m pairs of strings $(x_1^0, x_1^1), \ldots, (x_m^0, x_m^1)$ and the receiver has an m-bit string $(\sigma_1, \ldots, \sigma_m)$.
- **Auxiliary inputs:** Both parties have the security parameter 1^n and the description of a group $\mathbb{G}$ of *prime order*, including a generator g for the group and its order q. In addition, they have a probabilistic polynomial-time algorithm V that checks membership in $\mathbb{G}$ (i.e., for every h, $V(h) = 1$ if and only if $h \in \mathbb{G}$). The group can be chosen by R if not given as auxiliary input; in this case, it needs to be possible for S to check that all values are chosen correctly.
- **The protocol:**

  1. R chooses a single α_0, α_1 and computes h_0, h_1 as above. Then, R proves that it knows the discrete log of h_0, using a zero-knowledge proof of knowledge for $\mathcal{R}_{\mathrm{DL}}$.
  2. Next, for every $j = 1, \ldots, m$, the receiver R chooses a random r_j and computes $a^j = g^{r_j}$, $b_0^j = h_0^{r_j} \cdot g^{\sigma_j}$ and $b_1^j = h_1^{r_j} \cdot g^{\sigma_j}$. R sends all these values to S.
  3. S chooses random $\rho_1, \ldots, \rho_m \leftarrow_R \{1, \ldots, q\}$ and sends them to R.
  4. Both parties locally compute

     $$a = \prod_{j=1}^m (a^j)^{\rho_j} \quad \text{and} \quad b = \prod_{j=1}^m \left(\frac{b_0^j}{b_1^j}\right)^{\rho_j}$$

     and then R proves that $(\mathbb{G}, q, g, h_0/h_1, a, b)$ is a Diffie-Hellman tuple.
  5. S computes e_0^j, e_1^j as in the basic protocol π_{OT}, for every j.
  6. R decrypts $e_{\sigma_j}^j$ as in the basic protocol π_{OT}, for every j.

Before proceeding to the proof, we observe that if R is honest then

$$a = g^{\sum r_j \rho_j} \quad \text{and} \quad b = \frac{\prod_{j=1}^n (b_0^j)^{\rho_j}}{\prod_{j=1}^n (b_1^j)^{\rho_j}} = \frac{\prod_{j=1}^n (h_0)^{r_j \rho_j}}{\prod_{j=1}^n (h_1)^{r_j \rho_j}} = \left(\frac{h_0}{h_1}\right)^{\sum r_j \rho_j}.$$

Therefore, $(\mathbb{G}, q, g, h_0/h_1, a, b)$ is a Diffie-Hellman tuple, as required. We now prove that the protocol is secure.

Theorem 7.4.4 *Assume that the DDH problem is hard in $\mathbb{G}$ with generator g. Then, Protocol 7.4.3 securely computes $\mathcal{F}_{\mathrm{BOT}}$ in the presence of malicious adversaries.*

Proof (sketch). As we have seen, if the fact that $(\mathbb{G}, q, g, h_0/h_1, a, b)$ is a Diffie-Hellman tuple implies that all $(\mathbb{G}, q, g, h_0/h_1, b^j, b_0^j/b_1^j)$ are Diffie-Hellman tuples except with negligible probability, then the only difference between the proof of security of the batch protocol and the original one is that the simulator does not know the individual r values and so cannot compute each σ_j. However, for this reason, R proves that it knows the discrete log of h_0; this enables $\mathcal{S}_{\mathrm{REC}}$ to extract α_0 and therefore learn σ_j from h_0^j. In all other ways, the simulator remains identical. It therefore remains to show that if $(\mathbb{G}, q, g, h_0/h_1, a, b)$ is a Diffie-Hellman tuple, then this implies that all $(\mathbb{G}, q, g, h_0/h_1, a^j, b_0^j/b_1^j)$ are Diffie-Hellman tuples except with negligible probability. Now, assume that there exists an *index* k ($1 \leq k \leq m$) such that $(\mathbb{G}, q, g, h_0/h_1, a^k, b_0^k/b_1^k)$ is not a Diffie-Hellman tuple. Denote $h = h_0/h_1$ and let r_k be such that $a^k = g^{r_k}$. Then, this implies that there exists a value $t \neq 1$ such that $b_0^k/b_1^k = h^{r_k} \cdot t$ (this follows immediately from the fact that $h = h_0/h_1$ and the assumption that there exists a non-Diffie-Hellman tuple). Now, fix ρ_ℓ for all $\ell \neq k$. This fixes the values $a' = \prod_{j \neq k}(b^j)^{\rho_j}$, $b_0' = \prod_{j \neq k}(b_0^j)^{\rho_j}$ and $b_1' = \prod_{j \neq k}(b_0^j)^{\rho_j}$. We have two cases:

1. *Case 1 – $(\mathbb{G}, q, g, h, a', b_0'/b_1')$ is a Diffie-Hellman tuple:*
   We prove that $(\mathbb{G}, q, g, h, a, b_0/b_1)$ is *not* a Diffie-Hellman tuple in this case as follows. Note first that $a = a' \cdot (a^k)^{\rho_k} = a' \cdot g^{r_k \rho_k}$ (where r_k is such that $a^k = g^{r_k}$ as above). Next, $b_0 = b_0' \cdot (b_0^k)^{\rho_k}$ and $b_1 = b_1' \cdot (b_1^k)^{\rho_k}$. Therefore, we have that $b_0/b_1 = b_0'/b_1' \cdot (b_0^k)^{\rho_k}/(b_1^k)^{\rho_k} = b_0'/b_1' \cdot (h^{r_k} \cdot t)^{\rho_k}$. Letting r be such that $a' = g^r$ and $b_0'/b_1' = h^r$ we have that $(\mathbb{G}, q, g, h, a, b_0/b_1) = (\mathbb{G}, q, g, h, g^{r+r_k \cdot \rho_k}, h^{r+r_k \cdot \rho_k} \cdot t^{\rho_k})$. Since $\mathbb{G}$ is a group of prime order and $t \in \mathbb{G}$ we have that t is of order q and so $t^{\rho_k} \neq 1$, implying that the tuple is not of the Diffie-Hellman type, as required.
2. *Case 2 – $(\mathbb{G}, q, g, h_0/h_1, a', b_0'/b_1')$ is not a Diffie-Hellman tuple:*
   In this case, $(\mathbb{G}, q, g, h_0/h_1, a, b_0/b_1)$ can only be a Diffie-Hellman tuple if there exists a value r such that $a = a' \cdot (a^k)^{\rho_k} = g^r$ and

   $$b_0/b_1 = b_0'/b_1' \cdot (h^{r_k} \cdot t)^{\rho_k} = h^r.$$

Let α and β be such that $a' = g^\alpha$ and $b_0'/b_1' = h^\beta$ (by the assumption in this case, $\alpha \neq \beta$). This implies that $(\mathbb{G}, q, g, h_0/h_1, a, b_0/b_1)$ can only be a Diffie-Hellman tuple if $g^r = g^{\alpha + r_k \cdot \rho_k}$ and $h^r = h^{\beta + r_k \cdot \rho_k} \cdot t^{\rho_k}$. Let γ be such $t = h^\gamma$ (such a γ exists because $h \in \mathbb{G}$ and is thus a generator). We have that $(\mathbb{G}, q, g, h_0/h_1, a, b_0/b_1)$ is a Diffie-Hellman tuple if and only if $r = \alpha + r_k \cdot \rho_k$ and $r = \beta + r_k \cdot \rho_k + \gamma \cdot \rho_k$ and thus if and only if $0 = \alpha - \beta - \gamma \cdot \rho_k$ (where the operations here are in $\mathbb{Z}_q$). Since ρ_k is uniformly chosen *after* α, β and γ are fixed (because these values are a function of the values sent by R initially), we have that equality occurs with probability $1/q$ only, which is negligible.

∎

Exact efficiency. We conclude by analyzing the efficiency of the batch oblivious transfer. For m executions, we have the following:

1. *Rounds of communication:* there are still six rounds of communication because R's initial values can still be sent with the first round of the zero-knowledge protocol and S's random values ρ_j can be sent together with the second round of the zero-knowledge protocol. Furthermore, the proof of discrete log of h_0 can be run in parallel with the proof of the Diffie-Hellman tuple. In order to do this, R sends h_0, h_1 and all of the a^j, b_0^j, b_1^j values to S in the first round, together with the first message of both zero-knowledge protocols (this can be done because the first two rounds of the zero-knowledge proofs are independent of the statements). Next, the sender S sends R all of the ρ_j values together with the second message of the zero-knowledge proofs. The parties then conclude the proofs, which takes three more rounds, and finally S sends its last message.
2. *Exponentiations:* the number of exponentiations is equal to $14m + 23$ (counting the number of exponentiations for all operations plus the number of exponentiations in each zero-knowledge proof).
3. *Bandwidth:* Overall $8m + 15$ group elements are sent.

Implementation. In [58], Protocol 7.4.3 was implemented over an elliptic curve group, as part of a general implementation of the protocol of Chapter 4. An interesting result of this implementation is the observation that when oblivious transfers are batch, the savings made by using a random oracle are insignificant. In this implementation, 320 oblivious transfers took an overall time of 20 seconds per party and 480 oblivious transfers took an overall time of 34 seconds per party (this is the total amount of time from the beginning to end, which upper-bounds the actual computation of each side). We remark that a fully simulatable oblivious transfer was needed in this implementation, and due to the efficiency of Protocol 7.4.1, the oblivious transfers were not the main bottleneck of the overall protocol.

7.5 Another Oblivious Transfer – Full Simulation

In this section, we present another oblivious transfer protocol that is based on [70]. The protocol is similar to that of Section 7.4 and is also based on the private oblivious transfer of [62]. However, the protocol here is more efficient in batch mode. Due to its similarity to the protocol above, the presentation of the protocol here is more concise and we only sketch the proof of security. A full description appears in Protocol 7.5.1 below.

PROTOCOL 7.5.1 (Another Fully Simulatable Oblivious Transfer)

- **Inputs:** The sender's input is a pair (x_0, x_1) and the receiver's input is a bit σ.
- **Auxiliary input:** Both parties hold a security parameter 1^n and $(\mathbb{G}, q, g_0)$, where $\mathbb{G}$ is an efficient representation of a group of order q with a generator g_0, and q is of length n.
- **The protocol:**

  1. The receiver R chooses random values $y, \alpha_0 \leftarrow \mathbb{Z}_q$ and sets $\alpha_1 = \alpha_0 + 1$. R then computes $g_1 = (g_0)^y$, $h_0 = g_0^{\alpha_0}$ and $h_1 = g_1^{\alpha_1}$ and sends (g_1, h_0, h_1) to the sender S.
  2. R proves, using a zero-knowledge proof of knowledge, that $(g_0, g_1, h_0, \frac{h_1}{g_1})$ is a DH tuple; see Protocol 6.2.4.
  3. R chooses a random value r and computes $g = (g_\sigma)^r$ and $h = (h_\sigma)^r$, and sends (g, h) to S.
  4. The sender operates in the following way:
     - Define the function $RAND(w, x, y, z) = (u, v)$, where $u = (w)^s \cdot (y)^t$ and $v = (x)^s \cdot (z)^t$, and the values $s, t \leftarrow \mathbb{Z}_q$ are random.
     - S computes $(u_0, v_0) = RAND(g_0, g, h_0, h)$, and $(u_1, v_1) = RAND(g_1, g, h_1, h)$.
     - S sends the receiver the values (u_0, w_0) where $w_0 = v_0 \cdot x_0$, and (u_1, w_1) where $w_1 = v_1 \cdot x_1$.
  5. The receiver computes $x_\sigma = w_\sigma/(u_\sigma)^r$.

In order to see that the protocol works, observe that

$$\frac{w_\sigma}{(u_\sigma)^r} = \frac{v_\sigma \cdot x_\sigma}{(u_\sigma)^r} = \frac{g^s \cdot h^t \cdot x_\sigma}{((g_\sigma)^s \cdot (h_\sigma)^t)^r} = \frac{g^s \cdot h^t \cdot x_\sigma}{((g_\sigma)^r)^s \cdot ((h_\sigma)^r)^t} = \frac{g^s \cdot h^t \cdot x_\sigma}{g^s \cdot h^t} = x_\sigma$$

and so the receiver obtains the desired output.

Regarding security, if (g_0, g_1, h_0, h_1) is *not* a DH tuple, then the receiver can learn only one of the sender's inputs, since in that case one of the two pairs (u_0, w_0), (u_1, w_1) is uniformly distributed and therefore reveals no information about the corresponding input of the sender. This is proven in the same way as Claim 7.2.3. The nice observation is that in order to prove that (g_0, g_1, h_0, h_1) is not a DH tuple, it suffices to prove that $(g_0, g_1, h_0, \frac{h_1}{g_1})$ *is* a DH tuple, which can be efficiently proven in zero knowledge. The proof of security takes advantage of the fact that a simulator can extract R's input-bit σ because it can extract the value α_0 from the zero-knowledge proof of knowledge proven by R. Given α_0, the simulator can compute $\alpha_1 = \alpha_0 + 1$ and then check if $h = g^{\alpha_0}$ (in which case $\sigma = 0$) or if $h = g^{\alpha_1}$ (in which case $\sigma = 1$). For simulation in the case where S is corrupted, the simulator sets $\alpha_0 = \alpha_1$ and cheats in the zero-knowledge proof, enabling it to extract both sender inputs. We have the following theorem:

Theorem 7.5.2 *Assume that the DDH problem is hard in $\mathbb{G}$ with generator g. Then, Protocol 7.5.1 securely computes $\mathcal{F}_{\mathrm{OT}}$ in the presence of malicious adversaries.*

Exact efficiency. In the oblivious transfer without the zero-knowledge proof, the sender computes eight exponentiations and the receiver computes six. The zero-knowledge proof adds an additional five exponentiations for the prover (which is played by the receiver) and seven for the verifier (which is played by the sender). In addition, the parties exchange fourteen group elements (including the zero-knowledge proof), and the protocol takes six rounds of communication (three messages are sent by each party). In summary, there are 26 exponentiations, fourteen group elements sent and six rounds of communication.

Batch oblivious transfer and efficiency. The main observation here is that Steps 1 and 2 are independent of the parties' inputs (x_0, x_1) and σ. Furthermore, the tuple (g_0, g_1, h_0, h_1) can be reused for many exchanges. Thus, a batch oblivious transfer is obtained by running Steps 1 and 2 once, and then repeating Steps 3–5 for every execution. Observe also that the executions do not need to be run at the same time, and so this optimization can be utilized when oblivious transfers are carried out over time. This yields an extraordinarily efficient batch protocol. Specifically, the cost of carrying out m transfers is as follows:

1. *Rounds of communication:* Steps 3–5 can be run in parallel and so there are still six rounds of communication.
2. *Exponentiations:* the number of exponentiations is equal to $11m + 15$.
3. *Bandwidth:* overall $4m + 8$ group elements are sent.

7.6 Secure Pseudorandom Function Evaluation

Intuitively, the secure pseudorandom function evaluation problem is as follows. One party holds a secret key k for a pseudorandom function and the other party has an input x. The aim is for the second party to learn $F(k, x)$, where F is a pseudorandom function, without the first party learning anything about x.

Formally, let $(I_{\mathrm{PRF}}, F_{\mathrm{PRF}})$ be an ensemble of pseudorandom functions, where $I_{\mathrm{PRF}}(1^n)$ is a probabilistic polynomial-time algorithm that generates keys (or more exactly, that samples a function $k \leftarrow I_{\mathrm{PRF}}(1^n)$ from the ensemble), and $F_{\mathrm{PRF}}(k, \cdot)$ is a keyed function that is indistinguishable from a truly random function $H_{\mathrm{Func}}(\cdot)$. This is formalized by giving the distinguisher oracle access to either $F_{\mathrm{PRF}}(k, \cdot)$ or $H_{\mathrm{Func}}(\cdot)$ and comparing its output in both cases; for more details see [30]. The pseudorandom function evaluation functionality $\mathcal{F}_{\mathrm{PRF}}$ is defined by

$$(k, x) \mapsto \begin{cases} (\lambda, F_{\mathrm{PRF}}(k, x)) & \text{if } k \text{ is a valid key} \\ (\perp, \lambda) & \text{otherwise} \end{cases} \qquad (7.5)$$

where $k \leftarrow I_{\mathrm{PRF}}(1^n)$ and $x \in \{0, 1\}^m$.[5] Observe that the length of the input x is m, whereas the security parameter is n.

We show how to compute this functionality for the Naor-Reingold [64] pseudorandom function ensemble (with some minor modifications). For every n, the function's key is the tuple $k = (\mathbb{G}, q, g, g^{a_0}, a_1, \ldots, a_m)$, where $\mathbb{G}$ is the description of a cyclic group of prime order q (of length n), g is a generator of $\mathbb{G}$, and $a_0, a_1, \ldots, a_m \leftarrow_R \mathbb{Z}_q^*$. Note that the function is defined over inputs of length m. (This is slightly different from the description in [64] but makes no difference to the pseudorandomness of the ensemble.) The validity of the key k can be verified as follows. First, check that q is prime and that g is a generator of $\mathbb{G}$. Next, the values $a_1, \ldots, a_m$ should be strings from $\mathbb{Z}_q^*$, and g^{a_0} should also be of order q. (Note that for all $a \in \mathbb{Z}_q^*$, g^a is of order q because the group is of prime order, and so all elements are generators. Thus, if g is correctly chosen then this check will pass.) The function itself is defined by

$$F_{\mathrm{PRF}}(k, x) = g^{a_0 \cdot \prod_{i=1}^m a_i^{x_i}}.$$

We remark that this function is not pseudorandom in the classical sense of it being indistinguishable from a random function whose range is composed of all strings of a given length. Rather, it is indistinguishable from a random function whose range is the group $\mathbb{G}$. Nevertheless, this is sufficient for most applications.

A protocol for securely computing the pseudorandom function evaluation functionality for the Naor-Reingold function was presented in [28] and involves the parties running an oblivious transfer execution for every bit of the input x. Consequently, the security level of this protocol relies heavily on the security of the oblivious transfer subprotocol. Furthermore, by using OT with different levels of security (privacy only, one-sided simulation and so on), we obtain a set of protocols for $\mathcal{F}_{\mathrm{PRF}}$ with different levels of security.

7.6.1 Pseudorandom Function – Privacy Only

We begin by defining privacy for the pseudorandom function functionality. This definition follows the same intuition of private oblivious transfer that was given in Section 2.6.1. Specifically, privacy in the presence of a malicious

---

[5] An alternative formulation is to take any arbitrary value if k is not a valid key in the range of $I_{\mathrm{PRF}}(1^n)$.

P_1^* is formalized by showing that P_1^* cannot distinguish between the case where P_2 has input x and the case where it has input x', for any two strings x, x' of the same size. Since P_1^* does not obtain any output, this suffices. In contrast, party P_2^* does receive output; in particular, it learns $F_{\mathrm{PRF}}(k, x)$ for some value x of its choice. Thus, we must somehow require that it learn $F_{\mathrm{PRF}}(k, x)$ for one value of x and only one value of x. This is problematic to define for the case of general protocols because the value x used by P_2^* is not necessarily well defined; see the discussion regarding oblivious transfer in Section 2.6.1. In the case of oblivious transfer, we defined privacy specifically for the case of a two-message protocol (see Section 2.6.1 as to why this is beneficial). We therefore use the same methodology and define privacy for the case of pseudorandom function evaluation specifically for the case of two-message protocols.

Intuitively, in a two-message protocol (in which P_2^* receives output), the input used by P_2^* is *fully determined* by its first message. Therefore, we wish to formalize privacy by saying that for every first message from P_2^* there exists an input x such that P_2^* knows $F_{\mathrm{PRF}}(k, x)$, but for every other $x' \neq x$ it cannot distinguish $F_{\mathrm{PRF}}(k, x')$ from random. As in the case of oblivious transfer, we do not quantify over possible first messages. Rather, we observe that when considering deterministic non-uniform adversaries (which are no weaker than PPT adversaries), the first message is fully defined by the adversary's code and its auxiliary input. Then, a first attempt at defining privacy is to consider a game where P_2^* is given an oracle that it can query on every point except for x (as determined by its first message), and then to require that it cannot distinguish between the case where its oracle is a random function from the case where its oracle is $F_{\mathrm{PRF}}(k, \cdot)$, where k is the same key as used by P_1 in the protocol. The reason P_2^*'s oracle is restricted to not computing the function on x is to prevent P_2^* from comparing the output of the protocol execution with P_1 (which is always $F_{\mathrm{PRF}}(k, x)$) to the computation of the oracle on x. Although this definition is intuitively appealing, we were not able prove security according to it.

We therefore define privacy in the case of a malicious P_2^* in a different way. First, we provide P_2^* with an oracle (which is either a random function or F_{PRF}) and do not limit its access to the oracle; in particular, it may query the oracle on x. Second, we compare a real protocol execution between P_1 with input key k and P_2^* with oracle $F_{\mathrm{PRF}}(k, \cdot)$ to an execution between an imaginary party $\hat{P}_1$ given $H_{\mathrm{Func}}(x)$ and P_2^* given oracle access to H_{Func}, where H_{Func} is a random function whose range is $\mathbb{G}$. Intuitively, $\hat{P}_1$ is a "mini-simulator" that is given $H_{\mathrm{Func}}(x)$ where x is the determined input used by P_2^*. Then, the requirement is that the view of P_2^* generated by the execution with $\hat{P}_1$ be indistinguishable from its view in a real execution with P_1. This means that P_2^*'s view in a world with a random function is indistinguishable from its view in a world with F_{PRF}, and thus all of the intuition stating that the pseudorandom function behaves like a random function follows.

Definition 7.6.1 *Let $m = m(n)$ be a polynomial. A two-message two-party probabilistic polynomial-time protocol (P_1, P_2) is a* private pseudorandom function evaluation *for messages of length m if the following holds:*

- NON-TRIVIALITY: *If P_1 and P_2 follow the protocol then after an execution in which P_1 has for input a key k and P_2 has for input any string $x \in \{0,1\}^m$, the output of P_2 is $F_{\mathrm{PRF}}(k, x)$.*
- PRIVACY IN THE CASE OF A MALICIOUS P_1^*: *For every non-uniform probabilistic polynomial-time P_1^*, for every pair of inputs $x, x' \in \{0,1\}^m$ and every auxiliary input $z \in \{0,1\}^*$, it holds that*

$$\big\{\mathrm{VIEW}_{P_1^*}(P_1^*(1^n, z), P_2(1^n, x))\big\}_{n \in \mathbb{N}} \stackrel{c}{\equiv} \big\{\mathrm{VIEW}_{P_1^*}(P_1^*(1^n, z), P_2(1^n, x'))\big\}_{n \in \mathbb{N}}.$$

- PRIVACY IN THE CASE OF A MALICIOUS P_2^*: *There exists a probabilistic polynomial-time $\hat{P}_1$, such that for every non-uniform deterministic polynomial-time receiver P_2^* and every auxiliary input $z \in \{0,1\}^*$, there exists an input $x \in \{0,1\}^m$ such that*

$$\left\{\mathrm{VIEW}_{P_2^*}(P_1(1^n, k); P_2^{*F_{\mathrm{PRF}}(k, \cdot)}(1^n, z))\right\}_{n \in \mathbb{N}}$$
$$\stackrel{c}{\equiv} \left\{\mathrm{VIEW}_{P_2^*}(\hat{P}_1(1^n, H_{\mathrm{Func}}(x)); P_2^{*H_{\mathrm{Func}}}(1^n, z))\right\}_{n \in \mathbb{N}}$$

where $k \leftarrow I_{\mathrm{PRF}}(1^n)$ is randomly chosen.

Constructing private pseudorandom function evaluation. Recall first that the protocol for $\mathcal{F}_{\mathrm{PRF}}$, as presented in [28], requires an oblivious transfer evaluation for every bit in P_2's input. Moreover, as in private oblivious transfer, P_2^*'s first message must determine its input x and the value it learns from the execution. This can be achieved by having the parties run all of the oblivious transfer executions (for all bits of P_2's input) simultaneously, thereby having P_2's first message in the protocol determine its *entire* input x. We use batch oblivious transfer, denoted by $\pi_{\mathrm{BOT}}^{\mathrm{P}}$, for this task (see Section 7.4.2). Such a protocol can be achieved by simply running the private oblivious transfer protocol $\pi_{\mathrm{OT}}^{\mathrm{P}}$ (see Section 7.2) many times in parallel. Formally, we define private batch oblivious transfer as follows:

Definition 7.6.2 *Let $m = m(n)$ be a polynomial. A two-message two-party probabilistic polynomial-time protocol (S, R) is said to be a* private batch oblivious transfer *for m executions if the following holds:*

- NON-TRIVIALITY: *If S and R follow the protocol then after an execution in which S has for input any vector $\bar{x}$ of pairs of strings $x_0^i, x_1^i \in \{0,1\}^*$ $(1 \leq i \leq m)$, and R has for input any string $\bar{\sigma} = \sigma_1, \ldots, \sigma_m \in \{0,1\}$, the output of R is the vector $(x_{\sigma_1}^1, \ldots, x_{\sigma_m}^m)$.*
- PRIVACY IN THE CASE OF A MALICIOUS S^*: *For every non-uniform probabilistic polynomial-time S^*, every auxiliary input $z \in \{0,1\}^*$, and every pair of inputs $\bar{\sigma}, \bar{\sigma}' \in \{0,1\}^m$ for R it holds that*

$$\{\text{VIEW}_{S^*}(S^*(1^n, z), R(1^n, \overline{\sigma}))\}_{n \in \mathbb{N}} \overset{c}{\equiv} \{\text{VIEW}_{S^*}(S^*(1^n, z), R(1^n, \overline{\sigma}'))\}_{n \in \mathbb{N}}.$$

- PRIVACY IN THE CASE OF A MALICIOUS R^*: *For every non-uniform deterministic polynomial-time receiver R^* and every auxiliary input $z \in \{0, 1\}^*$, there exists a vector $\overline{\sigma} \in \{0, 1\}^m$ such that for every triple of vectors $\overline{x}_0, \overline{x}_1, \overline{x} \in \{0, 1\}^*$ such that for every i $|x_0^i| = |x_1^i| = |x^i|$ it holds that*

$$\{\text{VIEW}_{R^*}(S(1^n, (\overline{x}_0, \overline{x}_1)); R^*(1^n, z))\}_{n \in \mathbb{N}}$$
$$\overset{c}{\equiv} \{\text{VIEW}_{R^*}(S(1^n, (\overline{y}_0, \overline{y}_1)); R^*(1^n, z))\}_{n \in \mathbb{N}}$$

where for every i, $y_{\sigma_i}^i = x_{\sigma_i}^i$ and $y_{1-\sigma_i}^i = x^i$.

We now proceed to describe the protocol, using $\pi_{\text{BOT}}^{\text{P}}$ as defined above.

PROTOCOL 7.6.3 (Private Pseudorandom Function Evaluation $\pi_{\text{PRF}}^{\text{P}}$)

- **Inputs:** The input of P_1 is $k = (g^{a_0}, a_1, \ldots, a_m)$ and the input of P_2 is a value x of length m.
- **Auxiliary inputs:** Both parties have the security parameter 1^n and are given $\mathbb{G}$, q and g, as described above.
- **The protocol:**

  1. P_1 chooses m random values $r_1, \ldots, r_m \leftarrow_R \mathbb{Z}_q^*$.
  2. The parties engage in a 1-out-2 private batch oblivious transfer protocol $\pi_{\text{BOT}}^{\text{P}}$. In the ith iteration, P_1 inputs $y_0^i = r_i$ and $y_1^i = r_i \cdot a_i$ (with multiplication in $\mathbb{Z}_q^*$), and P_2 enters the bit $\sigma_i = x_i$ where $x = x_1, \ldots, x_m$. If the output of any of the oblivious transfers is $\perp$, then both parties output $\perp$ and halt.
  3. Otherwise, P_2's output from the m executions is a series of values $y_{x_1}^1, \ldots, y_{x_m}^m$. If any value $y_{x_i}^i$ is not in $\mathbb{Z}_q^*$, then P_2 redefines it to equal 1.
  4. P_1 computes $\tilde{g} = g^{a_0 \cdot \prod_{i=1}^m \frac{1}{r_i}}$ and sends it to P_2.
  5. P_2 computes $y = \tilde{g}^{\prod_{i=1}^m y_{x_i}^i}$ and outputs y.

Observe that Protocol 7.6.3 is a two-message protocol. In the first message, P_2 sends its receiver-message in $\pi_{\text{BOT}}^{\text{P}}$, and in the second message P_1 sends its sender-message in $\pi_{\text{BOT}}^{\text{P}}$ together with $\tilde{g}$. We now prove security:

Theorem 7.6.4 *Assume that $\pi_{\text{BOT}}^{\text{P}}$ privately computes $\mathcal{F}_{\text{BOT}}$ as in Definition 7.6.2 and that the DDH assumption holds in $\mathbb{G}$. Then, Protocol 7.6.3 privately computes $\mathcal{F}_{\text{PRF}}$ as in Definition 7.6.1.*

Proof. The first requirement of Definition 7.6.1 is that of non-triviality, and we prove this first. Let k be P_1's input and let x be P_2's input. Non-triviality follows from the fact that

$$y = \tilde{g}^{\prod_{i=1}^m y_{x_i}^i} = g^{a_0 \cdot \prod_{i=1}^m \frac{y_{x_i}^i}{r_i}} = g^{a_0 \cdot \prod_{i=1}^m a_i^{x_i}} = F_{\text{PRF}}(k, x)$$

where the second last equality is due to the fact that for $x_i = 0$ it holds that $a_i^{x_i} = 1$ and $y_{x_i}^i / r_i = y_0^i / r_i = 1$, and for $x_i = 1$ it holds that $a_i^{x_i} = a_i$ and $y_{x_i}^i / r_i = y_1^i / r_i = a_i$.

Next, we prove the requirement of privacy in the case of a malicious P_1^*. Recall that this requirement is that P_1^*'s view when P_2 has input x is indistinguishable from its view when P_2 has input x', for any two strings $x, x' \in \{0,1\}^m$. Now, the view of an adversarial P_1^* in Protocol 7.6.3 consists merely of P_2's first message within $\pi_{\text{BOT}}^{\text{P}}$. By the privacy of this protocol we have that for every pair of inputs $\overline{\sigma}, \overline{\sigma}' \in \{0,1\}^m$

$$\{\text{VIEW}_{S^*}(S^*(1^n, z), R(1^n, \overline{\sigma}))\}_{n \in \mathbb{N}} \stackrel{c}{\equiv} \{\text{VIEW}_{S^*}^n(S^*(1^n, z), R(1^n, \overline{\sigma}'))\}_{n \in \mathbb{N}}.$$

Replacing S^* with P_1^*, R with P_2, and $\overline{\sigma}$ and $\overline{\sigma}'$ with x and x' respectively, we obtain that

$$\{\text{VIEW}_{P_1^*}(P_1^*(1^n, z), P_2(1^n, x))\}_{n \in \mathbb{N}} \stackrel{c}{\equiv} \{\text{VIEW}_{P_1^*}(P_1^*(1^n, z), P_2(1^n, x'))\}_{n \in \mathbb{N}}$$

as required.

We now prove privacy in the case of a malicious P_2^*. We begin by describing the party $\hat{P}_1$ from Definition 7.6.1. Party $\hat{P}_1$ receives a value $w = H_{\text{Func}}(x)$ and runs the following protocol. For every i it sets both of its inputs x_0^i, x_1^i to the ith oblivious transfer to be the same random value $r_i \leftarrow_R \mathbb{Z}_q^*$, and thus $x_0^i = x_1^i = r_i$. Then, it sends P_2^* the value $\hat{g} = w^{\prod_{i=1}^m \frac{1}{r_i}}$.

We now reduce the security of Protocol 7.6.4 to the privacy of the batch oblivious transfer $\pi_{\text{BOT}}^{\text{P}}$ and the pseudorandomness of F_{PRF}. In order to do this, we first define a hybrid game H_1 with the same parties $\hat{P}_1$ and P_2^*, except that instead of $\hat{P}_1$ receiving $H_{\text{Func}}(x)$ and P_2^* having an oracle H_{Func}, party $\hat{P}_1$ is given $F_{\text{PRF}}(k, x)$ and P_2^* is given oracle $F_{\text{PRF}}(k, \cdot)$ for a random $k \leftarrow I(1^n)$. (Note that $\hat{P}_1$ is the same in both games and thus computes $\hat{g} = w^{\prod_{i=1}^m \frac{1}{r_i}}$ in both games). The indistinguishability of the view of P_2^* in an execution with $\hat{P}_1$ with a random oracle H_{Func} from its view in an execution with $\hat{P}_1$ and an oracle F_{PRF} follows from a standard reduction to the pseudorandomness of F_{PRF}. (Observe that there is no difference in the instructions of the parties and so a distinguisher for F_{PRF} can just run $\hat{P}_1$ and P_2^* using its oracle and output P_2^*'s view at the end. If the oracle is random, then the view is as with a random function, and otherwise the view is as in H_1.)

Next, we define a game H_2 between P_2^* and a party P_1' that is given x and the key $k = (g^{a_0}, a_1, \ldots, a_m)$ of F_{PRF} and works by setting the input vectors to the oblivious transfers to be $y_0^i = y_1^i = r_i \cdot a_i^{x_i}$ (that is, if $x_i = 0$ then $y_0^i = y_1^i = r_i$, and otherwise $y_0^i = y_1^i = r_i \cdot a_i$). In addition, P_1' computes $\tilde{g}' = w^{\prod_{i=1}^m \frac{1}{r_i \cdot a_i^{x_i}}}$. The view of P_2^* in game H_2 is *identical* to its view in game H_1 because the r_i values are random, and thus $r_i \cdot a_i^{x_i}$ is identically distributed to r_i.

It remains to show that the view of P_2^* in H_2 is computationally indistinguishable from its view in a real execution of $\pi_{\mathrm{PRF}}^{\mathrm{P}}$ with P_1. We prove this via a reduction to the privacy of the batch oblivious transfer. Observe that there are two differences between H_2 and $\pi_{\mathrm{PRF}}^{\mathrm{P}}$:

1. For every i, the inputs to the oblivious transfers in H_2 are $y_0^i = y_1^i = r_i \cdot a_i^{x_i}$, whereas in $\pi_{\mathrm{PRF}}^{\mathrm{P}}$ we have that $y_0^i = r_i$ and $y_1^i = r_i \cdot a_i$. (Note that when P_2^* uses input x_i in the ith execution, it obtains $r_i \cdot a_i^{x_i}$ in both cases.)

2. P_1' in H_2 computes $\tilde{g}' = w^{\prod_{i=1}^m \frac{1}{r_i \cdot a_i^{x_i}}}$, whereas P_1 in $\pi_{\mathrm{PRF}}^{\mathrm{P}}$ computes $\tilde{g} = g^{a_0 \cdot \prod_{i=1}^m \frac{1}{r_i}}$. (Note that when $w = g^{a_0 \cdot \prod_{i=1}^m a_i^{x_i}}$ it follows that $\tilde{g}' = \tilde{g}$.)

Let D be a distinguisher between the view of P_2^* in H_2 and its view in $\pi_{\mathrm{PRF}}^{\mathrm{P}}$. We construct a distinguisher D_{OT} for the batch oblivious transfer with receiver P_2^* that works as follows. D_{OT} is given x and $k = (g^{a_0}, a_1, \ldots, a_m)$. Then, given the view of P_2^* from the batch OT, it computes $\tilde{g}' = w^{\prod_{i=1}^m \frac{1}{r_i \cdot a_i^{x_i}}}$ where $w = F_{\mathrm{PRF}}(k, x)$. D_{OT} then invokes D on the view of P_2^* from the batch OT concatenated with the message $\tilde{g}'$. Finally D_{OT} outputs whatever D outputs.

We now define vectors of inputs to the batch OT, and show that if D successfully distinguishes between H_2 and $\pi_{\mathrm{PRF}}^{\mathrm{P}}$, then D_{OT} successfully distinguishes between these vectors when used in batch OT. Specifically, we define $\overline{\alpha}_0 = (r_1, \ldots, r_m)$ and $\overline{\alpha}_1 = (r_1 \cdot a_1, \ldots, r_m \cdot a_m)$. In addition, we define $\overline{\alpha} = (r_1 \cdot a_1^{x_1}, \ldots, r_1 \cdot a_1^{x_1})$. (Since x is already used in this context for P_2^*'s input, we replace the vectors $\overline{x}_0, \overline{x}_1, \overline{x}$ in Definition 7.6.1 with $\overline{\alpha}_0, \overline{\alpha}_1, \overline{\alpha}$.) Now, let $x = x_1, \ldots, x_m$ be the input to all the oblivious transfers, defined by $P_2^*(1^n, z)$. We then have that in the real execution of $\pi_{\mathrm{PRF}}^{\mathrm{P}}$, the vectors of inputs of P_1 to the oblivious transfers are exactly $\overline{\alpha}_0, \overline{\alpha}_1$. Furthermore, in H_2 the pair of inputs in the ith OT are both the same value $r_i \cdot a_i^{x_i}$. Observe that when we define a pair of vectors $\overline{\beta}_0, \overline{\beta}_1$ (instead of $\overline{y}_0, \overline{y}_1$ in Definition 7.6.1) so that for every i, $\beta_{x_i}^i = \alpha_{x_i}^i$ and $\beta_{1-x_i}^i = \alpha^i$, it follows that for every i both inputs are the same value $r_i \cdot a_i^{x_i}$. Finally, as we have already explained above, the computed value $\tilde{g}'$ in H_2 is exactly the same as the computed value $\tilde{g}$ in $\pi_{\mathrm{PRF}}^{\mathrm{P}}$. Therefore, we conclude that the view generated by D_{OT} (and given to D) when D_{OT} runs batch OT with a sender that inputs vectors $\overline{\beta}_0, \overline{\beta}_1$ is identical to the view of P_2^* in H_2. Furthermore, the view generated by D_{OT} (and given to D) when D_{OT} runs batch OT with a sender that inputs vectors $\overline{\alpha}_0, \overline{\alpha}_1$ is identical to the view of P_2^* in $\pi_{\mathrm{PRF}}^{\mathrm{P}}$. Thus, D_{OT} distinguishes with the same probability as D. We conclude that the view of P_2^* in H_2 is indistinguishable from its view in $\pi_{\mathrm{PRF}}^{\mathrm{P}}$, and thus the view of P_2^* in an execution with $\hat{P}_1$ is indistinguishable from its view in $\pi_{\mathrm{PRF}}^{\mathrm{P}}$, completing the proof. ∎

Efficiency. The protocol has only two rounds and consists of running a single batch OT with parameter m, with the addition of sending a single group element and computing a single exponentiation. Thus, using the private OT protocol of Section 7.2 we have that the number of exponentiations is $13m+1$ (this can be reduced to $12m+2$ by reusing the same α in all executions).

7.6.2 Pseudorandom Function – Full Simulation

We now present a protocol for computing the pseudorandom function evaluation functionality that is fully secure in the presence of malicious adversaries.

PROTOCOL 7.6.5 (Fully Simulatable PRF Evaluation π_{PRF})

- **Inputs:** The input of P_1 is $k = (\mathbb{G}, q, g, g^{a_0}, a_1, \ldots, a_m)$ and the input of P_2 is a value x of length m.
- **Auxiliary inputs:** Both parties have the security parameter 1^n and are given $\mathbb{G}$, q and g as above.
- **The protocol:**

    1. P_1 chooses m random values $r_1, \ldots, r_m \leftarrow_R \mathbb{Z}_q^*$.
    2. The parties engage in a 1-out-2 batch oblivious transfer protocol π_{BOT}. In the ith iteration, P_1 inputs $y_0^i = r_i$ and $y_1^i = r_i \cdot a_i$ (with multiplication in $\mathbb{Z}_q^*$), and P_2 enters the bit $\sigma_i = x_i$ where $x = x_1, \ldots, x_m$. If the output of any of the oblivious transfers is $\perp$, then both parties output $\perp$ and halt.
    3. Otherwise, P_2's output from the m executions is a series of values $y_{x_1}^1, \ldots, y_{x_m}^m$. If any value $y_{x_i}^i$ is not in $\mathbb{Z}_q^*$, then P_2 redefines it to equal 1.
    4. P_1 computes $\tilde{g} = g^{a_0 \cdot \prod_{i=1}^m \frac{1}{r_i}}$ and sends it to P_2.
    5. P_2 aborts if the order of $\tilde{g}$ is different than q. Otherwise, P_2 computes $y = \tilde{g}^{\prod_{i=1}^m y_{x_i}^i}$ and outputs y.

We remark that the only differences between Protocols 7.6.5 and 7.6.3 are that (1) the batch OT here is fully secure whereas in Protocol 7.6.3 it only achieved privacy, and (2) we require P_2 to check $\tilde{g}$ in the last step because we need correctness here (and this was not required for Protocol 7.6.3).

Theorem 7.6.6 *Assume that π_{BOT} securely computes $\mathcal{F}_{\text{BOT}}$ in the presence of malicious adversaries. Then Protocol 7.6.5 securely computes $\mathcal{F}_{\text{PRF}}$ in the presence of malicious adversaries.*

Proof. We separately analyze the case where P_1 is corrupted and the case where P_2 is corrupted. We prove the theorem in the hybrid model where a trusted party is used to compute $\mathcal{F}_{\text{BOT}}$; see Section 2.7.

Party P_1 is corrupted. Let $\mathcal{A}$ be an adversary controlling P_1. We construct a simulator $\mathcal{S}_1$ as follows.

1. $\mathcal{S}_1$ receives $\mathcal{A}$'s inputs for all the m iterations of the oblivious transfers for $\mathcal{F}_{\text{BOT}}$ (recall our analysis is in the hybrid model). Let y_{i_0} and y_{i_1} denote

the inputs that $\mathcal{A}$ handed $\mathcal{S}_1$ in the ith iteration. In addition, $\mathcal{S}_1$ receives from $\mathcal{A}$ the message $\tilde{g}$.

2. In case $\mathcal{A}$ does not send a valid message (where $\mathcal{S}_1$ conducts the same check as the honest P_2 does), $\mathcal{S}_1$ simulates P_2 aborting and sends $\perp$ to the trusted party for $\mathcal{F}_{\text{PRF}}$. Otherwise, $\mathcal{S}_1$ checks the validity of all the y_0^i and y_1^i values and modifies them to 1 if necessary (as an honest P_2 would). $\mathcal{S}_1$ defines $g_0 = \tilde{g}$. Then, for every $i = 1, \ldots, m$, $\mathcal{S}_1$ defines:

$$a_i = \frac{y_1^i}{y_0^i} \quad \text{and} \quad g_i = (g_{i-1})^{y_0^i}.$$

3. $\mathcal{S}_1$ defines the key used by $\mathcal{A}$ to be $k = (p, q, g_m, a_1, \ldots, a_m)$ and sends it to $\mathcal{F}_{\text{PRF}}$.

This completes the description of $\mathcal{S}_1$. It is immediate that the view of $\mathcal{A}$ is identical in a real and simulated execution because it receives no messages in a hybrid execution where the oblivious transfers are run by a trusted party. It thus remains to show that the output received by P_2 in a real execution is the same as in the ideal model. In order to see this, first note that $\mathcal{S}_1$ and P_2 replace any invalid y_0^i or y_1^i values in the same way. Next, note that for any $x \in \{0, 1\}^m$

$$F_{\text{PRF}}(k, x) = g_m^{\prod_{i=1}^m a_i^{x_i}} = \tilde{g}^{\prod_{i=1}^m y_0^i \cdot a_i^{x_i}} = \tilde{g}^{\prod_{i=1}^m y_0^i \cdot \left(\frac{y_1^i}{y_0^i}\right)^{x_i}} = \tilde{g}^{\prod_{i=1}^m \frac{y_0^i}{(y_0^i)^{x_i}} \cdot (y_1^i)^{x_i}}$$

where the first equality is by the definition of the key by $\mathcal{S}_1$, the second equality is by the fact that $g_m = \tilde{g}^{\prod_{i=1}^m y_0^i}$, and the third equality is by the fact that $a_i = y_1^i/y_0^i$. Notice now that if $x_i = 0$ we have that $y_0^i/(y_0^i)^{x_i} \cdot (y_1^i)^{x_i} = y_0^i$, whereas if $x_i = 1$ we have that $y_0^i/(y_0^i)^{x_i} \cdot (y_1^i)^{x_i} = y_1^i$. Thus,

$$\prod_{i=1}^m \frac{y_0^i}{(y_0^i)^{x_i}} \cdot (y_1^i)^{x_i} = \prod_{i=1}^m y_{x_i}^i,$$

exactly as computed by P_2 in a real execution. That is, the computation of $F_{\text{PRF}}(k, x)$ as carried out by the trusted party using the key supplied by $\mathcal{S}_1$ is the same as that obtained by P_2 in a real execution. Formally,

$$\{\text{IDEAL}_{\mathcal{F}_{\text{PRF}}, \mathcal{S}_1(z), 1}((\mathbb{G}, q, g, g^{a_0}, a_1, \ldots, a_m), x, n)\}$$
$$\equiv \{\text{HYBRID}_{\pi_{\text{PRF}}, \mathcal{A}(z), 1}^{\text{BOT}}((\mathbb{G}, q, g, g^{a_0}, a_1, \ldots, a_m), x, n)\}.$$

This completes the case where P_1 is corrupted.

Party P_2 is corrupted. In this case, $\mathcal{S}_2$ learns the full input x of $\mathcal{A}$ controlling P_2 (through the oblivious transfer inputs). In each oblivious transfer, $\mathcal{S}_2$ hands $\mathcal{A}$ a random value $r_i \leftarrow_R \mathbb{Z}_q^*$ (as if coming from the trusted third party computing the batch OT). After all of the oblivious transfers have concluded, $\mathcal{S}_2$ sends x to $\mathcal{F}_{\text{PRF}}$ and receives back a value $y = F_{\text{PRF}}(k, x)$. $\mathcal{S}_2$ then sets $\tilde{g} = y^{\prod_{i=1}^m \frac{1}{r_i}}$ and sends it to $\mathcal{A}$. This completes the simulation.

We claim that the view of $\mathcal{A}$ in an execution of π_{PRF} with P_1 (using a trusted party for the oblivious transfers) is identical to its view in an ideal execution with $\mathcal{S}_2$. This is true because all of the r_i values are distributed identically to the messages sent in a real execution (note that r_i and $r_i \cdot a_i$ have the same distribution). Furthermore, in a real execution, it holds that $\tilde{g}^{\Pi_{i=1}^m y_{x_i}^i} = g^{a_0 \Pi_{i=1}^m a_i^{x_i}} = y$, where the x_i values are those used by P_2 in the oblivious transfers. Likewise, in the simulation it holds that

$$\tilde{g}^{\Pi_{i=1}^m y_{x_i}^i} = \left(y^{\Pi_{i=1}^m \frac{1}{r_i}} \right)^{\Pi_{i=1}^m r_i} = y = g^{a_0 \Pi_{i=1}^m a_i^{x_i}}$$

where the first equality is due to the fact that the simulator sets each y^i value received by $\mathcal{A}$ to r_i, and the last equality is by the fact that y is computed correctly by the trusted party. Thus, the joint distribution over the values received by $\mathcal{A}$ and $\tilde{g}$ in the hybrid and ideal executions are exactly the same. Formally,

$$\left\{ \text{IDEAL}_{\mathcal{F}_{\text{PRF}}, \mathcal{S}_2(z), 2}((\mathbb{G}, q, g, g^{a_0}, a_1, \ldots, a_m), x, n) \right\}$$
$$\equiv \left\{ \text{HYBRID}_{\pi_{\text{PRF}}, \mathcal{A}(z), 2}^{\text{BOT}}((\mathbb{G}, q, g, g^{a_0}, a_1, \ldots, a_m), x, n) \right\},$$

and the proof is concluded.   ∎

Remark. Observe that when proving the security of Protocol 7.6.3 in Theorem 7.6.4 we assumed that the DDH assumption holds, whereas when proving the security of Protocol 7.6.5 in Theorem 7.6.6 we did not. This may seem strange at first sight. However, our definition of privacy only compares the output received by P_2 to a random function, and this only holds if the Naor-Reingold function is pseudorandom, which in turn relies on the DDH assumption. In contrast, the definition of security for the pseudorandom function evaluation functionality which is based on simulation merely states that the view of a corrupted party in a real execution is indistinguishable from its view when a trusted party computes the specified Naor-Reingold function. The important point to note is that Protocol 7.6.5 securely computes the Naor-Reingold function even if it is not pseudorandom. (One could argue that this should be called Naor-Reingold evaluation and not pseudorandom function evaluation. Nevertheless, under the DDH assumption, this is a pseudorandom function evaluation.)

7.6.3 Covert and One-Sided Simulation

As shown above, protocol π_{PRF} is carried out by m parallel invocations of oblivious transfers. Therefore, we conclude that by employing a subprotocol for $\mathcal{F}_{\text{OT}}$ that is secure in the presence of covert adversaries, we derive a protocol for $\mathcal{F}_{\text{PRF}}$ that is also secure in the presence of covert adversaries.

Theorem 7.6.7 *Assume that $\pi_{\mathrm{BOT}}^{\mathrm{CO}}$ securely computes the batch oblivious transfer functionality in the presence of covert adversaries with deterrent ϵ, and let $\pi_{\mathrm{PRF}}^{\mathrm{CO}}$ be the same as Protocol 7.6.5 except that $\pi_{\mathrm{BOT}}^{\mathrm{CO}}$ is used instead of π_{BOT}. Then $\pi_{\mathrm{PRF}}^{\mathrm{CO}}$ securely computes $\mathcal{F}_{\mathrm{PRF}}$ in the presence of covert adversaries with deterrent ϵ.*

Similarly, by replacing the batch oblivious transfer in protocol π_{PRF} with batch OT that is secure under one-sided simulation we obtain a protocol for π_{PRF} that achieves one-sided simulation. In this case, the proof of security when P_1^* is malicious can be taken from protocol $\pi_{\mathrm{PRF}}^{\mathrm{P}}$ (because in both cases privacy only is needed), and the proof of security when P_2^* is malicious can be taken from protocol π_{PRF} (because in both cases simulation is needed).

Theorem 7.6.8 *Assume that $\pi_{\mathrm{BOT}}^{\mathrm{OS}}$ securely computes the batch oblivious transfer functionality with one-sided simulation, and let $\pi_{\mathrm{PRF}}^{\mathrm{OS}}$ be the same as Protocol 7.6.5 except that $\pi_{\mathrm{BOT}}^{\mathrm{OS}}$ is used instead of π_{BOT}. Then $\pi_{\mathrm{PRF}}^{\mathrm{OS}}$ securely computes $\mathcal{F}_{\mathrm{PRF}}$ with one-sided simulation.*

7.6.4 Batch Pseudorandom Function Evaluation

In a similar way to batch oblivious transfer, we define the batch pseudorandom function evaluation functionality as follows:

$$\mathcal{F}_{\mathrm{BPRF}} : (k, (x_1, \ldots, x_\ell)) \to (\lambda, (F_{\mathrm{BPRF}}(k, x_1), \ldots, F_{\mathrm{BPRF}}(k, x_\ell))). \qquad (7.6)$$

Observe that the definition requires that the PRF evaluations be computed relative to the same key. Thus, it is not possible to simply run π_{PRF} many times, because a malicious P_1^* may use different keys in different executions. Thus, correctness – in the sense that the output of the honest party is correctly computed as in the functionality definition – is not achieved. (This is in contrast to batch OT where the parties may use any input in any execution and the only issue is to make the batch execution more efficient than just running the basic protocol many times.) Due to the above, we do not present a protocol for batch pseudorandom function evaluation that is fully secure in the presence of malicious adversaries. However, note that in the models of privacy only and one-sided simulation, there is no requirement of correctness. Thus, simultaneously running many executions of protocols $\pi_{\mathrm{PRF}}^{\mathrm{P}}$ and $\pi_{\mathrm{PRF}}^{\mathrm{OS}}$ *does* yield batch PRF evaluation that achieves privacy and one-sided simulation, respectively. We have the following proposition (for the sake of clarity, we state it for one-sided simulation only):

Proposition 7.6.9 *Assume that $\pi_{\mathrm{BOT}}^{\mathrm{OS}}$ securely computes $\mathcal{F}_{\mathrm{BOT}}$ with one-sided simulation, and let $\pi_{\mathrm{BPRF}}^{\mathrm{OS}}$ be many parallel executions of $\pi_{\mathrm{PRF}}^{\mathrm{OS}}$, which uses $\pi_{\mathrm{BOT}}^{\mathrm{OS}}$. Then Protocol $\pi_{\mathrm{BPRF}}^{\mathrm{OS}}$ securely computes $\mathcal{F}_{\mathrm{BPRF}}$ with one-sided simulation.*

Chapter 8
The kth-Ranked Element

In this chapter we describe the construction of Aggarwal et al. [1] for securely computing the kth-ranked element of the union of two distributed and sorted data sets in the presence of semi-honest and malicious adversaries. An important special case of this problem is that of securely computing the *median*. The construction of [1] is based on the iterative protocol of [74] for computing the kth-ranked element with low communication complexity. The secure implementation of this fundamental problem has been looked at by researchers in the last few years, mostly due to its potential applications. One particular setting is where the data sets contain sensitive data, yet the particular kth element is of mutual interest. For instance, two health insurance companies may wish to compute the median life expectancy of their insured smokers, or two companies may wish to compute the median salary of their employees. By running secure protocols for these tasks, sensitive information is not unnecessarily revealed. In this chapter we show how to securely compute the two-party kth element functionality $\mathcal{F}_{\mathrm{Kth}}$, where each party enters a *set* of values from some predetermined domain and the output is the kth element of the sorted list comprised of the union of these sets.

8.1 Background

The functionality computing the kth-ranked element is defined as follows:

Definition 8.1.1 *Let X and Y be subsets of a domain $\{0,1\}^{p(n)}$ for some known polynomial $p(n)$. Then, the functionality $\mathcal{F}_{\mathrm{Kth}}$ is as follows:*

$$((X,k),(Y,k)) \mapsto \begin{cases} ((X \cup Y)_k, (X \cup Y)_k) & \text{if } |X \cup Y| \geq k \\ (\bot, \bot) & \text{otherwise} \end{cases}$$

where $(\Gamma)_i$ denotes the ith element within the sorted set Γ.

We remark that the protocol of [1] presented here in Section 8.3 that is secure in the presence of malicious adversaries is a rare example of a protocol that achieves this level of security without having the parties commit to their inputs at the onset of the protocol (as in the general construction in Chapter 4). Rather, the simulator is able to construct the inputs used by the

C. Hazay, Y. Lindell, *Efficient Secure Two-Party Protocols*,
Information Security and Cryptography, DOI 10.1007/978-3-642-14303-8_8,
© Springer-Verlag Berlin Heidelberg 2010

adversary dynamically, by continually rewinding the adversary. This is feasible in this context because the communication complexity of the protocol is logarithmic in the input length, and so continued rewinding of the adversary (which can be exponential in the communication complexity) yields a polynomial-time strategy.

8.1.1 A Protocol for Finding the Median

We begin by providing a detailed description of a protocol for computing the median with only logarithmic communication; this protocol is due to [74]. This protocol is not supposed to be secure; however, it forms the basis of the secure protocols presented below. We focus our attention on the case in which the parties compute the *median* of two *equally-sized disjoint sets*, where the size of each set is *a power of 2*. (By definition, the median of a set X with an even number of elements is often taken to be the mean of the two middle values. Yet, in this chapter we consider the element ranked $\lceil |X|/2 \rceil$.) We stress that the general case can be reduced to this simpler instance; see below for more details. The protocol is repeated in rounds where in each round the parties compare the medians of their updated sets, and then remove elements from their sets accordingly. Specifically, if, for instance, the median of party P_1 is larger than the median of P_2, then P_1 removes from its set the elements that are larger than or equal to its median, whereas P_2 removes the elements that are smaller than its median, and vice versa. The protocol is concluded when the data sets are of size 1, yielding that the number of iterations is logarithmic in the size of the sets. Let Γ be a sorted list with ℓ items. We write $\Gamma^\top$ to denote elements $\gamma_{\ell/2+1}, \ldots, \gamma_\ell \subset \Gamma$ and by $\Gamma^\perp$ denote the elements $\gamma_1, \ldots, \gamma_{\ell/2} \subset \Gamma$. The formal details of the protocol are given in Protocol 8.1.2.

PROTOCOL 8.1.2 (A Protocol for Computing the Median − FindMed)

- **Input:** Disjoint data sets X for party P_1 and Y for party P_2, with $|X| = |Y| = 2^\ell$ for some ℓ.
- **Output:** The median m of $X \cup Y$.
- **The protocol:**

  1. If $|X| = |Y| = 1$, then output $\mathsf{max}(X, Y)$.
  2. Else (if $|X|, |Y| > 1$):
      a. P_1 computes the median m_X of X and sends it to P_2.
      b. P_2 computes the median m_Y of Y and sends it to P_1.
      c. If $m_X > m_Y$, then P_1 sets $X = X^\perp$ and P_2 sets $Y = Y^\top$.
      d. Else (if $m_X < m_Y$), then P_1 sets $X = X^\top$ and P_2 sets $Y = Y^\perp$.
      e. Return FindMed(X,Y).

A protocol for computing the median with logarithmic communication

The main idea behind the correctness of the protocol is that an equal number of elements is removed from two sides of the median of $X \cup Y$, and so it does not change during the execution. That is, the median of $X \cup Y$ is the median of $(X^{\perp}, Y^{\top})$ when $m_x \geq m_Y$ and vice versa. This can be seen from the following concrete example. Assume that in some iteration, $m_X \geq m_Y$ and that $|X| = |Y| = 2^i$ for some i. Now, recall that by definition the median is the $(2^i + 1)$th-ranked element. Therefore, each item that P_1 removes is larger than at least $2^i + 1$ elements in the combined sets, including the final median (meaning that, the median must be smaller than m_X). Likewise, the elements removed by P_2 are smaller than at least $2^i + 1$ elements of the combined sets, including the final median (meaning that the median must be larger than m_Y). We conclude that the number of removed elements that are larger than the final median matches the number of removed elements that are smaller than the final median. Formally,

Claim 8.1.3 *Let X and Y be disjoint sets such that $|X| = |Y| = 2^i$ for some $i > 0$, and let m_X and m_Y be the medians of X and Y, respectively. If $m_X > m_Y$ then $m_X > m \geq m_Y$, where m is the median of $X \cup Y$.*

Proof. Within the set X there are exactly 2^{i-1} elements that are strictly smaller than m_X (because m_X is the median and we define it to be the $(2^{i-1} + 1)$th element). Likewise, within the set Y there are exactly 2^{i-1} elements that are strictly smaller than m_Y. Since $m_X > m_Y$, every element that is smaller than m_Y in Y is also smaller than m_X. In addition, m_Y itself is smaller than m_X. This implies that within the set $X \cup Y$ there are *at least* $2^i + 1$ elements that are strictly smaller than m_X. Therefore, the median m of $X \cup Y$ must be strictly less than m_X. That is, $m < m_X$.

Next, within the set Y there are exactly 2^{i-1} elements that are greater than or equal to m_Y (because m_Y is the median). Likewise, within the set X there are exactly 2^{i-1} elements that are greater than or equal to m_X. Since $m_Y < m_X$, every element that is greater than or equal to m_X is greater than m_Y. This implies that within the set $X \cup Y$ there are *at least* 2^i elements that are greater than or equal to m_Y. Therefore, the median m of $X \cup Y$ must be greater than or equal to m_Y. That is, $m \geq m_Y$. ∎

We now use the above to prove correctness of the algorithm.

Corollary 8.1.4 *Let X and Y be disjoint sets such that $|X| = |Y| = 2^i$ for some i, and let m_X and m_Y be the medians of X and Y, respectively. If $m_X > m_Y$ then the median of $X \cup Y$ is the median of $X^{\perp} \cup Y^{\top}$.*

Proof. Observe that $X^{\perp}$ is obtained by removing 2^{i-1} elements that are greater than or equal to m_X from X. Likewise $Y^{\top}$ is obtained by removing 2^{i-1} elements that are strictly smaller than m_Y from Y. Since $m_X > m \geq m_Y$ it follows that in $X^{\perp} \cup Y^{\top}$ we remove exactly 2^{i-1} elements that are strictly smaller than m and 2^{i-1} elements that are strictly greater than m. Therefore, the median value m remains the same as in $X \cup Y$. ∎

We remark that the above deals with the case where $m_X > m_Y$; the proof for the case where $m_Y < m_X$ is symmetrical.

Protocol efficiency. Clearly, the number of iterations is $\log |X| + 1$ (since each set is reduced by half in each round). Moreover, in each iteration the parties exchange $p(n)$ bits (for comparing their medians). Therefore, the total communication cost is $O(p(n) \cdot \log |X|)$. The computation costs are dominated by the sorting algorithm that the parties employ at the beginning of the protocol, yielding $O(|X| \log |X| + |Y| \log |Y|)$ symmetric operations.

8.1.2 Reducing the kth-Ranked Element to the Median

As we have described, Protocol 8.1.2 finds the *median* on simplified input instances X and Y where

1. X and Y are disjoint sets (i.e., $X \cap Y = \phi$),
2. $|X| = |Y|$, and
3. there exists an integer ℓ such that $|X| = 2^{\ell}$ for some ℓ.

We now show that the ability to compute the median in the above simplified case suffices for computing the kth-ranked element of arbitrary lists X and Y. In order to show this, we first describe a series of modifications that each party needs to carry out locally on its own list.

Input modification. Let X and Y be two lists of arbitrary length for P_1 and P_2, respectively. In order for the parties to compute the kth-ranked element of their list they each carry out the following local manipulation on their lists:

1. Each party sorts its list in ascending order.
2. Each party constructs a list of size exactly k from its input:

   a. *Party P_1:* If $|X| \geq k$, P_1 takes the k smallest elements from its sorted list. If $|X| < k$, then P_1 adds $k - |X|$ elements to its list that are all guaranteed to be larger than all elements in X and Y (we denote such elements by $+\infty$).
      If k is not a power of 2, let ℓ be the smallest integer for which $k \leq 2^{\ell}$. P_1 then adds an additional $2^{\ell} - k$ elements that are equal to $+\infty$, thereby obtaining a list of size 2^{ℓ}.

   b. *Party P_2:* If $|Y| \geq k$, P_1 takes the k smallest elements from its sorted list. If $|Y| < k$, P_1 adds $k - |Y|$ elements $+\infty$ to its list; each element being equal to $+\infty$.
      In the same way as P_1, if k is not a power of 2, P_1 adds an additional $2^{\ell} - k$ elements to its list (where ℓ is as above). However, for P_2 these elements all equal $-\infty$ (i.e., are guaranteed to be smaller than all the elements in X and Y).

3. The final modifications that the parties make to their lists are to ensure that all elements in all lists are distinct. In order to do this, each party concatenates a unique index to each element, and a bit to distinguish inputs of P_1 from inputs of P_2. This concatenation is carried out as follows:

   a. *Party P_1:* Party P_1 concatenates to each element the bit 0 followed by the index of the element in its sorted list X. (This concatenation is to the least significant bits of the elements.)

   b. *Party P_2:* Party P_2 concatenates to each element the bit 1 followed by the index of the element in its sorted list Y.

4. We denote the modified lists of the parties by X' and Y', respectively.

It is easy to see that the result of the above manipulation is two disjoint lists X and Y of distinct elements (i.e., sets), of size exactly 2^ℓ each, for some integer ℓ.

Computing the kth-ranked element. We now show that the result of applying Protocol 8.1.2 for finding the median of the modified lists above is the kth-ranked element of $X \cup Y$, as long as $|X \cup Y| \geq k$ (observe that if this does not hold, then there is no kth-ranked element in any case). That is, the median of $X' \cup Y'$ equals the kth-ranked element of $X \cup Y$. In order to see this, we separately look at the cases where the original input lists are larger and smaller than k:

1. *Case 1 – $|X| > k$ and $|Y| > k$:* In this case, both parties take their k smallest elements. The result is that both parties hold lists of exactly size k, and so the median of the combined lists is exactly the kth-ranked element of the original lists. In addition, if k is not a power of 2, party P_1 adds exactly $2^\ell - k$ values $+\infty$ to its list, and party P_2 adds exactly $2^\ell - k$ values $-\infty$ to its list. Since they add the same number of values each, the median of the combined lists remains unchanged. Finally, adding a bit and an index clearly has no effect on the median.

2. *Case 2 – $|X| \leq k$ or $|Y| \leq k$:* If a party's original list is smaller than k, it works exactly as above except that it adds an additional $k - |X|$ values $+\infty$ to its list. Since the size of the combined original input lists $X \cup Y$ is greater than k, adding $+\infty$ to one or both lists does not change the result. In particular, the median of the lists after adding these $k - |X|$ or $k - |Y|$ values is the kth-ranked element of $X \cup Y$. The rest of the analysis remains the same as in the previous case.

Finally, we remark that the inputs X' and Y' fulfill the assumptions on the input sets as required by Protocol 8.1.2 (i.e., they are disjoint sets of the same size, which is a power of 2). Thus, the algorithm correctly computes the median of $X' \cup Y'$, which is the kth-ranked element of $X \cup Y$, as required.

Security in the semi-honest and malicious cases. It remains to argue that the result of applying a *secure protocol* for computing the median of combined lists $X' \cup Y'$ is a *secure protocol* for computing the kth-ranked element of $X \cup Y$. First observe that the modifications to the input sets are all carried out locally and therefore do not reveal any information. Furthermore, since the output when computing the median of $X' \cup Y'$ *equals* the output when computing the kth-ranked element of $X \cup Y$, and a secure protocol only reveals the output, security is preserved. One important subtlety however is to deal with the case where a party may not correctly modify its input. This does not arise in the semi-honest case. However, it must be dealt with in the malicious case. Specifically, we must consider the possibility that a malicious party replaces some of its original input values also with $+\infty$ or $-\infty$. This is seemingly not allowed because $+\infty$ and $-\infty$ are outside of the domain. However, this makes no difference because in an ideal execution of the kth-ranked element a malicious party can use the maximum and minimum values of the domain in order to achieve exactly the same effect.

From now on, we will consider only the problem of finding the median of two disjoint sets of size 2^i for some i. We denote the functionality that computes this by $\mathcal{F}_{\mathrm{MED}}$.

8.2 Computing the Median – Semi-honest

In this section we present a protocol for securely computing the median of two distributed sets that achieves security in the presence of semi-honest adversaries. The protocol follows the outline of protocol FindMed from Section 8.1.1 for finding the median with low communication complexity. Intuitively, a secure protocol can be derived from FindMed by having the parties use a secure subprotocol to compare their local medians m_X and m_Y, instead of just sending them to each other. At first sight, this seems to reveal more information than allowed. Specifically, in a secure protocol the parties are only allowed to learn the final output. However, here they seemingly learn more because in each iteration they learn whether their local median is greater or less than the other party's local median. Nevertheless, as we will see, this information can actually be obtained from each party's input and output alone. Thus, surprisingly, this actually yields a secure protocol.

The protocol $\pi_{\mathrm{MED}}^{\mathrm{SH}}$ below for securely computing the median in the presence of semi-honest adversaries uses a subprotocol $\pi_{\mathrm{GT}}^{\mathrm{SH}}$ for securely computing the *greater-than* functionality $\mathcal{F}_{\mathrm{GT}}$ in the presence of semi-honest adversaries. The greater-than functionality $\mathcal{F}_{\mathrm{GT}}$ is defined by $\mathcal{F}_{\mathrm{GT}}(x, y) = (1, 1)$ if $x > y$, and $\mathcal{F}_{\mathrm{GT}}(x, y) = (0, 0)$ otherwise. In addition, it also uses a subprotocol $\pi_{\mathrm{MAX}}^{\mathrm{SH}}$ for securely computing the *maximum* functionality $\mathcal{F}_{\mathrm{MAX}}(x, y) = \mathsf{max}\{x, y\}$ in the presence of semi-honest adversaries.

PROTOCOL 8.2.1 (Secure Median – Semi-Honest Adversaries $\pi_{\text{MED}}^{\text{SH}}$)

- **Inputs:** The inputs of P_1 and P_2 are disjoint sets $X, Y \subseteq \{0,1\}^{p(n)}$ of size exactly 2^ℓ each for some integer ℓ.
- **Auxiliary inputs:** A security parameter 1^n.
- **The protocol:**

  1. Denote $X^1 = X$ and $Y^1 = Y$.
  2. FOR $j = 1, \ldots, \ell$ DO:
     a. P_1 and P_2 locally compute the respective medians m_X^j and m_Y^j of their respective sets X^j and Y^j.
     b. P_1 and P_2 run subprotocol $\pi_{\text{GT}}^{\text{SH}}$; party P_1 inputs m_X^j and party P_2 inputs m_Y^j. Let b denote the result of the computation.
        - If $b = 1$ (i.e., $m_X^j > m_Y^j$), then P_1 sets $X^{j+1} = (X^j)^\perp$ (i.e., X^{j+1} is derived from X^j by removing all of the elements that are greater than or equal to m_X^j). Moreover, P_2 sets $Y^{j+1} = (Y^j)^\top$ (i.e., Y^{j+1} is derived from Y^j by removing all of the elements that are strictly smaller than m_Y^j).
        - If $b = 0$ (i.e., $m_X^j \leq m_Y^j$), then P_1 sets $X^{j+1} = (X^j)^\top$ and P_2 sets $Y^{j+1} = (Y^j)^\perp$.
     c. The parties continue to the next iteration with their new data sets.
  3. OBTAIN OUTPUT: In this stage, $|X^{\ell+1}| = |Y^{\ell+1}| = 1$ and the parties run a subprotocol $\pi_{\text{MAX}}^{\text{SH}}$ in order to obtain the maximum of the final elements. The parties both output the result of this computation.

We now prove the security of Protocol 8.2.1.

Theorem 8.2.2 *Assume that $\pi_{\text{GT}}^{\text{SH}}$ and $\pi_{\text{MAX}}^{\text{SH}}$ securely compute the greater-than and maximum functionalities in the presence of semi-honest adversaries. Then Protocol 8.2.1 securely computes $\mathcal{F}_{\text{MED}}$ in the presence of semi-honest adversaries.*

Proof. We prove the security using the simpler formulation in Equations (2.1) and (2.2) in Section 2.2. As such, we first need to prove correctness; however, this follows immediately from the proof of correctness of Protocol FindMed above. We now proceed to prove security in the case where P_1 is corrupted; the case where P_2 is corrupted is identical due to the symmetry of the roles of P_1 and P_2 in the protocol. We present the proof in a hybrid model in which a trusted party is used to compute $\mathcal{F}_{\text{GT}}$ and $\mathcal{F}_{\text{MAX}}$.

We construct a simulator S_1 that generates the view of P_1, as in (2.1). Observe first that the only messages that P_1 sees when running Protocol 8.2.1 in the hybrid setting are the outputs from $\mathcal{F}_{\text{GT}}$ and $\mathcal{F}_{\text{MAX}}$. Therefore, given P_1's input X and the median m of $X \cup Y$, the simulator S_1 needs to emulate these hybrid outputs by playing the role of the trusted party for $\mathcal{F}_{\text{GT}}$ and $\mathcal{F}_{\text{MAX}}$. S_1 works as follows:

1. S_1 receives the input set X and the median m of $X \cup Y$.

2. S_1 SIMULATES P_1'S VIEW IN ITERATION j: Let m_X^j denote the median value of X^j as P_1 would compute it in the protocol. Then, if $m_X^j \leq m$, S_1 simulates the output of $\mathcal{F}_{\mathrm{GT}}$ in this iteration to be 0 (as in the case where $m_X^j \leq m_Y^j$ in the protocol). Otherwise, S_1 simulates the output to be 1. S_1 sets X^{j+1} in the same way as P_1 would in the protocol (i.e., if $b = 0$ then $X^{j+1} = X^\top$, and if $b = 1$ then $X^{j+1} = X^\perp$).

3. S_1 SIMULATES THE OBTAIN OUTPUT PHASE: At this point $|X^{\ell+1}| = 1$. S_1 simulates the output of $\mathcal{F}_{\mathrm{MAX}}$ to be the median m that it received above.

We claim that the view generated by S_1 is identical to the view of P_1 in a hybrid execution of the protocol where a trusted party is used to compute $\mathcal{F}_{\mathrm{GT}}$ and $\mathcal{F}_{\mathrm{MAX}}$. We first prove that in every iteration j, $m_X^j > m$ if and only if $m_X^j > m_Y^j$ (where m_X^j is the median of X^j, m is the median of $X \cup Y$, and m_Y^j is the median of Y^j based on the real input set Y of P_2 that S_1 is not given). This follows immediately from Claim 8.1.3 that states that if $m_X^j > m_Y^j$ then $m_X^j > m \geq m_Y^j$. We prove the two directions:

1. *Direction 1 – if $m_X^j > m_Y^j$ then $m_X^j > m$:* this is written directly in the claim.
2. *Direction 2 – if $m_X^j > m$ then $m_X^j > m_Y^j$:* in order to see this, observe that Claim 8.1.3 is symmetric and thus if $m_X^j < m_Y^j$ then $m_X^j \leq m < m_Y^j$ (by simply exchanging the names X and Y). In other words, $m_X^j < m_Y^j$ implies that $m_X^j \leq m$, and so it cannot be that $m_X^j > m$ *and* $m_X^j < m_Y^j$.

This proves that the view of P_1 in every iteration generated by S_1 is identical to its view in a real execution. Regarding the last phase, this also follows from the correctness of Protocol FindMed because the correct median of the sets is the maximum of the two remaining elements at the end. ∎

Efficiency. The number of iterations in Protocol 8.2.1 is exactly ℓ. By using a constant-round subprotocol for computing $\mathcal{F}_{\mathrm{GT}}$ and $\mathcal{F}_{\mathrm{MAX}}$ we have that the total number of rounds of communication is $O(\ell)$. Specifically, if the protocol of Chapter 3 is used, then this number is exactly $3\ell+3$ (three rounds for every iteration and another three for the output phase). However, since there are three rounds in the protocol (a message from P_2 to P_1 followed by a message from P_1 to P_2 and finally a reply from P_2 to P_1) and party P_2 receives its output after the second round, P_2 can send its input for the next execution together with its round 3 message that it sends to P_1. Therefore, we obtain execution $2\ell + 2$ rounds, which is logarithmic in the size of the input sets. Furthermore, the computation and communication complexity is inherited from the cost of computing $\pi_{\mathrm{GT}}^{\mathrm{SH}}$ and $\pi_{\mathrm{MAX}}^{\mathrm{SH}}$. See Section 3.5 for an exact analysis (e.g., given that the circuit for computing greater than and maximum is of linear size, the total computation cost is $O(\ell \cdot p(n))$ exponentiations, where $p(n)$ is the number of bits used to represent each element in the input sets).

8.3 Computing the Median – Malicious

In order to make the protocol for computing the median secure also for the case of malicious adversaries, it is necessary to prevent a malicious adversary from effectively deviating from the protocol specification. In addition to using subprotocols π_{GT} and π_{MAX} that are secure in the presence of malicious adversaries, we also have to ensure that a corrupted party uses "consistent inputs" in every iteration. In order to understand this consistency requirement observe that if the local median of P_1 in the first iteration is m_X^1 and $m_X^1 > m_Y^1$, then in all later iterations the median used by P_1 must be strictly smaller than m_X^1. This is because P_1 sets $X^2 = X^\perp$ and because all elements are distinct. Furthermore, if $m_X^1 < m_Y^1$, then in all later iterations the median used by P_1 must be greater than m_X^1 (because P_1 sets $X^2 = X^\top$). More generally, the local median m_X^j must be less than m_X^i where i ($1 \leq i < j$) is the last iteration in which P_1 received output $b = 1$, and must be greater than $m_X^{i'}$ where i' ($1 \leq i' < j$) is the last iteration in which P_1 received output $b = 0$. A similar argument holds with respect to P_2. The protocol for the case of malicious adversaries enforces this consistency requirement between iterations by using a reactive greater-than functionality, described below. As we will see, it turns out that enforcing this suffices for obtaining full simulation.

8.3.1 The Reactive Greater-Than Functionality

In the protocol for the semi-honest setting, the parties used a simple greater-than functionality $\mathcal{F}_{\mathrm{GT}}$. As we have already discussed, in the case of malicious adversaries it is necessary to enforce that the inputs used by the parties are within the "allowed bounds", based on the inputs they used in previous iterations. We achieve this by making the functionality reactive, meaning that it keeps state between executions. Specifically, the functionality stores local "lower" and "upper" bounds for each party, based on the inputs and outputs of previous iterations. More formally, let m_X^j and m_Y^j denote the medians received from P_1 and P_2 respectively in iteration j. Then, if $m_X^j > m_Y^j$ (meaning that P_1 should remove all values that are greater than or equal to m_X^j from its input set, and P_2 should remove all values that are less than m_Y^j from its input set), the upper bound for P_1 is set to m_j^X and the lower bound for P_2 is set to m_j^Y. The functionality then outputs an error if at any later stage a party enters a value that is greater than or equal to its upper bound or less than its lower bound.

In addition to the above, it is also necessary to make sure that the values used by the parties in the last phase (where maximum is computed) are consistent with the values that they sent in the past. Observe that an honest

P_1's input to the last phase (where the maximum is computed) equals the last value that it input to $\mathcal{F}_{\mathrm{GT}}$ for which the output it received was 0. This is due to the fact that in an iteration j where it receives back output 1, it defines $X^{j+1} = (X^j)^{\perp}$, meaning that it throws out all elements from m_X^j and above, *including* m_X^j. Thus, m_X^j is no longer in the list. This implies that the last value in the list is the last value for which it received back output 0. Regarding an honest P_2, the argument is the same when reversing 0 and 1, specifically, the last value that an honest P_2 inputs to $\mathcal{F}_{\mathrm{GT}}$ for which the output it received was 1. (An exception to the above rule is the case for P_1 where all outputs received were 1; in this case, an honest P_1 can input any value to the maximum computation that is smaller than all values sent until this point. Likewise, if P_2 received 0 in all outputs then it can input any value to the maximum computation that is smaller than all values sent until this point.) The reactive greater-than functionality forces the parties to use an input in the maximum computation which is equal to the value defined above that an honest party would input.

Due to the above, we cannot use a separate functionality for computing maximum, because the functionality needs to know the bounds that are defined by all previous iterations. We therefore incorporate the computation of the maximum into the greater-than functionality, and distinguish between greater-than and maximum computations via an additional string sent with the values. A formal description of the reactive greater-than functionality $\mathcal{F}_{\mathrm{RGT}}$ is presented in Figure 8.3.1.

FIGURE 8.3.1 (The Reactive Greater-Than Functionality $\mathcal{F}_{\mathrm{RGT}}$)

Functionality $\mathcal{F}_{\mathrm{RGT}}$ works with parties P_1 and P_2 as follows (the variables $\mathsf{lower}_1, \mathsf{lower}_2$ are initially set to $-\infty$, whereas the variables $\mathsf{upper}_1, \mathsf{upper}_2$ are initially set to $+\infty$):

1. Upon receiving from P_1 a message (GT, m_1) and from P_2 a message (GT, m_2), functionality $\mathcal{F}_{\mathrm{RGT}}$ checks whether $\mathsf{lower}_1 < m_1 < \mathsf{upper}_1$ and $\mathsf{lower}_2 < m_2 < \mathsf{upper}_2$. If these conditions are not met it sends error to both parties and aborts. Otherwise:

   a. If $m_1 > m_2$ it sets $\mathsf{upper}_1 = m_1$ and $\mathsf{lower}_2 = m_2$, and returns 1 to the parties.

   b. If $m_1 \leq m_2$ it sets $\mathsf{lower}_1 = m_1$ and $\mathsf{upper}_2 = m_2$ and returns 0 to the parties.

2. Upon receiving from P_1 a message (MAX, m_1) and from P_2 a message (MAX, m_2): If m_1 (respectively m_2) is the smallest item sent by P_1 (resp., P_2) in the computation, the functionality $\mathcal{F}_{\mathrm{RGT}}$ checks that $m_1 < \mathsf{upper}_1$ (resp., that $m_2 < \mathsf{upper}_2$). Otherwise, $\mathcal{F}_{\mathrm{RGT}}$ checks that $m_1 = \mathsf{lower}_1$ and $m_2 = \mathsf{lower}_2$. If the checks fail, $\mathcal{F}_{\mathrm{RGT}}$ sends error to the parties. Otherwise, it sends them the value $\max(m_1, m_2)$.

Before proceeding we remark that the value lower_1 in the last phase of the median computation equals the last value m_X^j input by party P_1 for which the output was 0. Thus, in light of the discussion above regarding the input used in the computation of the maximum, $\mathcal{F}_{\mathrm{RGT}}$ checks that the input equals lower_1 and this is equivalent to a check that it is the last value input for which the output was 0. The same argument holds for P_2.

Observe that a binary circuit for computing $\mathcal{F}_{\mathrm{RGT}}$ can be constructed similarly to the circuit for $\mathcal{F}_{\mathrm{GT}}$, while additionally storing state between iterations. Such a circuit is of size that is linear in the input, and a protocol for securely computing $\mathcal{F}_{\mathrm{RGT}}$ based on such a circuit can be found in Chapter 4 (with explicit reference to reactive functionalities in Section 2.5.3).

8.3.2 The Protocol

In order to transform Protocol 8.2.1 for semi-honest parties into a protocol for malicious parties, it suffices to replace the subprotocols $\pi_{\mathrm{GT}}^{\mathrm{SH}}$ and $\pi_{\mathrm{MAX}}^{\mathrm{SH}}$ with a subprotocol π_{RGT} that securely computes the reactive greater-than functionality $\mathcal{F}_{\mathrm{RGT}}$ in the presence of malicious adversaries. Let π_{MED} denote this modified version of Protocol 8.2.1.

Theorem 8.3.2 *Assume that π_{RGT} securely computes the functionality $\mathcal{F}_{\mathrm{RGT}}$ in the presence of malicious adversaries. Then Protocol π_{MED} securely computes $\mathcal{F}_{\mathrm{MED}}$ in the presence of malicious adversaries.*

Proof. We present the proof in the hybrid model in which a trusted party is used to compute $\mathcal{F}_{\mathrm{RGT}}$. As in the proof of Theorem 8.2.2, the proof is identical for both corruption cases due to symmetry. Thus we only consider the corruption case of party P_1.

Let $\mathcal{A}$ be an adversary controlling party P_1. The main challenge in constructing a simulator is to extract the input used by $\mathcal{A}$. This is non-trivial because $\mathcal{A}$ does not commit to its input or even implicitly send it during the protocol. Nevertheless, as we will see, $\mathcal{S}$ can construct a list $\tilde{X}$ of length 2^ℓ that is consistent with the values used by $\mathcal{A}$ in a real execution. Specifically, when $\mathcal{S}$ sends the list $\tilde{X}$ to the trusted party, the resulting median obtained by $\mathcal{S}$ and the honest P_2 is the same value that they would obtain in a real execution of π_{MED} when P_2 uses its private input list Y. Informally speaking, the simulator $\mathcal{S}$ works as follows. All possible executions of the protocol can be viewed as a binary "execution tree", where the root is the input of $\mathcal{A}$ (playing P_1) in the first execution of π_{RGT}. Each internal node corresponds to the input used by $\mathcal{A}$ in some iteration. Specifically, the left child of a node is the input used by $\mathcal{A}$ in the case where the output of the previous execution of π_{RGT} was $b = 1$ (in which an honest P_1 would take the smaller half of its remaining list), and the right child of a node is the input used by $\mathcal{A}$ when the output was $b = 0$. The leaves of the tree are the inputs used by $\mathcal{A}$ in the last phase where the maximum function is computed. As we will see, the leaves defined in this way actually define an input list for P_1 that is consistent with $\mathcal{A}$'s actions in the protocol.

Before proceeding further, note that a single execution of the protocol between $\mathcal{A}$ and an honest P_2 corresponds to a single path in the tree from the root to a leaf. Furthermore, the actual path taken depends on the set Y used by an honest P_2. We remark that although $\mathcal{A}$ may be probabilistic, for any given random tape there exists a *single* binary tree that corresponds to *all* possible executions between $\mathcal{A}$ and P_2, for *all* possible inputs Y to P_2. A crucial observation is that although the tree corresponds to all possible executions with all possible inputs Y (and there are exponentially many such inputs), the tree has only 2^ℓ leaves (and so $2^{\ell+1} - 1$ nodes overall). This is due to the fact that there are only ℓ iterations in the protocol, which is logarithmic in the size of the input sets.

The simulator $\mathcal{S}$ constructs the binary tree described above, and works as follows:

1. $\mathcal{S}$ receives X, z and 1^n, and invokes $\mathcal{A}$ on this input.
2. $\mathcal{S}$ plays the trusted party for the reactive greater-than functionality, with $\mathcal{A}$ playing party P_1. $\mathcal{S}$ traverses over all executions paths in the execution tree by rewinding $\mathcal{A}$ and handing it $b = 0$ and $b = 1$ at each node. $\mathcal{S}$ records the input values sent by $\mathcal{A}$ at every node in the tree, including the leaves. If $\mathcal{A}$ aborts at a leaf then $\mathcal{S}$ records the value that $\mathcal{A}$ had to provide at that leaf (recall that there is only a single value that $\mathcal{A}$ can provide, namely lower_1, unless this is the leftmost leaf, in which case $\mathcal{A}$ could use any value smaller than upper_1, and so $\mathcal{S}$ records any such value). If $\mathcal{A}$ aborts at an internal node, then $\mathcal{S}$ records input values that $\mathcal{A}$ *could have* legally sent (i.e., values that are within the bounds of $\mathcal{F}_{\mathrm{RGT}}$) and continues to complete the subtree of that node by itself, without interacting with $\mathcal{A}$. (Note that an "abort" can be the refusal of $\mathcal{A}$ to send anything, or the event that it sends an invalid value that would cause $\mathcal{F}_{\mathrm{RGT}}$ to send error.)
3. $\mathcal{S}$ defines the set $\tilde{X}$ to be the set of values in all leaves of the execution tree.
4. $\mathcal{S}$ sends $\tilde{X}$ to the trusted party for $\mathcal{F}_{\mathrm{MED}}$ and receives back m (which is the median of $\tilde{X} \cup Y$).
5. $\mathcal{S}$ completes the simulation as the semi-honest simulator in the proof for Theorem 8.2.2 by creating a view for $\mathcal{A}$ that is consistent with $\tilde{X}$ and m, as follows. In iteration j:

   a. If $\mathcal{A}$ sends m_X^j and $\mathsf{lower}_1 < m_X^j < up_1$ and it sends continue to the trusted party computing $\mathcal{F}_{\mathrm{RGT}}$ (meaning that $\mathcal{A}$ did not abort in this iteration and P_2 would receive its output from $\mathcal{F}_{\mathrm{RGT}}$ in the iteration), then $\mathcal{S}$ simulates the answer from $\mathcal{F}_{\mathrm{RGT}}$ to equal 0 if and only if $m_X^j \leq m$.

   b. If $\mathcal{A}$ sends m_X^j that is not within the bounds or sends abort$_1$ to the trusted party computing $\mathcal{F}_{\mathrm{RGT}}$, then $\mathcal{S}$ simulates $\mathcal{F}_{\mathrm{RGT}}$ sending error or abort$_1$, respectively, to the parties. In addition, $\mathcal{S}$ sends abort$_1$ to its trusted party computing the median. Finally, it halts outputting whatever $\mathcal{A}$ outputs.

6. If $\mathcal{S}$ has not halted (i.e., in the case where all values sent by $\mathcal{A}$ are within the bounds and all computations of $\mathcal{F}_{\mathrm{RGT}}$ conclude successfully), then this corresponds to an execution in which the parties reach the last phase where maximum is computed. If $\mathcal{A}$ inputs the correct value (i.e., lower$_1$) and sends continue instructing the trusted party computing $\mathcal{F}_{\mathrm{RGT}}$ to send the output to P_2, then $\mathcal{S}$ sends continue to its trusted party computing the median, indicating that P_2 should receive output. Otherwise, it sends abort$_1$. Finally, $\mathcal{S}$ outputs whatever $\mathcal{A}$ does.

Note first that the traversal of the tree takes polynomial time because it has logarithmic depth. We now prove that $\mathcal{A}$'s view in a hybrid execution of π_{MED} (where a trusted party is used to compute $\mathcal{F}_{\mathrm{RGT}}$) and in an ideal execution with $\mathcal{S}$ are identical. In order to see this, we claim that an honest party P_1 with input $\tilde{X}$ would input *exactly* the same values as $\mathcal{A}$ used throughout the computation, and in the same order. Thus, by the correctness of the FindMed protocol, the output received by a real P_2 running with $\mathcal{A}$ is the median of $\tilde{X} \cup Y$, exactly what it receives in the ideal model. In addition, the view of $\mathcal{A}$ in the simulation by $\mathcal{S}$ is identical to its view in a hybrid model where $\mathcal{F}_{\mathrm{RGT}}$ is computed by a trusted party. This follows from exactly the same argument as in the proof of security for the semi-honest protocol $\pi_{\mathrm{MED}}^{\mathrm{SH}}$; see Theorem 8.2.2. Note that the proof for the semi-honest case suffices because, as we have already stated, the behavior of $\mathcal{A}$ is identical to the behavior of an *honest party* with input $\tilde{X}$ and so the simulation of $\mathcal{S}$ with $\mathcal{A}$ is the same as the simulation of the semi-honest simulator for a semi-honest adversary with input $\tilde{X}$. (The only exception is when $\mathcal{A}$ aborts or sends an inconsistent value to $\mathcal{F}_{\mathrm{RGT}}$. However, the simulator sends abort$_1$ to the trusted party computing $\mathcal{F}_{\mathrm{MED}}$ whenever an honest P_2 would not receive output in its execution with $\mathcal{A}$. Thus the output distribution is also identical in this case.) ∎

Efficiency. The round complexity of π_{MED} equals $\ell + 1$ times the round complexity of a protocol computing $\mathcal{F}_{\mathrm{RGT}}$, and thus is logarithmic in the size of the sets when a constant-round protocol π_{RGT} is used. Such a protocol appears in Chapter 4. The communication and computation costs are dominated by the protocol used for computing $\mathcal{F}_{\mathrm{RGT}}$; see Section 4.5 for a detailed analysis.

Proof technique – discussion. In order to achieve security in the presence of malicious adversaries, most known protocols work by having the parties commit to their entire input and then forcing them to work consistently with these committed inputs. In the classic GMW protocol [35] these commitments are explicit. However, in the protocol in Chapter 4, the same effect is also achieved, albeit more implicitly. Specifically, P_2 is committed to its inputs through the oblivious transfers that it runs, and P_1 is committed to its inputs through the commitment sets $W_{i,j}, W'_{i,j}$ which define its ith input bit; the input is only fully defined once P_1 sends the decommitments, but this does define the entire input. In contrast, in the protocol π_{MED}, the parties do not even read (let alone use) their entire input in any given execution. This is because they only use a logarithmic number of values; all other input elements are essentially undefined. Due to this, it is very surprising that it is possible to define a simulator that can construct a full-length input $\tilde{X}$ that is equivalent to the input used by $\mathcal{A}$ (in every execution with every possible list Y used by P_2) without knowing anything about Y. This is possible because the number of elements sent is logarithmic and so $\mathcal{S}$ can exhaustively rewind $\mathcal{A}$ to traverse all possible computation paths in a protocol execution.

Chapter 9
Search Problems

Recently, there has been much interest in the data mining and other communities for secure protocols for a wide variety of tasks. This interest exists not only in academic circles, but also in industry, in part due to the growing conflict between the privacy concerns of citizens and the homeland security needs of governments. In this chapter we focus on search problems involving two parties; *a client* that is interested in performing a search in a database that is being held by *a server*. The (informal) security requirements for search problems assert that the server should not learn any useful information regarding the search conducted by the client, while the client should learn the search result but nothing more. This chapter includes protocols for the following three basic problems:

- *Secure database search:* In this problem, a client is able to search a database held by a server so that the client can only carry out a single search (or a predetermined number of searches authorized by the server). We remark that searches are as in the standard database setting: the database has a "key attribute" and each record has a unique key value; searches are then carried out by inputting a key value – if the key exists in the database then the client receives back the entire record; otherwise it receives back a "non-existent" reply. This problem has been studied in [16, 28] and has important applications to privacy. For example, consider the case of homeland security where it is sometimes necessary for one organization to search the database of another. In order to minimize information flow (or stated differently, in order to preserve the "need to know" principle), we would like the agency carrying out the search to have access only to the single piece of information it is searching for. Furthermore, we would like the value being searched for to remain secret. Another, possibly more convincing, application comes from the commercial world. The LexisNexis database is a paid service provided to legal professionals that enables them – among other things – to search legal research and public records for the purpose of case preparation. Now, the content of searches made for case preparation is *highly confidential*; this information reveals much about the

C. Hazay, Y. Lindell, *Efficient Secure Two-Party Protocols,*
Information Security and Cryptography, DOI 10.1007/978-3-642-14303-8_9,
© Springer-Verlag Berlin Heidelberg 2010

legal strategy of the lawyers preparing the case, and if it were exposed would allow the other side to prepare counter-arguments well ahead of time. It is even possible that revealing the content of some of these searches may breach attorney-client privilege. We conclude that the searches made to LexisNexis must remain confidential, and even LexisNexis should not learn them (either because they may be corrupted, or more likely, a breach to their system could be used to steal this confidential information). Secure database search can be used to solve this exact problem.

- *Secure document search:* A similar, but seemingly more difficult, problem to that of secure database search is that of secure document search. Here, the database is made up of a series of unstructured documents and a keyword query should return all documents that contain that query. This is somewhat more difficult than the previous problem because of the dependence between documents (the client should not know if different documents contain the same keyword if it has not searched them both). We remark that in many cases, including the LexisNexis example above, what is really needed is the unstructured document search.

- *Text search:* The basic problem of private *text search* is the following one: given a text T held by a server, the aim of the client is to learn all the locations in the text where a pattern p appears (and there may be many) while the server learns nothing about the pattern. This problem has been intensively studied and can be solved optimally in time that is linear in size of the text [10, 53], when security is not taken into account.

9.1 Background

The protocols presented in this chapter all employ a subprotocol that securely computes the committed pseudorandom permutation (PRP) functionality denoted by $\mathcal{F}_{\text{CPRP}}$; see Figure 9.1.1 for a formal definition (the initialization stage is a technicality that is used to ensure that a single secret key k is used throughout). As opposed to batch pseudorandom function evaluation (formally introduced in Section 7.6.4), which considers parallel repetition, the committed functionality is reactive. This means that the client may choose its current query based on the output it received from previous queries. Furthermore, each query must be authorized by the server. We stress that as in the case of batch pseudorandom function evaluation, all computations of the pseudorandom function must be with the same secret key.

The protocols presented here can use any subprotocol π_{CPRP} that securely computes the functionality $\mathcal{F}_{\text{CPRP}}$. One possibility is to use a generic construction like that in Chapter 4. This has actually been implemented in [71]; they constructed a circuit computing the AES function (with 30,000 gates), and then implemented the protocol of Chapter 4 (and the covert protocol of Chapter 5) for securely computing the AES function. The implementation that achieves security in the presence of malicious adversaries takes approx-

---

FIGURE 9.1.1 (The Committed PRP Functionality $\mathcal{F}_{\text{CPRP}}$)

Functionality $\mathcal{F}_{\text{CPRP}}$, parameterized by 1^n and F_{PRP}, works with parties P_1 and P_2 as follows (the variable init is initially set to 0):

Initialize: Upon receiving from P_1 a message (init, k), if init = false, functionality $\mathcal{F}_{\text{CPRP}}$ sets init = true, stores k and sends receipt to P_2. If init = true, then $\mathcal{F}_{\text{CPRP}}$ ignores the message.[1]

Evaluation: Upon receiving a message (retrieve, x) from P_2, functionality $\mathcal{F}_{\text{CPRP}}$ checks that init = 1, and if not it returns notInit. Otherwise, if $x \in \{0,1\}^n$ it sends retrieve to P_1. If P_1 replies with approve then $\mathcal{F}_{\text{CPRP}}$ forwards $F_{\text{PRP}}(k,x)$ to P_2. Otherwise, $\mathcal{F}_{\text{CPRP}}$ forwards P_2 reject.

---

imately four minutes to carry out. Thus, such a computation could be used for low latency applications. However, more improvements are needed before this approach is truly practical. Despite the above, in Section 9.4.1 we show that the committed pseudorandom permutation functionality can be securely realized in a novel setting where the parties have secure (standard) smartcards in addition to regular network communication. Instantiating the π_{CPRP} protocol in this model yields protocols for database and document search that are truly practical; indeed the time that it takes to carry out a single search is almost the same as when using a basic search scheme without any security.

We remark that in contrast to the database and document search applications which may use *any* protocol for securely computing $\mathcal{F}_{\text{CPRP}}$, and for any underlying pseudorandom permutation, the protocol for secure text search in Section 9.5 is based on the Naor-Reingold pseudorandom function [64] and utilizes specific properties of it. In addition, the protocol for secure text search achieves only one-sided simulation. In contrast, the protocols for database and document search achieve full simulation when a fully secure protocol is used for $\mathcal{F}_{\text{CPRP}}$ (more efficient versions of the protocol that achieve security for covert adversaries and one-sided simulation can also be derived by instantiating the protocol π_{CPRP} appropriately).

9.2 Secure Database Search

In this section we study the problem of secure database search. The aim here is to allow a client to search a database without the server learning the query (or queries) made by the client. Furthermore, the client should only be able to make a single query (or, to be more exact, the client should only be able to make a search query after receiving explicit permission from the server). This latter requirement means that the client cannot just download the en-

---

[1] We assume an efficient mapping $\phi : \{0,1\}^* \to I_{\text{PRP}}(1^n)$ (i.e., ϕ maps *any* string to a valid key for $\mathcal{F}_{\text{PRP}}$). This works well for block ciphers used below, where a key is any possible string of appropriate length.

tire database and run local searches (this naive solution perfectly preserves the privacy of the database searches because the server cannot learn anything, but enables the client to carry out an unlimited number of searches). In the solution presented here, the client therefore downloads the database in *encrypted* form. A search is then carried on the database by running an interactive protocol that enables the client to decrypt a single database record. Our starting point is the one-sided simulatable protocol of [28], where the use of secure pseudorandom function evaluation lies at its heart. Replacing this subprotocol with a protocol for $\mathcal{F}_{\mathrm{CPRP}}$ is the first step towards achieving full simulation against malicious adversaries.

We now provide an inaccurate description of our solution. Denote the ith database record by (p_i, x_i), where p_i is the value of the search attribute (as is standard, the values $p_1, \ldots, p_N$ are unique). We assume that each $p_i \in \{0,1\}^n$, and for some ℓ each $x_i \in \{0,1\}^{\ell n}$ (recall that the pseudorandom permutation works over the domain $\{0,1\}^n$; thus p_i is made up of a single "block" and x_i is made up of ℓ blocks). Then, the server chooses a key $k \leftarrow I_{\mathrm{PRP}}$ and computes $t_i = F_{\mathrm{PRP}}(k, p_i)$, $u_i = F_{\mathrm{PRP}}(k, t_i)$ and $c_i = E_{u_i}(x_i)$ for every $i = 1, \ldots, N$. The server sends the encrypted database (t_i, c_i) to the client. Now, since F_{PRP} is a pseudorandom function, the value t_i reveals nothing about p_i, and the "key" u_i is pseudorandom, implying that c_i is a cryptographically sound (i.e., secure) encryption of x_i that therefore reveals nothing about x_i. In order to search the database for attribute p, the client computes $t = F_{\mathrm{PRP}}(k, p)$ and $u = F_{\mathrm{PRP}}(k, t)$ using $\mathcal{F}_{\mathrm{CPRP}}$. If there exists an i for which $t = t_i$, then the client decrypts c_i using the key u, obtaining the record x_i as required. Note that the server has no way of knowing the search query of the client. Furthermore, the client cannot carry out the search without explicit approval from the server, and thus the number of searches can be audited and limited (if required for privacy purposes), or a charge can be issued (if a pay-per-search system is in place).

We warn that the above description is not a fully secure solution. To start with, it is possible for a client to use the key k to compute t and t' for two different values p and p'. Although this means that the client will not be able to obtain the corresponding records x and/or x', it does mean that it can see whether the two values p and p' are in the database (something which it is not supposed to be able to do, because just the existence of an identifier in a database can reveal confidential information). We therefore use two different keys k_1 and k_2; k_1 is used to compute t and k_2 is used to compute u. In addition, we do not use u to directly mask x and use a third key k_3 instead. The reason for this will become apparent in the proof.

9.2.1 Securely Realizing Basic Database Search

We begin by describing the ideal functionality for the problem of secure database search; the functionality is a *reactive* one where the server S first sends the database to the trusted party, and the client can then carry out searches. We stress that the client can choose its queries adaptively, meaning that it can choose what keywords to search for after it has already received the output from previous queries. However, each query must be explicitly approved by the server (this allows the server to limit queries or to charge per query). In this section, we show how to securely realize a basic database search functionality in which the database is completely static (i.e., no changes can be made to it throughout the lifetime of the functionality). In Section 9.2.2 we show how to extend the functionality and protocol to enable some database updates. See Figure 9.2.1 for a specification of the basic database functionality.

FIGURE 9.2.1 (The Database Search Functionality $\mathcal{F}_{\text{basicDB}}$)

Functionality $\mathcal{F}_{\text{basicDB}}$ works with a server S and a client C as follows (the variable init is initially set to 0):

Initialize: Upon receiving from S a message (init, $(p_1, x_1), \ldots, (p_N, x_N)$), if init = false, functionality $\mathcal{F}_{\text{basicDB}}$ sets init = true, stores all pairs and sends (init, N) to C. If init = true, then $\mathcal{F}_{\text{basicDB}}$ ignores the message.

Search: Upon receiving a message (retrieve, p) from C, functionality $\mathcal{F}_{\text{basicDB}}$ checks that init = 1, and if not it returns notInit. Otherwise, it sends retrieve to S. If S replies with approve then $\mathcal{F}_{\text{basicDB}}$ works as follows:

  1. If there exists an i for which $p = p_i$, functionality $\mathcal{F}_{\text{basicDB}}$ sends (retrieve, x_i) to C.
  2. If there is no such i, then $\mathcal{F}_{\text{basicDB}}$ sends notFound to C.

  If S replies with reject, then $\mathcal{F}_{\text{basicDB}}$ forwards reject to C.

The protocol for securely computing the basic functionality $\mathcal{F}_{\text{basicDB}}$ below uses an efficiently invertible pseudorandom permutation over $\{0,1\}^n$. Let F_{PRP} denote such a permutation, and let I_{PRP} denote the key generation algorithm. We define a keyed function $\hat{F}_{\text{PRP}}$ from $\{0,1\}^n$ to $\{0,1\}^{\ell n}$ by

$$\hat{F}_{\text{PRP}}(k,t) = \langle F_{\text{PRP}}(k, t+1), F_{\text{PRP}}(k, t+2), \ldots, F_{\text{PRP}}(k, t+\ell) \rangle$$

where addition is modulo 2^n. We remark that $\hat{F}_{\text{PRP}}(k, \cdot)$ is a pseudorandom function when the input t is uniformly distributed (this follows directly from the proof of security of the counter mode for block ciphers; for example, see [49]). We assume that all records in the database are exactly of length ℓn (and that this is known); if this is not the case, then padding can be used. See Protocol 9.2.2 for a full description.

PROTOCOL 9.2.2 (secure basic database search π_{basicDB})

- **Inputs:** The server has database (init, $(p_1, x_1), \ldots, (p_N, x_N)$).
- **Auxiliary inputs:** Both parties have the security parameter 1^n.
- **The protocol:**

  - INITIALIZATION:
    1. The server S chooses three keys $k_1, k_2, k_3 \leftarrow I_{\text{PRP}}$ for the pseudorandom permutation, and the parties employ the initialize phase within π_{CPRP} in which S initializes three distinct executions with keys k_1, k_2, k_3. The output of C from every execution is receipt.
    2. S randomly permutes the pairs (p_i, x_i) and for every i, computes $t_i = F_{\text{PRP}}(k_1, p_i)$, $u_i = F_{\text{PRP}}(k_2, t_i)$ and $c_i = \hat{F}_{\text{PRP}}(k_3, t_i) \oplus x_i$, where $\hat{F}$ is as defined above.
    3. S sends $(u_1, c_1), \ldots, (u_N, c_N)$ to the client C (these pairs denote an encrypted version of the database).
    4. Upon receiving $(u_1, c_1), \ldots, (u_N, c_N)$, C stores the pairs and outputs (init, N).
  - SEARCH: The parties employ three *sequential* invocations of the evaluation phase within π_{CPRP} as follows:
    1. For the first run which was initialized with a key k_1, C uses input (retrieve, p). If S approves the evaluation, the output of C is $t = F_{\text{PRP}}(k_1, p)$.
    2. Next, C uses input (retrieve, t) and learns $u = F_{\text{PRP}}(k_2, t)$ if the evaluation is approved by S. If there does not exist any i for which $u = u_i$, then C outputs notFound (but continues to the third execution anyway).
    3. The parties engage in a third execution, initialized with a key k_3, as follows:
       a. If there exists an i for which $u = u_i$, C uses input (retrieve, t) and learns $r = \hat{F}_{\text{PRP}}(k_3, t)$ (assuming S approves); this involves ℓ calls to π_{CPRP} for computing $F_{\text{PRP}}(k_3, \cdot)$. Then, C sets $x = r \oplus c_i$ and outputs (retrieve, x).
       b. Else, C enters an arbitrary value and ignores the outcome.
    If S does not approve the computation at any stage, C halts and outputs $\perp$.

It is easy to verify that if S and C follow the instructions of the protocol, then the client outputs the correct result of its search query. We proceed to prove security.

Theorem 9.2.3 *Assume that F_{PRP} is a strong pseudorandom permutation over $\{0,1\}^n$ and that π_{CPRP} securely computes $\mathcal{F}_{\text{CPRP}}$ in the presence of malicious adversaries. Then, Protocol 9.2.2 securely computes $\mathcal{F}_{\text{basicDB}}$ in the presence of malicious adversaries.*

Proof. We treat each corruption case separately. We note that the proof is in a hybrid model where a trusted party is used to compute an ideal functionality for $\mathcal{F}_{\text{CPRP}}$.

The server S is corrupted. The idea behind the proof is to show the existence of a simulator that can extract the input of a malicious server. Intuitively we construct such a simulator by having it extract the server's keys to the pseudorandom permutations via the calls to $\mathcal{F}_{\mathrm{CPRP}}$, thus enabling it to decrypt the entire database and send it to the trusted party computing $\mathcal{F}_{\mathrm{basicDB}}$. Since pseudorandom permutations are used, the database is fully defined by the pairs (u_i, c_i) and the keys to the pseudorandom permutations. Thus, a malicious server cannot do anything beyond approving or rejecting a query, and the simulation therefore proceeds by approving and rejecting queries of the client (in the ideal model) whenever a malicious server (in the real model) approves or rejects queries. Formally, let $\mathcal{A}$ be an adversary controlling S; we construct a simulator $\mathcal{S}_{\mathrm{SERV}}$ as follows:

1. $\mathcal{S}_{\mathrm{SERV}}$ obtains the keys k_1, k_2, k_3 that $\mathcal{A}$ sends to the trusted party for $\mathcal{F}_{\mathrm{CPRP}}$. If $\mathcal{A}$ does not send three valid pseudorandom permutation keys, then $\mathcal{S}_{\mathrm{SERV}}$ sends $\perp$ to $\mathcal{F}_{\mathrm{basicDB}}$.
2. Upon receiving $(u_1, c_1), \dots, (u_N, c_N)$ from $\mathcal{A}$, simulator $\mathcal{S}_{\mathrm{SERV}}$ computes $t_i = F_{\mathrm{PRP}}^{-1}(k_2, u_i)$, $p_i = F_{\mathrm{PRP}}^{-1}(k_1, t_i)$ and $x_i = \hat{F}_{\mathrm{PRP}}(k_3, t_i) \oplus c_i$, for every i. Then, $\mathcal{S}_{\mathrm{SERV}}$ sends $(\mathsf{init}, (p_1, x_1), \dots, (p_N, x_N))$ to $\mathcal{F}_{\mathrm{basicDB}}$.
3. Upon receiving a message $\mathsf{retrieve}$ from $\mathcal{F}_{\mathrm{basicDB}}$, simulator $\mathcal{S}_{\mathrm{SERV}}$ simulates C sending three $\mathsf{retrieve}$ requests to $\mathcal{A}$ for the three sequential calls to $\mathcal{F}_{\mathrm{CPRP}}$. Simulator $\mathcal{S}_{\mathrm{SERV}}$ sends $\mathsf{approve}$ to $\mathcal{F}_{\mathrm{basicDB}}$ if and only if $\mathcal{A}$ answers $\mathsf{approve}$ in all three calls; otherwise $\mathcal{S}_{\mathrm{SERV}}$ sends reject to $\mathcal{F}_{\mathrm{basicDB}}$.

This completes the simulation. The output distribution from the simulation is identical to that from a hybrid execution. This is due to the fact that F_{PRP} is a pseudorandom permutation and thus k_1, k_2, k_3 together with a pair (u_i, c_i) define a unique (p_i, x_i) that is sent to $\mathcal{F}_{\mathrm{basicDB}}$. In addition, C can carry out a search if and only if $\mathcal{S}_{\mathrm{SERV}}$ sends $\mathsf{approve}$ to $\mathcal{F}_{\mathrm{basicDB}}$. Finally, we note that $\mathcal{A}$'s view is identical in the simulation and in a hybrid execution because it has no incoming messages in the hybrid protocol.

The client C is corrupted. We now proceed to the case where C is corrupted. Intuitively, C can only learn by querying π_{CPRP} in the specified way. This is due to the fact that if C does not first compute $t = F_{\mathrm{PRP}}(k, p)$ for some keyword p, then the probability that it queries π_{CPRP} with a value t such that $F_{\mathrm{PRP}}(k_2, t) = u_i$ for some u_i in the encrypted database is negligible. This follows from the pseudorandomness of F_{PRP}. Thus, C can only learn information by querying π_{CPRP} as specified in the protocol. Again, let $\mathcal{A}$ be an adversary controlling C; we construct $\mathcal{S}_{\mathrm{CL}}$ as follows:

1. Upon receiving input (init, N) from $\mathcal{F}_{\mathrm{basicDB}}$, simulator $\mathcal{S}_{\mathrm{CL}}$ constructs N tuples $(t_1, u_1, c_1), \dots, (t_N, u_N, c_N)$ where each $t_i, u_i \leftarrow_R \{0, 1\}^n$ and $c_i \leftarrow_R \{0, 1\}^{\ell n}$ (recall that ℓ is known to $\mathcal{S}_{\mathrm{CL}}$). If there exist t_i, t_j with $i \neq j$ such that
$$\{t_i + 1, \dots, t_i + \ell\} \cap \{t_j + 1, \dots, t_j + \ell\} \neq \phi,$$
then $\mathcal{S}_{\mathrm{CL}}$ outputs fail_1 and halts. Otherwise, $\mathcal{S}_{\mathrm{CL}}$ hands $\mathcal{A}$ the pairs $(u_1, c_1), \dots, (u_N, c_N)$.

2. Upon receiving a new query (retrieve, p) from $\mathcal{A}$ to the ideal execution for $\mathcal{F}_{\mathrm{CPRP}}$ that was initialized by k_1, $\mathcal{S}_{\mathrm{CL}}$ forwards (retrieve, p) to $\mathcal{F}_{\mathrm{basicDB}}$. If it receives back reject then it hands reject to $\mathcal{A}$. Otherwise,

   a. If this is the first time that $\mathcal{A}$ has queried p, then:
      i. If $\mathcal{F}_{\mathrm{basicDB}}$ replies with (retrieve, x), then $\mathcal{S}_{\mathrm{CL}}$ chooses a random index $i \in \{1, \ldots, N\}$ that has not yet been chosen, hands t_i to $\mathcal{A}$, and stores the association (i, p, x).
      ii. If $\mathcal{F}_{\mathrm{basicDB}}$ replies with notFound, then $\mathcal{S}_{\mathrm{CL}}$ chooses a random $t_p \leftarrow_R \{0,1\}^n$ that was not chosen before, stores the pair (p, t_p), and hands t_p to $\mathcal{A}$.
   b. If this is not the first time that $\mathcal{A}$ queried p, then $\mathcal{S}_{\mathrm{CL}}$ returns the same reply as the last time (either t_p or t_i, appropriately).

3. When $\mathcal{A}$ queries the trusted party for $F_{\mathrm{PRP}}(k_2, \cdot)$ with some t, simulator $\mathcal{S}_{\mathrm{CL}}$ works as follows:

   a. If there exists an i and a tuple (t_i, u_i, c_i) where $t = t_i$, then $\mathcal{S}_{\mathrm{CL}}$ checks that there is a recorded tuple (i, p_i, x_i). If no, $\mathcal{S}_{\mathrm{CL}}$ outputs fail$_2$. Otherwise, it hands $\mathcal{A}$ the value u_i from the tuple (t_i, u_i, c_i).
   b. If there does not exist such an i, then $\mathcal{S}_{\mathrm{CL}}$ chooses a random $u \leftarrow_R \{0,1\}^n$ and hands u to $\mathcal{A}$. $\mathcal{S}_{\mathrm{CL}}$ also stores the pair (t, u) so that if t is queried again, then $\mathcal{S}_{\mathrm{CL}}$ will reply with the same u.

4. When $\mathcal{A}$ queries the trusted party for $F_{\mathrm{PRP}}(k_3, \cdot)$ with some value t, simulator $\mathcal{S}_{\mathrm{CL}}$ checks if there exists an i and a tuple (t_i, u_i, c_i) where $t = t_i + j$ for some $j \in \{1, \ldots, \ell\}$.

   a. If so, $\mathcal{S}_{\mathrm{CL}}$ checks that there is a recorded tuple (i, p_i, x_i). If not, $\mathcal{S}_{\mathrm{CL}}$ outputs fail$_2$. Otherwise, it hands $\mathcal{A}$ the n-bit string $r_j = c_i^j \oplus x_i^j$ (as the output of the jth block in $\hat{F}_{\mathrm{PRP}}(k_3, t_i)$) where c_i^j is the jth n-bit block of c_i, and x_i^j is the jth n-bit block of x_i.
   b. If not, then $\mathcal{S}_{\mathrm{CL}}$ returns a random value ($\mathcal{S}_{\mathrm{CL}}$ stores a set to maintain consistency, meaning that if in the future the same t' is queried, it returns the same random value).

This completes the simulation. We begin by showing that in the simulation, the probability that $\mathcal{S}_{\mathrm{CL}}$ outputs fail$_1$ or fail$_2$ is negligible. Regarding fail$_1$, this follows from the fact that ℓ is polynomial in n, and the values t are chosen randomly within a range of size 2^n. Thus, the probability that there exist t_i, t_j where $i \neq j$ such that $\{t_j + 1, \ldots, t_j + \ell\} \cap \{t_i + 1, \ldots, t_i + \ell\} \neq \phi$ is negligible. (For a detailed proof that this probability is negligible, see the proof of security of counter mode for encryption; e.g., see [49].) Regarding fail$_2$, recall that $\mathcal{S}_{\mathrm{CL}}$ outputs fail$_2$ if $\mathcal{A}$ sends a value $t \in \{t_i, t_i + 1, \ldots, t_i + \ell\}$ for some tuple (t_i, u_i, c_i) but there is no stored tuple (i, p_i, x_i). Now, if no tuple (i, p_i, x_i) is stored, then this means that $\mathcal{S}_{\mathrm{CL}}$ never gave $\mathcal{A}$ the value t_i from the ith tuple (t_i, u_i, c_i). However, t_i is uniformly distributed and so the probability that $\mathcal{A}$ sends $t \in \{t_i, t_i + 1, \ldots, t_i + \ell\}$ is negligible.

To prove that $\mathcal{A}$'s output in the hybrid and simulated executions are computationally close we construct a sequence of hybrid games and show that the corresponding random variables $H_\ell^{\mathcal{A}(z)}(\{p_i, x_i\}_i, n)$ that consist of the output of $\mathcal{A}$ in hybrid game H_ℓ are computationally close. In each hybrid game, we modify the simulator (and game definition), and we denote by $\mathcal{S}_i$ the simulator used in hybrid game H_i. Here and below, we denote by F_{Perm} an ensemble of truly random permutations.

Game H_1: In the first game the simulator $\mathcal{S}_1$ has access to an oracle $\mathcal{O}_{F_{\mathrm{Perm}}}$ computing a truly random permutation F_{Perm}; more exactly, the oracle computes three distinct random permutations. Whenever $\mathcal{S}_{\mathrm{CL}}$ chooses a random value t_i (resp., u_i, c_i), $\mathcal{S}_1$ chooses t_i (resp., u_i, c_i) by querying the first (resp. second, third) permutation oracle on input i. We stress that the execution in this game still involves a trusted party that computes $\mathcal{F}_{\mathrm{basicDB}}$. The output distribution of the current and original simulation are statistically close since $\mathcal{S}_{\mathrm{CL}}$ uses truly random values, which is equivalent to using a truly random function, and $\mathcal{S}_1$ uses a random permutation.

Game H_2: The next game is identical to the previous one except that the simulator $\mathcal{S}_2$ knows the real input $\{p_i, x_i\}_i$ of the server and uses it for the computation of $\{t_i, u_i, c_i\}_i$ instead of the input i used by $\mathcal{S}_1$ (note that $\mathcal{S}_2$ still interacts with a trusted party that computes $\mathcal{F}_{\mathrm{basicDB}}$). Since the oracle is a truly random permutation, the distribution here is identical unless fail_1 or fail_2 occurs; if either fail event occurs it is possible to distinguish between the case where the real input is used and a fake input i is used. Thus, the output distributions of H_1 and H_2 are statistically close.

Game H_3: In this game, the simulator $\mathcal{S}_3$ has access to an oracle $\mathcal{O}_{F_{\mathrm{PRP}}}$ computing a pseudorandom permutation, instead of an oracle $\mathcal{O}_{F_{\mathrm{Perm}}}$ computing a truly random permutation. The computational indistinguishability of H_2 from H_3 follows from the pseudorandomness of F_{PRP}. The reduction is straightforward and we therefore only briefly sketch it. We show that a distinguisher that distinguishes between the above distributions can be transformed into a distinguisher for F_{PRP}. Assume that there exist an adversary $\mathcal{A}$, a distinguisher D and infinitely many inputs $(1^n, \{p_i, x_i\}_i)$ such that D distinguishes between $\mathcal{A}$'s outputs in the above games whenever the server's input is $(1^n, \{p_i, x_i\}_i)$. Then a distinguisher D_{PRP} with oracle access to either $\mathcal{O}_{F_{\mathrm{PRP}}}$ or $\mathcal{O}_{F_{\mathrm{Perm}}}$, and an auxiliary input $(\{p_i, x_i\}_i, z)$, is constructed the following way. On input 1^n, D_{PRP} invokes $\mathcal{A}(1^n, z)$ and plays the roles of the simulator and the trusted party computing $\mathcal{F}_{\mathrm{CPRP}}$. Note that if the oracle of D_{PRP} computes $\mathcal{O}_{F_{\mathrm{Perm}}}$, then the adversary's view is identical to its view in game H_2, whereas if D_{PRP}'s oracle computes $\mathcal{O}_{F_{\mathrm{PRP}}}$, then it is identical to its view in game H_3. Thus D_{PRP} distinguishes F_{PRP} from F_{Perm} with the same probability that D distinguishes game H_2 from game H_3.

Game H_4: Finally, we let the simulator $\mathcal{S}_4$ perfectly emulate the role of the server S instead of working with a trusted party computing $\mathcal{F}_{\mathrm{basicDB}}$. In particular, $\mathcal{S}_4$ carries out the computations of the pseudorandom permutations by itself, using keys k_1, k_2, k_3. This does not affect the outputs of these computations and yields an identical distribution.

Observe that H_4 yields exactly the same distribution as in a hybrid execution of Protocol 9.2.2. This concludes the proof. ∎

Efficiency. The computational cost of Protocol 9.2.2 is dominated by the cost incurred by the $\ell + 2$ invocations of π_{CPRP}. We note that if π_{CPRP} is secure under parallel composition, then the number of rounds of communication is constant; otherwise $O(\ell)$ rounds are needed.

9.2.2 Securely Realizing Full Database Search

The main drawback with $\mathcal{F}_{\text{basicDB}}$ is that the database is completely static and updates cannot be made by the server. We therefore modify $\mathcal{F}_{\text{basicDB}}$ so that *inserts* and *updates* are included. An insert operation adds a new record to the database, while an update operation makes a change to the x portion of an existing record. We stress that we define an update by concatenating the new x portion to the record without deleting the previous x value. We define the functionality in this way because it affords greater efficiency. In order to see why this is more efficient, recall that in the protocol for computing $\mathcal{F}_{\text{basicDB}}$, and thus also in the protocol that we will present for the full database functionality, the client holds the entire database in encrypted form and retrievals are carried out obtaining a decryption key based on the keyword p. Since p does not change, and we do not wish to change the pseudorandom permutation keys (because this would require re-encrypting the entire database), we have that the old and new x portions are encrypted with the same key. Thus, if the client does not erase an old encrypted x value, it can decrypt it at the same time that it is able to decrypt the new x value. We model this capability by defining an update to be a concatenation.

Another subtlety that arises is that since inserts are carried out over time, and the client receives encrypted records when they are inserted, it is possible for the client to know when a decrypted record was inserted. However, a naive definition of the database functionality (that merely notifies the client whenever an insert takes place) would not leak this information. In order to model this information leakage, we include unique identifiers to records; when a record is inserted, the ideal functionality hands the client the identifier of the inserted record. Then, when a search succeeds, the client receives the identifier together with the x portion. This allows the client in the ideal model to track when a record was inserted (of course, without revealing anything about its content). Finally, we remark that our solution does not efficiently support delete commands (this is for the same reason that updates are modeled as concatenations). We therefore include a reset command that deletes all records. This requires the server to re-encrypt the entire database from scratch and send it to the client. Thus, such a command cannot be issued at too frequent intervals. See Figure 9.2.4 for the full definition of $\mathcal{F}_{\mathrm{DB}}$.

FIGURE 9.2.4 (The Database Functionality $\mathcal{F}_{\mathrm{DB}}$)

Functionality $\mathcal{F}_{\mathrm{DB}}$ works with a server S and client C as follows:

Insert: Upon receiving a message (insert, p, x) from S, functionality $\mathcal{F}_{\mathrm{DB}}$ checks that there is no recorded tuple (id_i, p_i, x_i) for which $p = p_i$. If there is such a tuple it ignores the message. Otherwise, it assigns an identifier id to (p, x), sends (insert, id) to C, and records the tuple (id, p, x).

Update: Upon receiving a message (update, p, x) from S, functionality $\mathcal{F}_{\mathrm{DB}}$ checks whether there is a recorded tuple (id_i, p_i, x_i) for which $p = p_i$. If there is no such tuple it ignores the message. Otherwise it updates the tuple, by *concatenating* x to x_i.

Retrieve: Upon receiving a query (retrieve, p) from the client C, functionality $\mathcal{F}_{\mathrm{DB}}$ sends retrieve to S. If S replies with approve then:

1.   If there exists a recorded tuple (id_i, p_i, x_i) for which $p = p_i$, $\mathcal{F}_{\mathrm{DB}}$ sends (id_i, x_i) to C.

2.   If there does not exist such a tuple, $\mathcal{F}_{\mathrm{DB}}$ sends notFound to C.

Reset: Upon receiving a message reset from S, functionality $\mathcal{F}_{\mathrm{DB}}$ sends reset to C and erases all entries.

A protocol for securely computing the more sophisticated functionality $\mathcal{F}_{\mathrm{DB}}$ can be derived directly from Protocol 9.2.2. Specifically, instead of sending all the pairs (u_i, c_i) at the onset, S sends a new pair every time an insert is carried out, and C verifies that u_i does not already appear in its encrypted database. In addition, an update just involves S re-encrypting the new x_i value and sending the new ciphertext c_i' together with u_i (so that the client can associate it with the correct record). Finally, a reset is carried out by choosing new keys k_1, k_2, k_3 and rejecting all $\mathcal{F}_{\mathrm{CPRP}}$ requests for the old keys. Then, any future inserts are computed using these new keys. We denote by π_{DB} the modified construction of Protocol 9.2.2 and state the following,

Theorem 9.2.5 *Assume that F_{PRP} is a strong pseudorandom permutation over $\{0,1\}^n$ and that π_{CPRP} securely computes $\mathcal{F}_{\mathrm{CPRP}}$ in the presence of malicious adversaries. Then, Protocol π_{DB} securely computes $\mathcal{F}_{\mathrm{DB}}$ in the presence of malicious adversaries.*

9.2.3 Covert and One-Sided Simulation

Our protocols π_{basicDB} and π_{DB} only use calls to $\mathcal{F}_{\mathrm{CPRP}}$. Thus, if a subprotocol for computing $\mathcal{F}_{\mathrm{CPRP}}$ is used that achieves security in the presence of covert adversaries, the result is a protocol for computing $\mathcal{F}_{\text{basicDB}}$ or $\mathcal{F}_{\text{DB}}$ that achieves security in the presence of covert adversaries. As shown in [71], $\mathcal{F}_{\text{CPRP}}$ can be securely computed much more efficiently in the presence of covert adversaries. Furthermore, when considering one-sided simulation, it suffices to use a subprotocol π_{PRF} that securely computes the ordinary secure pseudorandom function evaluation, rather than the more expensive committed pseudorandom permutation functionality. This is due to the fact that in the setting of one-sided simulation there is no need to extract the server's input, and correctness need not be guaranteed. See Section 7.6.3 for an efficient protocol that securely computes the pseudorandom function evaluation functionality with one-sided simulation. We have the following theorem:

Theorem 9.2.6 *Assume that F_{PRP} is a strong pseudorandom permutation over $\{0,1\}^n$, that $\pi_{\text{PRF}}^{\text{OS}}$ securely computes $\mathcal{F}_{\text{PRF}}$ with one-sided simulation, and that $\pi_{\text{CPRP}}^{\text{CO}}$ securely computes $\mathcal{F}_{\text{CPRP}}$ in the presence of covert adversaries. Then, Protocol $\pi_{\text{DB}}^{\text{OS}}$, which is identical to π_{DB} except that $\pi_{\text{PRF}}^{\text{OS}}$ is used, securely computes $\mathcal{F}_{\text{DB}}$ with one-sided simulation. Furthermore, Protocol $\pi_{\text{DB}}^{\text{CO}}$, which is identical to π_{DB} except that $\pi_{\text{CPRP}}^{\text{CO}}$ is used, securely computes $\mathcal{F}_{\text{DB}}$ in the presence of covert adversaries.*

9.3 Secure Document Search

In Section 9.2 we showed how a database can be searched securely, where the search is based only on a key attribute. Here, we show how to extend this to a less structured database, and in particular to a corpus of texts. In this case, there are many keywords that are associated with each document and the user wishes to gain access to all of the documents that contain a specific keyword. A naive solution would be to define each record value so that it contains all the documents the keyword appears in. However, this would be horrifically expensive because the same document would have to be re-encrypted many times. We present a solution where each document is stored (encrypted) only once, as follows.

The following solution uses Protocol 9.2.2 as a subprotocol, and we model this by constructing a protocol for secure document search in a "hybrid" model where a trusted party is used to compute the ideal functionality $\mathcal{F}_{\text{basicDB}}$. The basic idea is for the parties to use $\mathcal{F}_{\text{basicDB}}$ to store an index to the corpus of texts as follows. The server chooses a random value s_i for every document D_i and then associates with a keyword p the values s_i where p appears in the document D_i. Then, this index is sent to $\mathcal{F}_{\text{basicDB}}$, enabling C to search it securely. In addition, S sends (s_i, D_i) to $\mathcal{F}_{\text{basicDB}}$, enabling the client to retrieve D_i if it knows s_i. Since C is only able to decrypt a document if it has the appropriate s_i value, it can only do this if it queried $\mathcal{F}_{\text{basicDB}}$ with a keyword p that is in document D_i. Observe that in this way, each document is only encrypted once. We remark that the server is allowed to associate p with a *subset* of the documents D_i for which $p \in D_i$ and not all of them. This makes sense because even in a regular non-secure search, upon query p the server can send any subset of documents containing p (and so can filter at will). Preventing this type of behavior is problematic and would be very expensive. Furthermore, it is not clear that it is even desirable; in some cases a word p is crucial to a document and thus should be viewed as a keyword for retrieval and in other cases it is completely inconsequential to the document. In this latter case, the server should be able to decide that the document should not be retrieved upon query p.

Let $\mathcal{P}$ be the space of keywords of size M, let $D_1, \ldots, D_N$ denote N text documents from some predetermined domain, and let $P_i = \{p_{i_j}\}$ be the set of keywords that appear in D_i (note $P_i \subseteq \mathcal{P}$) as defined by the server S. Using this notation, when a search is carried out for a keyword p, the client is supposed to receive the set of documents D_i for which $p \in P_i$. We now proceed to formally define the document search functionality $\mathcal{F}_{\text{DOC}}$ in Figure 9.3.1.

FIGURE 9.3.1 (The Document Search Functionality $\mathcal{F}_{\text{DOC}}$)

Functionality $\mathcal{F}_{\text{DOC}}$ works with a server S and client C as follows (the variable init is initially set to 0):

Initialize: Upon receiving from S a message $(\text{init}, \mathcal{P}, P_1, D_1, \ldots, P_N, D_N)$, if init $= 0$, functionality $\mathcal{F}_{\text{DOC}}$ sets init $= 1$, stores all documents and $\mathcal{P}$, and sends (init, N, M) to C, where N is the number of documents and M is the size of the keyword set $\mathcal{P}$. If init $= 1$, then $\mathcal{F}_{\text{DOC}}$ ignores the message.

Search: Upon receiving a message (search, p) from C, functionality $\mathcal{F}_{\text{DOC}}$ checks that init $= 1$, and if not it returns notInit. Otherwise, it sends search to S. If S replies with approve, $\mathcal{F}_{\text{DOC}}$ works as follows:

1.   If there exists an i for which $p \in P_i$, then functionality $\mathcal{F}_{\text{DOC}}$ sends $(\text{search}, \{D_i\}_{p \in P_i})$ to C.
2.   If there is no such i, then $\mathcal{F}_{\text{DOC}}$ sends notFound to C.

If S replies with reject, then $\mathcal{F}_{\text{DOC}}$ forwards reject to C.

Our protocol makes use of a perfectly-hiding commitment scheme, denoted by $\text{com}_h(\cdot, \cdot)$. We let $\text{com}_h(m; r)$ denote the commitment to a message m using random coins r. For efficiency, we instantiate $\text{com}_h(\cdot; \cdot)$ with Pedersen's commitment scheme [69]; see Protocol 6.5.3 in Section 6.5.1.

We now present a protocol for securely computing $\mathcal{F}_{\text{DOC}}$ which uses a secure protocol π_{basicDB} to compute $\mathcal{F}_{\text{basicDB}}$. For the sake of clarity, we describe the protocol in the hybrid model using a trusted party computing $\mathcal{F}_{\text{basicDB}}$; a real protocol is obtained by replacing the ideal calls with an execution of π_{basicDB}.

PROTOCOL 9.3.2 (secure document search by keyword π_{DOC})

- **Inputs:** The server has database $(\mathcal{P}, D_1, \ldots, D_N)$.
- **Auxiliary inputs:** Both parties have the security parameter 1^n and a parameter ℓ that upper bounds the size of every keyword set P_i. Unless stated differently, $i \in \{1, \ldots, N\}$ and $j \in \{1, \ldots, M\}$.
- **The protocol:**

  - INITIALIZE:
    1. The server S chooses random values $s_1, \ldots, s_N \leftarrow_R \{0,1\}^n$ (one random value for each document), and computes the set of commitments $\{\text{com}_i = \text{com}_h(s_i; r_i)\}_{i=1}^N$ where $r_1, \ldots, r_N$ are random strings of appropriate length.
    2. S defines a database of M records (p_j, x_j) where $p_j \in \mathcal{P}$ is a keyword, and $x_j = \{(i, (s_i, r_i))\}_{p_j \in P_i}$ (i.e., x_j is the set of pairs $(i, (s_i, r_i))$ where i is such that p_j appears in the keyword list P_i associated with D_i). We assume that all x_js are of length ℓ; this can be achieved by padding if necessary.
    3. S sends the commitments $(\text{com}_1, \ldots, \text{com}_n)$ to the client C, and sends $(\text{init}, (p_1, x_1), \ldots, (p_M, x_M), (s_1, D_1), \ldots, (s_N, D_N))$ to $\mathcal{F}_{\text{basicDB}}$.
    4. Upon receiving $(\text{com}_1, \ldots, \text{com}_N)$ from the server S and (init, M) from $\mathcal{F}_{\text{basicDB}}$, the client C outputs (init, N, M).
  - SEARCH: Upon input (search, p) to C, the client C sends $(\text{retrieve}, p)$ to $\mathcal{F}_{\text{basicDB}}$, and receives the response from $\mathcal{F}_{\text{basicDB}}$.
    1. If $\mathcal{F}_{\text{basicDB}}$ sent notFound then C outputs notFound.
    2. If $\mathcal{F}_{\text{basicDB}}$ sent reject then C outputs reject.
    3. Otherwise, let $x = \{(i, (s_i, r_i))\}$ denote the set that C receives from $\mathcal{F}_{\text{basicDB}}$. Then, for every i in the set x, C verifies first that $\text{com}_i = \text{com}_h(s_i; r_i)$. If there exists an i in which the verification does not hold, then C outputs notFound.
    4. Finally, for every pair $(i, s_i) \in x$, the client C sends $(\text{retrieve}, s_i)$ to $\mathcal{F}_{\text{basicDB}}$ and receives back either notFound, reject or a document D_i. If it receives reject in any of the ℓ queries then it outputs reject. Otherwise, it outputs $(\text{search}, \{D_i\})$, including in the output every document D_i containing the keyword p that it received back from $\mathcal{F}_{\text{basicDB}}$.

We now prove the security of Protocol 9.3.2.

Theorem 9.3.3 *Assume that π_{basicDB} securely computes $\mathcal{F}_{\text{basicDB}}$ in the presence of malicious adversaries. Then, Protocol 9.3.2 securely computes $\mathcal{F}_{\text{DOC}}$ in the presence of malicious adversaries.*

Proof. We treat each corruption case separately. Our proof is in a hybrid model where a trusted party computes the ideal functionality $\mathcal{F}_{\text{basicDB}}$.

The server S is corrupted. Intuitively, the simulator $\mathcal{S}_{\text{SERV}}$ learns the set of documents and associated keywords from the init value sent by the server to the trusted party computing $\mathcal{F}_{\text{basicDB}}$. The only issue to deal with is the case of incorrectly generated inputs by S. This is dealt with by having the simulator verify the validity of each decommitment and include only the documents where the attributes that correspond to their keywords include only valid decommitments. Let $\mathcal{A}$ be an adversary controlling S; we construct a simulator $\mathcal{S}_{\text{SERV}}$ as follows:

1. Upon receiving from $\mathcal{A}$ the messages $(\text{com}_1, \ldots, \text{com}_N)$ and $(\text{init}, (p_1, x_1), \ldots, (p_M, x_M))$ (where the last message is for the trusted party for $\mathcal{F}_{\text{basicDB}}$), simulator $\mathcal{S}_{\text{SERV}}$ sets $\mathcal{P} = \{p_1, \ldots, p_M\}$, eliminating repetitions. Then, $\mathcal{S}_{\text{SERV}}$ records (s_i, D_i) if there exists a pair (i, j) for which $(i, (s_i, r_i)) \in x_j$, $\text{com}_i = \text{com}_h(s_i; r_i)$, and D_i includes the keyword p_j. Furthermore, for every such pair (i, j), $\mathcal{S}_{\text{SERV}}$ adds p_j to the list of keywords P_i associated with D_i. If there exists i such that $(i, (s_i, r_i)) \in x_j$, yet $\text{com}_i \neq \text{com}_h(s_i; r_i)$, $\mathcal{S}_{\text{SERV}}$ deletes p_j from $\mathcal{P}$. In addition, if there exist i, j_0, j_1, r_0, r_1 and $s_0 \neq s_1$, such that $(i, (s_{i_b}, r_{i_b})) \in x_{j_b}$ and $\text{com}_i = \text{com}_h(s_{i_b}; r_{i_b})$ for both $b \in \{0, 1\}$, $\mathcal{S}_{\text{SERV}}$ outputs fail.

   If $\mathcal{S}_{\text{SERV}}$ recorded less than N pairs (s_i, D_i) then it completes this set using random s_i and arbitrary documents D_i of appropriate size (and setting the set of keywords for the document to be empty). Finally, $\mathcal{S}_{\text{SERV}}$ sends $(\text{init}, \mathcal{P}, P_1, D_1, \ldots, P_N, D_N)$ to $\mathcal{F}_{\text{DOC}}$.

2. Upon receiving search from $\mathcal{F}_{\text{DOC}}$, $\mathcal{S}_{\text{SERV}}$ sends retrieve to $\mathcal{A}$ (as if it were sent by $\mathcal{F}_{\text{basicDB}}$). If $\mathcal{A}$ replies with reject, then $\mathcal{S}_{\text{SERV}}$ sends reject to $\mathcal{F}_{\text{DOC}}$. Otherwise, $\mathcal{S}_{\text{SERV}}$ simulates C sending ℓ queries s_i to $\mathcal{F}_{\text{basicDB}}$. If $\mathcal{A}$ sends reject for any of these queries, then $\mathcal{S}_{\text{SERV}}$ sends reject to $\mathcal{F}_{\text{DOC}}$. Otherwise, $\mathcal{S}_{\text{SERV}}$ sends approve to $\mathcal{F}_{\text{DOC}}$.

This completes the simulation. We claim that the output distribution of C in the simulation is statistically close to its output in the hybrid execution. The only difference is when $\mathcal{S}_{\text{SERV}}$ outputs fail. However, this happens with negligible probability by the binding property of the commitment scheme. Observe that when $\mathcal{S}_{\text{SERV}}$ does not output fail, the output distribution is identical. This is due to the way that $\mathcal{S}_{\text{SERV}}$ constructs the input from the server to $\mathcal{F}_{\text{basicDB}}$. In particular, the commitments guarantee that the keyword s_i used for storing D_i is unique. Furthermore, $\mathcal{S}_{\text{SERV}}$ defines P_i to be the set of keywords associated with D_i based on the sets x_j. Thus, $\mathcal{F}_{\text{basicDB}}$ returns a document D_i to the client C whenever a real client would receive that document in a real protocol execution.

The client C is corrupted. We now proceed to the case where C is corrupted. Let $\mathcal{A}$ be an adversary controlling C; we construct $\mathcal{S}_{\text{CL}}$ as follows:

1. Upon receiving (init, N, M) from $\mathcal{F}_{\text{DOC}}$, simulator $\mathcal{S}_{\text{CL}}$ chooses N random pairs (s_i, r_i) of appropriate length, and sends $\mathcal{A}$ the commitments $\text{com}_i = \text{com}_h(s_i; r_i)$.

2. Upon receiving a message $(\text{retrieve}, p)$ from $\mathcal{A}$, simulator $\mathcal{S}_{\text{CL}}$ sends (search, p) to its trusted party that computes $\mathcal{F}_{\text{DOC}}$.

   a. If $\mathcal{F}_{\text{DOC}}$ returns t documents $D_1, \ldots, D_t$, then $\mathcal{S}_{\text{CL}}$ chooses t random indices $i_1, \ldots, i_t \in \{1, \ldots, N\}$ that were not chosen before, and sets $x = \{(i', (s_{i'}, r_{i'}))\}$ for all $i' \in \{i_1, \ldots, i_t\}$ while associating $s_{i'}$ with D_i (if a document D' was already returned in a previous search, $\mathcal{S}_{\text{CL}}$

chooses the same index for D' that was previously associated with it). It then sends x to $\mathcal{A}$, emulating $\mathcal{F}_{\text{basicDB}}$.

   b. For every query (retrieve, $s_{i'}$) that $\mathcal{A}$ makes to $\mathcal{F}_{\text{basicDB}}$, if $s_{i'}$ was previously given to $\mathcal{A}$ by $\mathcal{S}_{\text{CL}}$ in a set x, then $\mathcal{A}$ returns the document D_i associated with $s_{i'}$ (as if coming from $\mathcal{F}_{\text{basicDB}}$). If $s_{i'}$ was not previously given to $\mathcal{A}$ by $\mathcal{S}_{\text{CL}}$, then $\mathcal{A}$ returns notFound.

   c. If $\mathcal{F}_{\text{DOC}}$ returns notFound, then $\mathcal{S}_{\text{CL}}$ forwards it to $\mathcal{A}$ as if it were coming from $\mathcal{F}_{\text{basicDB}}$.

The output distributions of the simulated execution in the ideal model and the protocol execution with $\mathcal{A}$ in the $\mathcal{F}_{\text{basicDB}}$-hybrid model are statistically close. The only difference occurs if $\mathcal{A}$ somehow queries $\mathcal{F}_{\text{basicDB}}$ with a value $s_{i'}$ that it did not receive from a previous keyword query p. However, since these $s_{i'}$ values are of length n, this can happen with only negligible probability. ∎

One-sided simulation and covert adversaries. As in previous examples, if a subprotocol $\pi_{\text{basicDB}}^{\text{OS}}$ that securely computes $\mathcal{F}_{\text{basicDB}}$ with one-sided simulation is used in Protocol 9.3.2, then the result is a protocol that securely computes $\mathcal{F}_{\text{DOC}}$ with one-sided simulation. Likewise, security in the presence of covert adversaries is achieved by using $\pi_{\text{basicDB}}^{\text{CO}}$, which achieves security in the presence of covert adversaries.

9.4 Implementing Functionality $\mathcal{F}_{\text{CPRP}}$ with Smartcards

In the protocols that we have seen above for computing database and document search, the expensive part of the computation is due to the secure implementation of the functionality $\mathcal{F}_{\text{CPRP}}$. Although relatively efficient for the case of covert adversaries, it is still quite expensive in the case of malicious adversaries.

In this section we consider a setting where in addition to standard network communication, the parties have access to smartcards. Namely, the participating parties may initialize smartcards in some way and send them to each other, in addition to sending messages over a network. As we describe below, a standard smartcard can be used to securely compute $\mathcal{F}_{\text{CPRP}}$ with extraordinary efficiency (i.e., approximately 50 ms per execution) and thus our protocols for database and document search when implemented in this way become practical even in settings where retrieval must be very fast.

Clearly, such a modus operandi is only reasonable it is not necessary to continually send smartcards between parties. In the uses described here, the server initializes a smartcard and sends it to the client, and that is all. Importantly, it is also sufficient to send a smartcard once, which can then be used for many executions of the protocol (and even for different protocols). This model is clearly not suitable for protocols that must be run by ad hoc participants over the Internet (e.g., for secure eBay auctions or secure Internet purchases). However, it is indeed suitable whenever parties with non-transient relationships need to run secure protocols. Thus, this model is suitable for the purpose of *privacy-preserving data mining* between commercial, governmental and security agencies.

9.4.1 Standard Smartcard Functionality and Security

A smartcard is a piece of secured hardware that carries out cryptographic computations on board. Smartcards are designed to withstand physical and logical attacks, while preserving the secrecy of their cryptographic keys and the integrity of the computations carried out. Smartcards are widely used today for the purposes of secure authentication, digital signatures and disk encryption. High-end smartcards are considered to provide a high level of security in practice; see the discussion regarding this at the end of this section.

One of the basic cryptographic operations of any smartcard is the computation of a block cipher using a secret key that was imported into the smartcard (and is never exported from it later). We assume that the block cipher in the smartcard behaves like a pseudorandom permutation. This is widely accepted for modern block ciphers, and in particular for 3DES and AES. Since smartcards preserve the secrecy of their cryptographic keys, it follows that a party holding a smartcard with a key k for block cipher operation is able to query the smartcard in order to compute $F_{\mathrm{PRP}}(k, x)$ without learning anything about k.

We now provide a description of standard smartcard functionality. We do not include an exhaustive list of all available functions on standard smartcards. Rather we describe the most basic functionality and some additional specific properties that we use.

1. *Onboard cryptographic operations:* Smartcards can store cryptographic keys for private and public-key operations. Private keys that are stored (for decryption or signing/MACing) can only be used according to their specified operation and *cannot* be exported. We note that symmetric keys are always generated outside of the smartcard and then imported, whereas asymmetric keys can either be imported or generated onboard (in which case, no one can ever know the private key). Two important operations that smartcards can carry out are basic block cipher operations and CBC-MAC computations. These operations may be viewed as pseudorandom function computations, and we will use them as such. The symmetric algorithms typically supported by smartcards use 3DES and/or AES, and the asymmetric algorithms use RSA (with some also supporting elliptic curve operations).

2. *Authenticated operations:* It is possible to "protect" a cryptographic operation by a logical test. In order to pass such a test, the user must either

present a password or pass a challenge/response test (in the latter case, the smartcard outputs a random challenge and the user must reply with a response based on some cryptographic operation using a password or key applied to the random challenge).

3. *Access conditions:* It is possible to define which operations on a key are allowed and which are not allowed. There is great granularity here. For all operations (e.g., use key, delete key, change key, and so on), it is possible to define that no one is ever allowed, anyone is allowed, or only a party passing some test is allowed. We stress that for different operations (such as use and delete) a different test (e.g., a different password) can also be defined.

4. *Special access conditions:* There are a number of special operations; we mention two here. The first is a *usage counter*; such a counter is defined when a key is either generated or imported and it says how many times the key can be used before it "expires". Once the key has expired it can only be deleted. The second is an *access-granted counter* and is the same as a usage counter except that it defines how many times a key can be used after passing a test, before the test must be passed again. For example, setting the access-granted counter to 1 means that the test (e.g., passing a challenge/response) must be passed every time the key is used.

5. *Secure messaging:* Operations can be protected by "secure messaging", which means that all data is encrypted and/or authenticated by a private (symmetric) key that was previously imported into the smartcard. An important property of secure messaging is that it is possible to receive a "receipt" testifying to the fact that the operation was carried out; when secure messaging with message authentication is used, this receipt cannot be tampered with by a man-in-the-middle adversary. Thus, it is possible for one party to initialize a smartcard and send it to another party, with the property that the first party can still carry out secure operations with the smartcard without the second party being able to learn anything or tamper with the communication in an undetected way. One example of how this may be useful is that the first party can import a secret key to the smartcard without the second party that physically holds the card learning the key. We remark that it is typically possible to define a different key for secure messaging that is applied to messages being sent *to* the smartcard and to messages that are received *from* the smartcard (and thus it is possible to have unidirectional secure messaging only). In addition to ensuring privacy, secure messaging can be used to ensure *integrity*. Thus, a message authentication code (MAC) can be used on commands to the smartcard and responses from the smartcard. This can be used, for example, to enable a remote user to verify that a command was issued to the smartcard by the party physically holding the smartcard. (In order to implement this, a MAC is applied to the smartcard response to the command and this MAC is forwarded to the remote user. Since it is not possible to forge a MAC without knowing the secret key, the party physically holding

the smartcard cannot forge a response and so must issue the command, as required.)

6. *Store files:* A smartcard can also be used to store files. Such files can either be public (meaning anyone can read them) or private (meaning that some test must be passed in order to read the file). We stress that private keys are not files because such a key can never be read out of a smartcard. In contrast a public key is essentially a file.

We stress that all reasonable smartcards have all of the above properties, with the possible exception of the special access conditions mentioned above in item 4.

Smartcard Security. It is important to note that smartcards provide a high level of *physical security*. They are not just regular microcontrollers with defined functionality. Rather, great progress has been made over the years to make it very hard to access the internal memory of a smartcard. Typical countermeasures against physical attacks on a smartcard include shrinking the size of transistors and wires to 200 nm (making them too small for analysis by optical microscopes and too small for probes to be placed on the wires), multiple layering (enabling sensitive areas to be buried beneath other layers of the controller), protective layering (a grid is placed around the smartcard and if this is cut, then the chip automatically erases all of its memory), sensors (if the light, temperature, etc. are not as expected then all internal memory is immediately destroyed), bus scrambling (obfuscating the communication over the data bus between different components to make it hard to interpret without full reverse engineering), and glue logic (mixing up components of the controller in random ways to make it hard to know which components hold which functionality). For more information, we refer the reader to [76]. Having said the above, there is no perfect security mechanism, and this includes smartcards. Nevertheless, we strongly believe that it is a reasonable assumption to trust the security of high-end smartcards (for example, smartcards that have FIPS 140-2, level 3 or 4 certification). Our belief is also supported by the computer security industry: smartcards are widely used today as an authentication mechanism to protect security-critical applications.

Standard smartcards – what and why. We stress that our use of smartcards is such that any standard smartcard can be used. It is important for us to use *standard* – rather than special-purpose – smartcards for the following reasons:

1. *Ease of deployment:* It is much easier to actually deploy a protocol that uses standard smartcard technology. This is due to the fact that many organizations have already deployed smartcards, typically for authenticating users. However, even if this is not the case, it is possible to purchase any smartcard from essentially any smartcard vendor.

2. *Trust:* If a special-purpose smartcard needs to be used for a secure protocol, then we need to trust the vendor that built the smartcard. This trust extends to believing that it did not incorrectly implement the smartcard functionality on purpose or unintentionally. In contrast, if standard smartcards can be used then it is possible to use smartcards constructed by a third-party vendor (and possibly constructed before the protocol design). In addition to reducing the chance of malicious implementation, the chance of an unintentional error is much smaller, because these cards have been tried and tested over many years.

We remark that Javacards can also be considered for the application that we are considering. Javacards are smartcards with the property that special-purpose Java applets can be loaded onto them in order to provide special-purpose functionality. We remark that such solutions are also reasonable. However, it does make deployment slightly more difficult as already-deployed smartcards (that are used for smartcard logon and VPN authentication, for example) cannot be used. Furthermore, it is necessary to completely trust whoever wrote the applet; this can be remedied by having an open source applet which can be checked before being loaded. Therefore, protocols that do need smartcards with some special-purpose functionality can be used, but are slightly less desirable.

For more discussion regarding this model of using smartcards in cryptographic protocols, see [44].

9.4.2 Implementing $\mathcal{F}_{\mathrm{CPRP}}$ with Smartcards

We model the smartcard as an additional entity that interacts with the parties, where in the ideal model, the ideal adversary/simulator plays the role of the smartcard for the adversary. This means that the simulator receives the messages that the adversary sends to the smartcard and returns the responses that the adversary expects to receive from the smartcard. This is the standard way of carrying out simulation, where the simulator plays the role of the honest parties interacting with the adversary. Importantly, we model this entity as an *incorruptible* trusted party (albeit with functionality that is limited to what standard smartcards can do). Alternatively, we model the smartcard as an ideal functionality $\mathcal{F}_{\mathrm{SC}}$ which receives commands from only one party at a time. Formally, the party that initializes the smartcard has access until it sends a transfer message to the functionality, after which only the other party can interact with the trusted party computing the smartcard functionality.

In order to securely compute $\mathcal{F}_{\mathrm{CPRP}}$ we need to show how to implement the INITIALIZATION and EVALUATION stages. An important feature of our implementation is that it suffices for P_1 (which owns the key k) to send a single smartcard to P_2, and this smartcard can be used for multiple invocations of $\mathcal{F}_{\text{CPRP}}$ with independent keys. In addition, this can be done before P_1 and P_2 have their inputs. This is crucial because although we argue that it is realistic for parties in non-transient relationships to send smartcards to each other, it is not very practical for them to do this every time they wish to run a protocol. Rather, they should be able to do this only once, and then run their protocol many times.

PROTOCOL 9.4.1 (Implementing $\mathcal{F}_{\text{CPRP}}$ with Smartcards)

- **Preprocessing:** P_1 initializes a smartcard so that a key for a pseudorandom permutation can be imported, while encrypted under a secure messaging key k_{sm}. P_1 sends the smartcard to P_2. This preprocessing step is run once in the smartcard lifetime.
- **The protocol:**

    - INITIALIZE: Party P_1 chooses a key $k \leftarrow \{0,1\}^n$ and imports it into the smartcard for use for a pseudorandom permutation. The imported key is encrypted and authenticated using the secure messaging key k_{sm}. In addition, P_1 imports a key k_{test} as a test object that protects the key k by challenge/response. Finally, P_1 sets the access-granted counter of k to 1. Again, these operations are carried out using secure messaging with encryption and message authentication.
    - EVALUATION:
        1. Upon input (retrieve, x), party P_2 queries the smartcard for a random challenge, and receives back r. P_2 sends (challenge, r) to P_1.
        2. Upon receiving (challenge, r), if party P_1 allows the computation, it computes $s = F_{\text{PRP}}(k_{\text{test}}, r)$ and sends (response, s) to P_2. Otherwise, it sends reject to P_2.
        3. If P_2 receives reject, then it outputs reject. Otherwise, upon receiving (response, s), party P_2 hands it to the smartcard in order to pass the test. If s is incorrect and thus the test is not passed, then P_2 outputs reject. Otherwise, P_2 uses the smartcard to compute $F_{\text{PRP}}(k, x)$ and outputs the result.

We prove the security of Protocol 9.4.1 by modeling the smartcard as an ideal functionality $\mathcal{F}_{\text{SC}}$.

Theorem 9.4.2 *Let $\mathcal{F}_{\text{SC}}$ be a functionality implementing the smartcard functionality described above. Then, Protocol 9.4.1 securely computes $\mathcal{F}_{\text{CPRP}}$ in the $\mathcal{F}_{\text{SC}}$-hybrid model.*

Proof (sketch). Let $\mathcal{A}$ be an adversary controlling P_1. Observe that P_1 receives no messages from P_2 in the protocol. Therefore, $\mathcal{A}$ can learn nothing. In addition, a simulator $\mathcal{S}$ can learn k because $\mathcal{A}$ imports k into the smartcard by sending it to $\mathcal{F}_{\text{SC}}$. Note that if $\mathcal{A}$ provides an illegal k (e.g., a string that is not of length n), then $\mathcal{S}$ can replace it with a default key k because a real smartcard will refuse to work with incorrect input and will output an error message that P_2 will see. Finally, observe that if $\mathcal{A}$ provides P_2 with an incorrect response, then this is equivalent to it sending reject, which is what P_2 outputs in this case.

Regarding the case where P_2 is corrupted and controlled by an adversary $\mathcal{A}$, the security stems from the fact that unless $\mathcal{A}$ can guess a correct response s to a challenge r, it is unable to access the smartcard and use $F_{\mathrm{PRP}}(k, \cdot)$ unless authorized by P_1. We stress that $\mathcal{A}$ cannot change the keys imported during the initialize step or learn their value except with negligible probability, because all messages sent by P_1 for the smartcard functionality $\mathcal{F}_{\mathrm{SC}}$ are encrypted and authenticated. Thus, $\mathcal{A}$ can only not import the keys at all, in which case it receives no information about k and so the execution can be easily simulated.

This completes the proof sketch. ∎

Efficiency. Protocol 9.4.1 achieves extraordinary efficiency. Only two rounds of communication are needed between the parties and then P_2 interacts with the smartcard only in order to carry out the computations.

9.5 Secure Text Search (Pattern Matching)

The basic problem of *text search* is the following: given a text T of length N (for simplicity we assume that N is a power of 2) and a pattern p of length m, find all the locations in the text where pattern p appears in the text. Stated differently, for every $i = 1, \ldots, N - m + 1$, let T_i be the substring of length m that begins at the ith position in T. Then, the basic problem of text search is to return the set $\{i \mid T_i = p\}$. This problem has been intensively studied and can be solved optimally in time that is linear in size of the text [10, 53].

In this section, we address the question of how to securely compute the above basic text search functionality. The functionality, denoted $\mathcal{F}_{\mathrm{TS}}$, is defined by

$$((T, m), p) \mapsto \begin{cases} (\lambda, \{i \mid T_i = p\}) & \text{if } |p| = m \\ (\bot, \bot) & \text{otherwise} \end{cases}$$

where T_i is defined as above, and T and p are binary strings. Note that S, which holds the text, learns nothing about the pattern held by C, and the only thing that C learns about the text held by S is the locations where its pattern appears.

In this section, we present a protocol for securely computing $\mathcal{F}_{\mathrm{TS}}$ with *one-sided simulation*. The basic idea of the protocol is for S and C to run a *single* execution of π_{PRF} for securely computing a pseudorandom function with one-sided simulatability; let $f = F_{\mathrm{PRF}}(k, p)$ be the output received by C. Then, S locally computes the pseudorandom function on T_i for every i and sends the results $\{F_{\mathrm{PRF}}(k, T_i)\}$ to C. C can then find all the matches by just seeing where f appears in the series sent by S. Unfortunately, within itself, this is insufficient because C can then detect repetitions in T. That is, if $T_i = T_j$,

C will learn this because it implies that $F_{\mathrm{PRF}}(k, T_i) = F_{\mathrm{PRF}}(k, T_j)$. However, if $T_i \neq p$, this should not be revealed. We therefore include the index i of the subtext T_i in the computation and have S send the values $F_{\mathrm{PRF}}(k, T_i \| \langle i \rangle)$ where $\langle i \rangle$ denotes the binary representation of i. This in turn generates another problem because now it is not possible for C to see where p appears, given only $F_{\mathrm{PRF}}(k, p)$. This is solved by having C obtain $F_{\mathrm{PRF}}(k, p \| \langle i \rangle)$ for every i, while ensuring that the same p appears in all of the $F_{\mathrm{PRF}}(k, p \| \langle i \rangle)$ computations. We utilize specific properties of the Naor-Reingold pseudorandom function, and the protocol π_{PRF} for computing it (see Section 7.6), in order to have C obtain all of these values, while running only a *single* execution of π_{PRF} and while ensuring that the same p appears in all the computations.

We remark that we cannot use a generic protocol for committed secure pseudorandom evaluation $\mathcal{F}_{\mathrm{CPRP}}$, and so in particular cannot use the smart-card methodology of Section 9.4.2. In order to see this, the $\mathcal{F}_{\mathrm{CPRP}}$ functionality does not ensure any consistency between evaluations, and so nothing prevents the client C from obtaining values $F_{\mathrm{PRF}}(k, p \| \langle i \rangle)$ and $F_{\mathrm{PRF}}(k, p' \| \langle j \rangle)$, where $p \neq p'$. We also remark that the above methodology does not seem to readily yield protocols that are fully secure with simulation even in the case where the server S is corrupted. This is due to the fact that in the text search problem the T_i substrings are *related*; in particular, T_i almost equals T_{i+1} (with most values overlapping). Thus, in order to obtain full simulation, the protocol must force a corrupt server S to use values $T_1, T_2, \dots$ that are consistent with each other, and with a single well-defined text T. This seems to be difficult to achieve efficiently.

9.5.1 Indexed Implementation for Naor-Reingold

Following our discussion above, we continue with a modified version of π_{PRF} for computing the Naor-Reingold function such that C's output is the *set* $\{F_{\mathrm{PRF}}(k, x \| \langle i \rangle)\}_{i=1}^{N-m+1}$, rather than just the single value $F_{\mathrm{PRF}}(k, x)$. We call this functionality *indexed pseudorandom function evaluation*, denoted by $\mathcal{F}_{\mathrm{IND}}$, and define it by

$$((k, 1^N), x) \mapsto \left(\lambda, \{F_{\mathrm{PRF}}(k, x \| \langle i \rangle)\}_{i=1}^{N-m+1} \right).$$

Observe that when computing $\mathcal{F}_{\mathrm{IND}}$, the input to F_{PRF} is of length $m + \log N$, where m is the length of the pattern p and $\log N$ is the length of the index i (because it is an index into the text of length N). Recall that the Naor-Reingold pseudorandom function is defined by

$$F_{\mathrm{PRF}}(k, x) = g^{a_0 \cdot \prod_{j=1}^m a_j^{x_j}}.$$

where the key k is defined by the vector $(g^{a_0}, a_1, \ldots, a_m)$ and the a_j values are random; see Section 7.6 for more details. Furthermore, the protocol for secure pseudorandom evaluation that is based on this function works by running an oblivious transfer for each bit x_i of the receiver's input. We assume familiarity with this construction here; see Section 7.6.1.

The idea behind securely computing $\mathcal{F}_{\mathrm{IND}}$ is due to the following observation. Let $x = (x_1 \| x_2)$ where $|x_1| = m$ and $|x_2| = \ell$ (i.e., $|x_1| + |x_2| = m + \ell$), and let $k = (g^{a_0}, a_1, \ldots, a_{m+\ell})$. Then, by the definition of the Naor-Reingold function, we have

$$
\begin{aligned}
F_{\mathrm{PRF}}(k, x) &= F_{\mathrm{PRF}}((g^{a_0}, a_1, \ldots, a_{m+\ell}), (x_1 \| x_2)) \\
&= F_{\mathrm{PRF}}((g^{a_0}, a_1, \ldots, a_m), x_1)^{\Pi_{j=m+1}^{m+\ell} a_j^{x_j}}.
\end{aligned}
$$

Thus, it follows that

$$
F_{\mathrm{PRF}}(k, x \, \| \langle i \rangle) = F_{\mathrm{PRF}}((g^{a_0}, a_1, \ldots, a_m), x)^{\Pi_{j=m+1}^{m+\log N} a_j^{\langle i \rangle_j}},
$$

where $\langle i \rangle_j$ denotes the jth bit in $\langle i \rangle$. We use this in order to securely compute $\mathcal{F}_{\mathrm{IND}}$ as follows. First, S uses a pseudorandom function key $k = (g^{a_0}, a_1, \ldots, a_{m+\log N})$ for inputs of length $m + \log N$, as we have in this case. Next, S and R essentially run Protocol π_{PRF} in Section 7.6 in order to compute $F_{\mathrm{PRF}}((g^{a_0}, a_1, \ldots, a_m), x)$. Recall that in this protocol, after the oblivious transfers the server S holds a value $\tilde{g} = g^{a_0 \cdot \Pi_{j=1}^m \frac{1}{r_j}}$ and the client C holds values $y_{x_1}^1, \ldots, y_{x_m}^m$ such that

$$
\tilde{g}^{\Pi_{j=1}^m y_{x_j}^j} = F_{\mathrm{PRF}}((g^{a_0}, a_1, \ldots, a_m), x).
$$

Thus, for every i, the server S can locally compute $\tilde{g}_{\langle i \rangle} = \tilde{g}^{\Pi_{j=m+1}^{m+\log N} a_j^{\langle i \rangle_j}}$ and it then follows that

$$
\tilde{g}_{\langle i \rangle}^{\Pi_{j=1}^m y_{x_j}^j} = F_{\mathrm{PRF}}(k, x \, \| \langle i \rangle).
$$

After having computed all of these values (for every i), S can send the set to C, which can then compute $F_{\mathrm{PRF}}(k, x \, \| \langle i \rangle)$ for every i, as required. See Protocol 9.5.1 for a full description.

We now prove that Protocol 9.5.1 securely computes the indexed pseudorandom function evaluation functionality. Due to the similarity to the proof of security of Protocol 7.6.5 in Section 7.6, we present a proof sketch only.

Proposition 9.5.2 *Assume that π_{BOT} securely computes $\mathcal{F}_{\mathrm{BOT}}$ with one-sided simulation. Then Protocol 9.5.1 securely computes $\mathcal{F}_{\mathrm{IND}}$ with one-sided simulation.*

Proof (sketch). Note first that the messages that S receives in Protocol 9.5.1 and $\pi_{\mathrm{PRF}}^{\mathrm{OS}}$ (see Theorem 7.6.8) are identical. Therefore, the proof of

---

PROTOCOL 9.5.1 (Indexed Pseudorandom Function Evaluation $\pi_{\text{IND}}^{\text{OS}}$)

- **Inputs:** The input of S is $k = (g^{a_0}, a_1, \ldots, a_{m+\log N})$ and a value 1^N, and the input of C is a value x of length m.
- **Auxiliary inputs:** Both parties have the security parameter 1^n and are given a description of the group $\mathbb{G}$, its order q and a generator g.
- **The protocol:**

  1. S chooses m random values $r_1, \ldots, r_m \leftarrow_R \mathbb{Z}_q^*$.
  2. The parties run a 1-out-2 batch oblivious transfer protocol π_{BOT}. In the ith iteration, S inputs $y_0^i = r_i$ and $y_1^i = r_i \cdot a_i$ (with multiplication in $\mathbb{Z}_q^*$), and C enters the bit $\sigma_i = x_i$ where $x = x_1, \ldots, x_n$. If the output of any of the oblivious transfers is $\perp$, then both parties output $\perp$ and halt. Otherwise:
  3. C's output from the m executions is a series of values $y_{x_1}^1, \ldots, y_{x_m}^m$. If any value $y_{x_i}^i$ is not in $\mathbb{Z}_q^*$, then C redefines it to equal 1.
  4. S sets $\tilde{g} = g^{a_0 \cdot \prod_{i=1}^n \frac{1}{r_i}}$ and sends C the set

$$\left\{ \left(i, g_{\langle i \rangle} = \tilde{g}^{\prod_{j=1}^{\log N} a_{m+j}^{\langle i \rangle_j}} \right) \right\}_{i=1}^{N-m+1}$$

  where $\langle i \rangle_j$ denotes the jth bit of the binary representation of i.
  5. C aborts if the order of any $g_{\langle i \rangle}$ is not equal to q. Otherwise, C computes and outputs the set

$$\left\{ \left(i, y_{\langle i \rangle} = g_{\langle i \rangle}^{\prod_{j=1}^n y_{x_j}^j} \right) \right\}_{i=1}^{N-m+1} = \{ (i, F_{\text{PRF}}(k, p \,\|\, \langle i \rangle)) \}_{i=1}^{N-m+1}.$$

---

privacy in the case where the server is corrupted follows from the proof in Section 7.6.

In the case where C is corrupted, we prove security in the π_{BOT}-hybrid model. Let $\mathcal{A}$ be an adversary that controls C; we construct a simulator $\mathcal{S}_{\text{CL}}$ as follows. $\mathcal{S}_{\text{CL}}$ receives $\mathcal{A}$'s input $x' \in \{0,1\}^m$ for the batch oblivious transfer execution, sends it to the trusted party computing $\mathcal{F}_{\text{IND}}$ and receives the output set $Z = \{z_i\}_{i=1}^{N-m+1}$. Then, in each oblivious transfer $\mathcal{S}_{\text{CL}}$ hands $\mathcal{A}$ a random value $r_i \leftarrow_R \mathbb{Z}_q^*$. Next, $\mathcal{S}_{\text{CL}}$ sets $\tilde{z}_i = z_i^{\prod_{j=1}^m \frac{1}{r_j}}$ for every $z_i \in Z$ and sends $(i, \tilde{z}_i)$ to $\mathcal{A}$. This completes the simulation.

The proof that $\mathcal{A}$'s view is identical in both the simulated and hybrid executions and that $\mathcal{A}$ returns the same value in both executions is the same as in the proof of Theorem 7.6.6. This is because the only difference between the executions is due to the fact that the intermediate value $\tilde{g}$ is raised to the power of $\prod_{j=1}^{\log N} a_{m+j}^{\langle i \rangle_j}$ for every i. Since the initial $\tilde{g}$ value is distributed identically in the hybrid and simulated executions, it follows that raising either value to the power of the same set of constants yields exactly the same distribution. ∎

Full simulation. Protocol 9.5.1 does not achieve full security with simulation in both corruption cases, even if a fully secure oblivious transfer subprotocol is used. This is due to the fact that a corrupt P_1 may not carry out the computation of all the $g_{\langle i \rangle}$ values from $\tilde{g}$ correctly. Thus, correctness is not guaranteed.

Efficiency. $\pi_{\text{IND}}^{\text{OS}}$ has six rounds of communication. In addition, using the batch oblivious transfer of Section 7.5, the number of exponentiations computed is $11(m + \log N) + 15 + 2(N - m)$ since the parties run $\pi_{\text{IND}}^{\text{OS}}$ with input length $m + \log N$ and the additional $2(N - m)$ values are needed to compute the set of $N - m$ values in the last message.

9.5.2 The Protocol for Secure Text Search

We are now ready to present our main result for this section, which uses $\pi_{\text{IND}}^{\text{OS}}$ as a subprotocol.

PROTOCOL 9.5.3 (Secure Text Search $\pi_{\text{TS}}^{\text{OS}}$)

- **Inputs:** The input of S is a binary string T of size N, and the input of C is a binary pattern p of size m.
- **Auxiliary Inputs:** The security parameter 1^n, the input sizes N and m, and $(\mathbb{G}, q, g)$ for the Naor-Reingold function.
- **The protocol:**
  1. S chooses a random key for computing the Naor-Reingold function on inputs of length $m + \log N$; let the key $k = (g^{a_0}, a_1, \ldots, a_{m+\log N})$.
  2. The parties execute $\pi_{\text{IND}}^{\text{OS}}$ where S enters the key k and C enters its pattern p of length m. The output of C from this execution is the set $\{(i, f_i)\}_{i=1}^{N-m+1}$.
  3. For every i, let $t_i = F_{\text{PRF}}(k, T_i \| \langle i \rangle)$, where T_i is the m-length substring of T beginning at location i. Then, S sends C the set $\{(i, t_i)\}_{i=1}^{N-m+1}$.
  4. C outputs the set of indices $\{i\}$ for which $f_i = t_i$.

Note that S can choose the parameters $(\mathbb{G}, q, g)$ as long as their validity can be verified (which is typically the case), and they do not need to be auxiliary inputs.

Theorem 9.5.4 *Let F_{PRF} denote the Naor-Reingold function and assume that the DDH assumption holds in $\mathbb{G}$. Furthermore, assume that protocol $\pi_{\text{IND}}^{\text{OS}}$ securely computes $\mathcal{F}_{\text{IND}}$ with one-sided simulation. Then Protocol 9.5.3 securely computes $\mathcal{F}_{\text{TS}}$ with one-sided simulation.*

Proof. We separately consider the cases that S and C are corrupted.

The server S is corrupted. Since we are only proving one-sided simulatability here, all we need to show is that S learns nothing about C's input.

This follows from the fact that the only messages received by S in the protocol execution are those received within a single execution of the subprotocol $\pi_{\mathrm{IND}}^{\mathrm{OS}}$. Thus, the security in this case follows directly from the one-sided simulatability of $\pi_{\mathrm{IND}}^{\mathrm{OS}}$.

The client C is corrupted. We prove security here in the $\mathcal{F}_{\mathrm{IND}}$-hybrid model. Let $\mathcal{A}$ be an adversary controlling C; we construct a simulator $\mathcal{S}_{\mathrm{CL}}$ as follows:

1. $\mathcal{S}_{\mathrm{CL}}$ receives input p and auxiliary input z and invokes $\mathcal{A}$ on this input.
2. $\mathcal{S}_{\mathrm{CL}}$ receives from $\mathcal{A}$ the input p' that it sends to $\mathcal{F}_{\mathrm{IND}}$. Simulator $\mathcal{S}_{\mathrm{CL}}$ sends p' to its trusted party computing $\mathcal{F}_{\mathrm{TS}}$ and receives back the set of text locations I for which there exists a match.
3. $\mathcal{S}_{\mathrm{CL}}$ chooses a random key $k \leftarrow I_{\mathrm{PRF}}(1^{m+\log N})$ and sends $\mathcal{A}$ the set $\{(i, F_{\mathrm{PRF}}(k, p' \| \langle i \rangle))\}_{i=1}^{N-m+1}$, as the trusted party that computes $\mathcal{F}_{\mathrm{IND}}$ would.
4. Let $k = (g^{a_0}, a_1, \ldots, a_{m+\log N})$. Then for every $1 \le i \le N - m + 1$, $\mathcal{S}_{\mathrm{CL}}$ continues as follows:

   - If $i \in I$, the simulator $\mathcal{S}_{\mathrm{CL}}$ defines $t_i = F_{\mathrm{PRF}}(k, p' \| \langle i \rangle)$.
   - Otherwise, $\mathcal{S}_{\mathrm{CL}}$ defines $t_i = F_{\mathrm{PRF}}(k, \hat{p} \| \langle i \rangle)$ where $\hat{p} \ne p'$ is an arbitrary string of length m.

5. $\mathcal{S}_{\mathrm{CL}}$ hands $\mathcal{A}$ the set $\{(i, t_i)\}$ and outputs whatever $\mathcal{A}$ outputs.

Observe that the only difference between the hybrid and the simulated executions is in the last step where for every text location i such that $T_i \ne p'$, the simulator $\mathcal{S}_{\mathrm{CL}}$ defines t_i based on a fixed $\hat{p} \ne p'$ instead of basing it on the substring T_i (which is unknown to the simulator). Intuitively, this makes no difference due to the pseudorandomness of the function. Formally, we prove this by going through a series of hybrid experiments.

Game H_1: We begin by modifying $\mathcal{S}_{\mathrm{CL}}$ so that it uses an oracle $\mathcal{O}_{F_{\mathrm{PRF}}}$ for computing the function F_{PRF}. The modified simulator $\mathcal{S}_{\mathrm{CL}}^1$ can use this oracle by handing $\mathcal{A}$ the set $F = \{\mathcal{O}_{F_{\mathrm{PRF}}}(p' \| \langle i \rangle)\}_{i=0}^{N-m+1}$ as its output from the trusted party computing $\mathcal{F}_{\mathrm{IND}}$. Furthermore, it can define $t_i = \mathcal{O}_{F_{\mathrm{PRF}}}(p' \| \langle i \rangle)$ if $i \in I$, and $t_i = \mathcal{O}_{F_{\mathrm{PRF}}}(\hat{p} \| \langle i \rangle)$ otherwise. By the definition of F_{PRF}, this is exactly the same distribution as generated by $\mathcal{S}_{\mathrm{CL}}$ above. We stress that $\mathcal{S}_{\mathrm{CL}}^1$ interacts with a trusted party that computes $\mathcal{F}_{\mathrm{TS}}$, as in the ideal world.

Game H_2: In this game, we replace $\mathcal{O}_{F_{\mathrm{PRF}}}$ with an oracle $\mathcal{O}_{H_{\mathrm{Func}}}$ computing a truly random function. Clearly, the resulting distributions in both games are computationally indistinguishable. This can be proven via a reduction to the pseudorandomness of the function F_{PRF}. Informally, let D_{PRF} denote a distinguisher that attempts to distinguish F_{PRF} from H_{Func}. Then D_{PRF}, playing the role of $\mathcal{S}_{\mathrm{CL}}^1$ above, invokes its oracle on the sets $\{p_i = p' \| \langle i \rangle\}_{i=0}^{N-m+1}$ and $\{t_i\}_{i=0}^{N-m+1}$, where $t_i = p_i \| \langle i \rangle$ when $i \in I$, and $t_i = \hat{p} \| \langle i \rangle$ otherwise (note that the difference is whether p' or $\hat{p}$ is used). Now, any distinguisher for the distributions of games H_1 and H_2 can be utilized by D_{PRF} to distinguish between F_{PRF} and H_{Func}.

Game H_3: In this game, we define a modified simulator $\mathcal{S}^3_{\mathrm{CL}}$ that computes all of the t_i values correctly using the honest S's text T, instead of invoking a trusted party. The resulting distribution is identical because the oracle computes a truly random function and all inputs are distinct in both cases (the fact that the inputs are distinct is due to the index i that is concatenated each time).

Game H_4: Here, we modify the oracle $\mathcal{O}_{H_{\mathrm{Func}}}$ back to an oracle $\mathcal{O}_{\mathrm{PRF}}$ computing F_{PRF}. Again, the fact that the distributions in games H_3 and H_4 are computationally indistinguishable follows from a straightforward reduction.

Game H_5: Finally, we compute the pseudorandom function instead of using an oracle. This makes no difference whatsoever for the output distribution.

Noting that the last game is exactly the distribution generated in a hybrid execution of Protocol 9.5.3 with a trusted party computing $\mathcal{F}_{\mathrm{IND}}$, we have that the hybrid and ideal executions are computationally indistinguishable, completing the proof. ∎

Efficiency. The cost of Protocol 9.5.3 equals a single execution of $\pi^{\mathrm{OS}}_{\mathrm{IND}}$ with the additional exponentiations needed for S to compute the set $\{(i, t_i)\}_{i=1}^{N-m+1}$.

One-sided versus full simulatability. Observe that Protocol 9.5.3 does not achieve correctness when S is corrupted because S may construct the t_i values in a way that is not consistent with any text T. Specifically, for every i, the last $m-1$ bits of T_i are supposed to be the first $m-1$ bits of T_{i+1}, but S is not forced to construct the values in this way. Protocol 9.5.3 is therefore not simulatable in this case, and it is not known how to enforce such behavior efficiently.

References

[1] G. Aggarwal, N. Mishra and B. Pinkas. Secure Computation of the k'th-ranked Element. In *EUROCRYPT'04*, Springer-Verlag (LNCS 3027), pages 40–55, 2004.

[2] W. Aiello, Y. Ishai and O. Reingold. Priced Oblivious Transfer: How to Sell Digital Goods. In *EUROCRYPT'01*, Springer-Verlag (LNCS 2045), pages 110–135, 2001.

[3] Y. Aumann and Y. Lindell. Security Against Covert Adversaries: Efficient Protocols for Realistic Adversaries. In *4th TCC*, Springer-Verlag (LNCS 4392), pages 137–156, 2007.

[4] D. Beaver. Multiparty Protocols Tolerating Half Faulty Processors. In *CRYPTO'89*, Springer-Verlag (LNCS 435), pages 560–572, 1990.

[5] D. Beaver. Foundations of Secure Interactive Computing. In *CRYPTO'91*, Springer-Verlag (LNCS 576), pages 377–391, 1991.

[6] D. Beaver and S. Goldwasser. Multiparty Computation with Faulty Majority. In *30th FOCS*, pages 468–473, 1989.

[7] M. Bellare and O. Goldreich. On Defining Proofs of Knowledge. In *CRYPTO'92*, Springer-Verlag (LNCS 740), pages 390–420, 1992.

[8] M. Bellare and O. Goldreich. On Probabilistic Versus Deterministic Provers in the Definition of Proofs of Knowledge. Manuscript, 2006.

[9] M. Ben-Or, S. Goldwasser and A. Wigderson. Completeness Theorems for Non-Cryptographic Fault-Tolerant Distributed Computations. In *20th STOC*, pages 1-10, 1988.

[10] R.S. Boyer and J.S. Moore. A Fast String Searching Algorithm. *Communications of the Association for Computing Machinery*, 20:762–772, 1977.

[11] R. Canetti. Security and Composition of Multiparty Cryptographic Protocols. *Journal of Cryptology*, 13(1):143–202, 2000.

[12] R. Canetti. Universally Composable Security: A New Paradigm for Cryptographic Protocols. In *42nd FOCS*, pages 136–145, 2001.

C. Hazay, Y. Lindell, *Efficient Secure Two-Party Protocols*,
Information Security and Cryptography, DOI 10.1007/978-3-642-14303-8,
© Springer-Verlag Berlin Heidelberg 2010

[13] R. Canetti and A. Herzberg. Maintaining Security in the Presences of Transient Faults. In *CRYPTO'94*, Springer-Verlag (LNCS 839), pages 425–438, 1994.

[14] R. Canetti, E. Kushilevitz and Y. Lindell. On the Limitations of Universal Composable Two-Party Computation Without Set-Up Assumptions. *Journal of Cryptology,* 19(2):135-167, 2006.

[15] D. Chaum, C. Crépeau and I. Damgård. Multiparty Unconditionally Secure Protocols. In *20th STOC*, pages 11-19, 1988.

[16] B. Chor, N. Gilboa and M. Naor. Private Information Retrieval by Keywords. *Technical Report TR-CS0917,* Department of Computer Science, Technion, 1997.

[17] R. Cleve. Limits on the Security of Coin Flips When Half the Processors Are Faulty. In *18th STOC*, pages 364–369, 1986.

[18] R. Cramer and I. Damgård. On the Amortized Complexity of Zero-Knowledge Protocols. In *CRYPTO'09*. Springer-Verlag (LNCS 5677), pages 177–191, 2009.

[19] R. Cramer, I. Damgård and B. Schoenmakers. Proofs of Partial Knowledge and Simplified Design of Witness Hiding Protocols. In *CRYPTO'94*, Springer-Verlag (LNCS 839), pages 174–187, 1994.

[20] I. Damgård. On Σ Protocols. http://www.daimi.au.dk/~ivan/Sigma.pdf.

[21] I. Damgård, T. P. Pedersen and B. Pfitzmann. On the Existence of Statistically Hiding Bit Commitment Schemes and Fail-Stop Signatures. In *CRYPTO'93*, Springer-Verlag (LNCS 773), pages 250–265, 1994.

[22] I. Damgård and T. Toft. Trading Sugar Beet Quotas – Secure Multiparty Computation in Practice. *ERCIM News* 2008(73), 2008.

[23] C. Dwork, M. Naor and O. Reingold. Immunizing Encryption Schemes from Decryption Errors. In *Eurocrypt'04*, Springer-Verlag (LNCS 3027), pages 342–360, 2004.

[24] T. El-Gamal. A Public-Key Cryptosystem and a Signature Scheme Based on Discrete Logarithms. In *CRYPTO'84*, Springer-Verlag (LNCS 196), pages 10–18, 1984.

[25] S. Even, O. Goldreich and A. Lempel. A Randomized Protocol for Signing Contracts. In *Communications of the ACM*, 28(6):637–647, 1985.

[26] U. Feige and A. Shamir. Zero Knowledge Proofs of Knowledge in Two Rounds. In *CRYPTO'89*, Springer-Verlag (LNCS 435), pages 526–544, 1989.

[27] U. Feige and A. Shamir. Witness Indistinguishability and Witness Hiding Protocols. In *22nd STOC*, pages 416–426, 1990.

[28] M. J. Freedman, Y. Ishai, B. Pinkas and O. Reingold. Keyword Search and Oblivious Pseudorandom Functions. In *2nd TCC*, Springer-Verlag (LNCS 3378), pages 303–324, 2005.

[29] Z. Galil, S. Haber and M. Yung. Cryptographic Computation: Secure Fault-Tolerant Protocols and the Public Key Model. In *CRYPTO'87*, Springer-Verlag (LNCS 293), pages 135–155, 1987.

[30] O. Goldreich. *Foundations of Cryptography: Volume 1 – Basic Tools*. Cambridge University Press, 2001.

[31] O. Goldreich. Concurrent Zero-Knowledge with Timing, Revisited. In *34th STOC*, pages 332–340, 2002.

[32] O. Goldreich. *Foundations of Cryptography: Volume 2 – Basic Applications*. Cambridge University Press, 2004.

[33] O. Goldreich. On Expected Probabilistic Polynomial-Time Adversaries: A Suggestion for Restricted Definitions and Their Benefits. In *4th TCC*, Springer-Verlag (LNCS 4392), pages 174–193, 2007.

[34] O. Goldreich and A. Kahan. How to Construct Constant-Round Zero-Knowledge Proof Systems for NP. *Journal of Cryptology*, 9(3):167–190, 1996.

[35] O. Goldreich, S. Micali and A. Wigderson. How to Play Any Mental Game – A Completeness Theorem for Protocols with Honest Majority. In *19th STOC*, pages 218–229, 1987.

[36] O. Goldreich, S. Micali and A. Wigderson. How to Prove all NP-Statements in Zero-Knowledge, and a Methodology of Cryptographic Protocol Design. In *CRYPTO'86*, Springer-Verlag (LNCS 263), pages 171–185, 1986.

[37] S. Goldwasser and L. Levin. Fair Computation of General Functions in Presence of Immoral Majority. In *CRYPTO'90*, Springer-Verlag (LNCS 537), pages 77–93, 1990.

[38] S. Goldwasser, S. Micali, and R. L. Rivest. A Digital Signature Scheme Secure Against Adaptive Chosen-Message Attacks. *SIAM Journal on Computing*, 17(2):281–308, 1988.

[39] S. D. Gordon, C. Hazay, J. Katz and Y. Lindell. Complete Fairness in Secure Two-Party Computation. In *40th STOC*, pages 413–422, 2008.

[40] S. D. Gordon and J. Katz. Partial Fairness in Secure Two-Party Computation. In *EUROCRYPT'10*, Springer-Verlag (LNCS 6110), 2010.

[41] S. Halevi and S. Micali. Practical and Provably-Secure Commitment Schemes from Collision-Free Hashing, In *CRYPTO'96*, Springer-Verlag (LNCS 1109), pages 201–215, 1996.

[42] S. Halevi and Y. Tauman-Kalai. Smooth Projective Hashing and Two-Message Oblivious Transfer. *Cryptology ePrint Archive*, Report 2007/118, 2007.

[43] S. Har-Peled. Lecture Notes on Approximation Algorithms in Geometry, Chapter 27, Excercise 27.5.3, 2010. Currently found at http://valis.cs.uiuc.edu/~sariel/teach/notes/aprx/.

[44] C. Hazay and Y. Lindell. Constructions of Truly Practical Secure Protocols Using Standard Smartcards. In *ACM CCS*, pages 491–500, 2008.

[45] Y. Ishai. Personal Communication, 2004.

[46] Y. Ishai, M. Prabhakaran and A. Sahai. Founding Cryptography on Oblivious Transfer – Efficiently. In *CRYPTO'08*, Springer-Verlag (LNCS 5157), pages 572–591, 2008.

[47] S. Jarecki and V. Shmatikov. Efficient Two-Party Secure Computation on Committed Inputs. In *EUROCRYPT'07*, Springer-Verlag (LNCS 4515), pages 97–114, 2007.

[48] J. Katz and Y. Lindell. Handling Expected Polynomial-Time Strategies in Simulation-Based Proofs. In *Journal of Cryptology,* 21(3):303-349, 2008.

[49] J. Katz and Y. Lindell. *Introduction to Modern Cryptography.* Chapman and Hall/CRC Press, 2007.

[50] J. Katz, R. Ostrovsky and A. Smith. Round Efficiency of Multiparty Computation with a Dishonest Majority. In *EUROCRYPT'03*, Springer-Verlag (LNCS 2656), pages 578–595, 2003.

[51] J. Katz and R. Ostrovsky. Round-Optimal Secure Two-Party Computation. In *CRYPTO'04,* Springer-Verlag (LNCS 3152), pages 35–354, 2004.

[52] J. Kilian. Improved Efficient Arguments. In *CRYPTO'95*, Springer-Verlag (LNCS 963), pages 311–324, 1995.

[53] D. E. Knuth, J. H. Morris and V. R. Pratt. Fast Pattern Matching in Strings. *SIAM Journal on Computing*, 6(2): 323–350, 1977.

[54] Y. Lindell. *Composition of Secure Multi-party Protocols – A Comprehensive Study.* Lecture Notes in Computer Science Vol. 2815, Springer-Verlag, 2003.

[55] Y. Lindell and B. Pinkas. An Efficient Protocol for Secure Two-Party Computation in the Presence of Malicious Adversaries. In *EURO-CRYPT'07*, Springer-Verlag (LNCS 4515), pages 52–78, 2007.

[56] Y. Lindell and B. Pinkas. A Proof of Security of Yao's Protocol for Two-Party Computation. *Journal of Cryptology*, 22(2):161–188, 2009.

[57] Y. Lindell and B. Pinkas. Secure Two-Party Computation via Cut-and-Choose Oblivious Transfer. Manuscript, 2010.

[58] Y. Lindell, B. Pinkas and N. Smart. Implementing Two-Party Computation Efficiently with Security Against Malicious Adversaries. In *Conference on Security and Cryptography for Networks*, pages 2–20, 2008.

[59] S. Micali and P. Rogaway. Secure Computation. Unpublished manuscript, 1992. Preliminary version in *CRYPTO'91*, Springer-Verlag (LNCS 576), pages 392–404, 1991.

[60] M. Naor. Bit Commitment Using Pseudorandomness. *Journal of Cryptology*, 4(2):151–158, 1991.

[61] M. Naor and K. Nissim. Communication Preserving Protocols for Secure Function Evaluation. In *33rd STOC*, pages 590–599, 2001.

[62] M. Naor and B. Pinkas. Efficient Oblivious Transfer Protocols. In *12th SODA*, pages 448–457, 2001.

[63] M. Naor, B. Pinkas and R. Sumner. Privacy Preserving Auctions and Mechanism Design. In the *ACM Conference on Electronic Commerce*, pages 129–139, 1999.

[64] M. Naor and O. Reingold. Number-Theoretic Constructions of Efficient Pseudo-Random Functions. In *38th FOCS*, pages 231–262, 1997.

[65] J. B. Nielsen and C. Orlandi. LEGO for Two-Party Secure Computation. In *6th TCC*, Springer-Verlag (LNCS 5444), pages 368–386, 2009.

[66] C. Orlandi. Personal communication, 2010.

[67] R. Ostrovsky and M. Yung. How to Withstand Mobile Virus Attacks. In *10th PODC*, pages 51–59, 1991.

[68] P. Paillier. Public-Key Cryptosystems Based on Composite Degree Residuosity Classes. In *EUROCRYPT'99*, Springer-Verlag (LNCS 1592), pages 223–238, 1999.

[69] T. P. Pedersen. Non-interactive and Information-Theoretical Secure Verifiable Secret Sharing. In *CRYPTO'91*, Springer-Verlag (LNCS 576) pp. 129–140, 1991.

[70] C. Peikert, V. Vaikuntanathan and B. Waters. A Framework for Efficient and Composable Oblivious Transfer. In *CRYPTO'08*, Springer-Verlag (LNCS 5157), pages 554–571, 2008.

[71] B. Pinkas, T. Schneider, N. P. Smart and S. C. Williams. Secure Two-Party Computation Is Practical. In *ASIACRYPT 2009*, Springer-Verlag (LNCS 5912), pages 250–267, 2009.

[72] M. Rabin. How to Exchange Secrets by Oblivious Transfer. Tech. Memo TR-81, Aiken Computation Laboratory, Harvard University, 1981.

[73] T. Rabin and M. Ben-Or. Verifiable Secret Sharing and Multi-party Protocols with Honest Majority. In *21st STOC*, pages 73–85, 1989.

[74] M. Rodeh. Finding the Median Distributively. *Journal of Computer and System Sciences*, 24(2): 162–166, 1982.

[75] P. Schnorr. Efficient Identification and Signatures for Smart Cards. In *CRYPTO'89*, Springer-Verlag (LNCS 435), pages 239–252, 1989.

[76] M. Witteman. Advances in Smartcard Security. In *Information Security Bulletin*, pages 11–22, July 2002.

[77] A. Yao. How to Generate and Exchange Secrets. In *27th FOCS*, pages 162–167, 1986.

Index